Automation in Garment Manufacturing

The Textile Institute Book Series

Incorporated by Royal Charter in 1925, The Textile Institute was established as the professional body for the textile industry to provide support to businesses, practitioners and academics involved with textiles and to provide routes to professional qualifications through which Institute Members can demonstrate their professional competence. The Institute's aim is to encourage learning, recognise achievement, reward excellence and disseminate information about the textiles, clothing and footwear industries and the associated science, design and technology; it has a global reach with individual and corporate members in over 80 countries.

The Textile Institute Book Series supersedes the former 'Woodhead Publishing Series in Textiles', and represents a collaboration between The Textile Institute and Elsevier aimed at ensuring that Institute Members and the textile industry continue to have access to high calibre titles on textile science and technology.

Books published in The Textile Institute Book Series are offered on the Elsevier web site at: store.elsevier.com and are available to Textile Institute Members at a substantial discount. Textile Institute books still in print are also available directly from the Institute's web site at: www.textileinstitute.org

To place an order, or if you are interested in writing a book for this series, please contact Matthew Deans, Senior Publisher: m.deans@elsevier.com

Recently Published and Upcoming Titles in The Textile Institute Book Series

Handbook of Technical Textiles, Volume 1, 2nd Edition, A. Richard Horrocks and Subhash C. Anand, 9781782424581

Handbook of Technical Textiles, Volume 2, 2nd Edition, A. Richard Horrocks and Subhash C. Anand, 9781782424659

Geotextiles, Robert Koerner, 9780081002216

Advances in Braiding Technology, Yordan Kyosev, 9780081009260

Antimicrobial Textiles, Gang Sun, 9780081005767

Active Coatings for Smart Textiles, Jinlian Hu, 9780081002636

Advances in Women's Intimate Apparel Technology, Winnie Yu, 9781782423690

Smart Textiles and Their Applications, Vladan Koncar, 9780081005743

Advances in Technical Nonwovens, George Kellie, 9780081005750

Activated Carbon Fiber and Textiles, Jonathan Chen, 9780081006603

Performance Testing of Textiles, Lijing Wang, 9780081005705

Colour Design, Janet Best, 9780081012703

Forensic Textile Science, Debra Carr, 9780081018729

Principles of Textile Finishing, Asim Kumar Roy Choudhury, 9780081006467

High-Performance Apparel, John McLoughlin and Tasneem Sabir, 9780081009048

The Textile Institute Book Series

Automation in Garment Manufacturing

Edited by

Rajkishore Nayak
Rajiv Padhye

The Textile Institute

ELSEVIER

WP
WOODHEAD PUBLISHING
An imprint of Elsevier

Woodhead Publishing is an imprint of Elsevier
The Officers' Mess Business Centre, Royston Road, Duxford, CB22 4QH, United Kingdom
50 Hampshire Street, 5th Floor, Cambridge, MA 02139, United States
The Boulevard, Langford Lane, Kidlington, OX5 1GB, United Kingdom

Library of Congress Cataloging-in-Publication Data
A catalog record for this book is available from the Library of Congress

British Library Cataloguing-in-Publication Data
A catalogue record for this book is available from the British Library

ISBN: 978-0-08-101211-6

For information on all Woodhead Publishing publications visit our website at https://www.elsevier.com/books-and-journals

Publisher: Mathew Deans
Acquisition Editor: David Jackson
Editorial Project Manager: Edward Payne
Production Project Manager: Surya Narayanan Jayachandran
Cover Designer: Victoria Pearson

Typeset by TNQ Books and Journals

Contents

List of contributors

Hein A.M. Daanen Vrije Universiteit Amsterdam, Amsterdam, The Netherlands

Ionuţ Dulgheriu "Gheorghe Asachi" Technical University of Iaşi, Iaşi, Romania

Hans-Christian Früh Institut fuer Textiltechnik of RWTH Aachen University, Aachen, Germany

Ashvani Goyal The Technological Institute of Textile & Sciences, Bhiwani, India

Thomas Gries Institut für Textiltechnik der RWTH Aachen University, Aachen, Germany

Irina Ionescu "Gheorghe Asachi" Technical University of Iaşi, Iaşi, Romania

Prabir Jana National Institute of Fashion Technology, New Delhi, India

Yamini Jhanji Technological Institute of Textile and Sciences, Bhiwani, India

Josef Klingele Institut fuer Textiltechnik of RWTH Aachen University, Aachen, Germany

Emil-Constantin Loghin "Gheorghe Asachi" Technical University of Iaşi, Iaşi, Romania

Maria-Carmen Loghin "Gheorghe Asachi" Technical University of Iaşi, Iaşi, Romania

Volker Lutz Institut fuer Textiltechnik of RWTH Aachen University, Aachen, Germany

Katerina Machova Hochschule Reutlingen, Reutlingen, Germany

Rajkishore Nayak RMIT University Vietnam, Vietnam

Volker Niebel Institut fuer Textiltechnik of RWTH Aachen University, Aachen, Germany

Kanwalpreet Nijhar RMIT University, Melbourne, VIC, Australia

Rajiv Padhye RMIT University, Melbourne, VIC, Australia

Asis Patnaik Cape Peninsula University of Technology, Cape Town, South Africa

Sweta Patnaik Cape Peninsula University of Technology, Cape Town, South Africa

Agnes Psikuta Empa - Swiss Federal Laboratories for Materials Science and Technology, St. Gallen, Switzerland

Marco Saggiomo Institut für Textiltechnik der RWTH Aachen University, Aachen, Germany

Kristina Simonis Institut für Textiltechnik der RWTH Aachen University, Aachen, Germany

Amanpreet Singh RMIT University, Melbourne, VIC, Australia

Ineta Vilumsone-Nemes University of Novi Sad, Novi Sad, Serbia

Marko Wischnowski Institut für Textiltechnik der RWTH Aachen University, Aachen, Germany

Introduction to automation in garment manufacturing

1

Rajkishore Nayak[1], *Rajiv Padhye*[2]
[1]RMIT University Vietnam, Vietnam; [2]RMIT University, Melbourne, VIC, Australia

1.1 Introduction

Automation is the process or technique of doing certain works by the use of automatic equipment in the place of human operators during a product manufacturing (Groover, 2007). Automation is achieved by the use of highly automatic tools and equipment embedded with sophisticated electronic devices. Although automation eliminates the human operators from a specific job, they create new jobs to assist the automatic tools and equipment (Hoos, 2000). Automation is widely used in several areas such as manufacturing industries, medicine, healthcare, engineering, supply chain, and distribution (Viswanadham, 2002). There are several areas where automation reduces human intervention to a minimum resulting in saving of labor and energy; improved precision, accuracy, and quality of products; and high productivity (Parasuraman and Riley, 1997; Paul and Becker, 1983; Stylios, 1996).

Before 1947, the concept of automation was not widely used. Although the knowledge of automation existed in some areas such as temperature regulation, automatic loom, automatic spinning mills, and automatic flour mills, the concept did not gain wide industrial acceptance. Automation became familiar only after 1947, when the automotive manufacturer Ford established an automation department (Jarvis, 2000). Feedback controllers were widely used during this time for automation in manufacturing. The developments in digital technology, controllers, relay switches, and sensors helped in the designing of automatic tools for various automation applications. Today, there have been wide applications of automation in various fields such as chemical plants, oil refineries, mining, textile industries, garment manufacturing, steel plants, plastic manufacturing, automotive components, aircraft production, and food processing (Ostrouh and Kuftinova, 2012; Risch et al., 2014; Aitken-Christie et al., 2013).

Clothing is the second most important need to human beings after food. This need is increasing around the world because of increased population and behavioral changes of consumers toward fast fashion. The global need for clothing is fulfilled by the production facilities in developing countries as it is not economically viable to produce cheaper clothes in developed countries (Gereffi and Frederick, 2010; Nayak and Padhye, 2015). The last few decades have witnessed the shifting of clothing production to countries such as Bangladesh, Vietnam, China, Indonesia, India, and Cambodia, where the wages are the lowest (Mani and Wheeler, 1998). This has helped to keep the price of final garment low because of cheap labor overhead. However, the recent garment production is suffering from stiff global competition, rising labor

Automation in Garment Manufacturing. https://doi.org/10.1016/B978-0-08-101211-6.00001-X

costs in many countries, lack of skilled workforce, and a change in consumer behavior influenced by fast fashion and social media (Nayak and Padhye, 2015). Furthermore, the consumers today expect high quality and trendy clothes at cheaper price delivered to their doorstep in a short time.

Clothing production starts from fiber and includes yarn, fabric, and garment manufacturing (Nayak and Padhye, 2015). In addition, other industries that produce trims and accessories for garments, leather industries, and fashion accessories industries are also considered as a part of the global fashion industry (Nayak et al., 2015b). The logistic providers for the supply chain management (SCM) of textile and clothing industries, retail stores, and the stores dealing with the recycling of end-of-life clothes are also considered as part of the fashion production process. Apparel manufacturing is labor intensive, but often there is a high demand on product quality. Hence, to fulfill the high-quality requirements, it is necessary that the labor-intensive processes are converted into automated processes accomplished by the use of computerized tools, digital components, and artificial intelligence (AI) (Nayak et al., 2016).

Although there is a wide scope for automation in all the above activities, automation has not been widely adopted because of reasons such as high cost, complexity of processes, and availability of cheap labor (Stylios, 1996). Inspite of several benefits, in many of the developing countries, the labor-intensive clothing production still use manual practices as it was many years ago, rather than automatic equipment. This can be attributed to the factors such as: (1) clothing production has not progressed to the same extent as it has done in other sectors such as automobile production, (2) availability of cheap labor in many developing countries, (3) high initial investment on the automatic tools and equipment, (4) complexities involved in the automation because of inherent nature of clothing production, (5) frequent style changes, and (6) production of a garment style in different sizes.

Several researches have been done on the automation and application of AI in garment manufacturing (Stylios, 1996; Wang et al., 2005; Fang and Ding, 2008; Stylios et al., 1995). During the preparation of the book, a gap was observed in the number of published articles reviewing the automation of garment manufacturing and the recent trends. Hence, an attempt was made to cover all the areas of automation in garment manufacturing in this chapter. This chapter discusses the global scenario of automation in garment manufacturing including the requirement and fundamental concepts. The major problems of automation lie in fabric handling, which has been covered in detail. Automation in various processes of garment manufacturing has been covered in detail. The other areas of automation such as spinning, weaving, and fabric inspection have also been covered. In addition, the advantages and disadvantages of automation and the future trends have also been discussed in this chapter.

1.1.1 Garment manufacturing: from concept to consumer

The garment manufacturing process starts from a concept or conceptualization stage and ends with the consumers. In the initial stage, a clothing style is conceptualized based on the forthcoming trends in silhouette, color, fabrics, and trims. These concepts are translated into the forms of "mood boards" and "inspiration boards." These concepts are converted into real garment shapes by the designers with the help of computer-aided design (CAD) software (Nayak and Padhye, 2015; Kim and Kang, 2003). Then, in the range planning a range of colors, fabrics and trims are finalized including the raw materials. The prices for the range of garment styles and their corresponding volume are finalized before moving into the production process.

The production process involves the selection and procurement of raw materials such as fibers, yarns, and fabrics (Fig. 1.1). A garment manufacturer can source the finished fabric and start manufacturing the garment or it can start from the initial phase of fiber selection, yarn manufacturing, fabric production, and then finally the garment manufacturing as a vertically integrated garment industry (Nayak and Padhye, 2015). In the fiber selection process the required fibers (natural and/or synthetic) are selected for spinning. In yarn manufacturing the fibers are converted into yarn of required fineness, strength, and uniformity by several spinning processes such as ring, rotor, and air-jet spinning. There are several automations done in the spinning process such as automatic yarn mixing, auto-doffing, auto splicing, and automatic bobbin change (Oxenham, 2003).

Fabric is produced by weaving or knitting processes. Weaving is performed by shuttle looms and shuttleless looms such as miniature gripper, rapier, water-jet, and air-jet looms, whereas knitting is performed by circular or flat knitting machines. Each process produces fabric with different properties and their suitability for specific end use application also varies. There are several automation in the weaving process, which involves automatic warp tension control, automatic pick repair, electronic warp and weft stop motion, and online fabric fault monitoring. Similarly, the automation in knitting involves seamless garment manufacturing, automatic yarn selection, and online fabric fault detection (Nawaz and Nayak, 2015). The details of automation in spinning, weaving, and knitting processes are discussed in Chapter 3.

The readers can refer to Fig. 5.2, which describes various steps followed during the process of garment manufacturing from receiving the fabric till the packaging. The major steps in garment manufacturing can be categorized into three groups such as (Nayak and Padhye, 2015):

1. **Preproduction processes**: Preproduction processes cover product planning, sample development, designing, approvals, raw material sourcing, preproduction meeting, and production scheduling (Stylios, 1996). Selection and procurement of trims, threads, and accessories are also covered in this step. These preproduction processes ensure that the garment manufacturing is performed on time so that the final garments are delivered within the lead time.
2. **Production processes**: The production process includes fabric spreading, cutting, bundling, and sewing. Fabrics are spread in flat tables and cut by tools such as knife cutter, laser cutter,

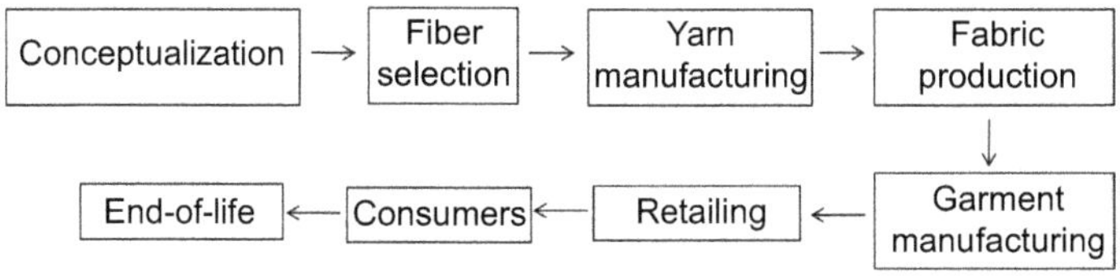

Figure 1.1 The process sequence of garment manufacturing.

or water-jet cutter. The cut components are separated, bundled and fixed with a bundle tickets, and moved to the sewing operation. A number of sewing operations are performed by different workers to finish the garment.
3. **Postproduction processes**: Postproduction processes involve thread trimming, pressing, inspection, folding, packaging, and shipment. Once the garments are manufactured, loose threads are trimmed, garments are pressed and inspected for quality, and packed and transported to the retail stores by the manufactures own logistic network or any third-party logistic providers. The consumers purchase their favorite clothes from the retail stores.

Once the garments are manufactured, they are transported to the retail stores, which link the suppliers in the upstream and the consumers at the downstream end. Consumers buy their required clothes from the retail stores and use it as desired. Once the service life of a garment is finished, it reaches its end-of-life stage. At the end-of-life stage, the garments can be reused, recycled, or else they go to the landfill. Numerous fashion brands are trying to reduce the amount of end-of-life garments going into the landfill by the concept of reduce, reuse, and recycle (Pui-Yan Ho and Choi, 2012; Farrant et al., 2010).

1.1.2 Global scenario of automation

The current scenario of automation in the developing countries where the garments are manufactured will be covered in this section. The production of garments has moved from developed countries to developing countries to keep low cost of production mainly because of low labor costs. In spite of the technological developments, garment production is still labor intensive in these countries. There are only few technologies that have been widely accepted as automation by garment manufacturers, which include button holing machine, button attaching machine, bar tacking machine, label attaching machine, and pocket sewer.

Technological advancements have helped the application of new concepts in garment manufacturing, which includes high sewing machine speed, CAD and computer-aided manufacturing (CAM) applications, new techniques in cutting, fusing, and pressing, and application of robotics (Nayak and Padhye, 2014; Kim and Kang, 2003; Yan and Fiorito, 2007). By introducing the new technologies into the process of garment production, a substantial increase in productivity and quality of work can be achieved. Consequently, the clothing industry is being transformed from a traditional, labor-intensive industry, into a highly automated and computer-aided industry. Garment production processes require, above all, the development and application of the computer-aided technologies as described in Table 1.1:

A garment manufacturer can have its own yarn and fabric manufacturing plants from where the fabric is brought for the garment production. This can help to produce the needed fabric within a short lead time with desired quality. However, majority of the clothing manufacturing companies procure finished fabric externally as per their requirement and convert them into garment. Some clothing manufacturers can also perform various other processes relating to garment manufacturing externally such as embroidery, patch work, or design printing from other producers and complete the remaining processes in-house.

Table 1.1 **Various automation systems and advanced tools in garment manufacturing**

Technology used in automation	Abbreviation	Description	Areas of application
Computer-aided design	CAD	Creation of design, drawing of garment components by the use of computers	Designing, patternmaking, digitizing, and grading
Computer-aided manufacturing	CAM	Manufacturing of garments by the use of machines controlled by software	Spreading, cutting, sewing, and material handling
Computer-aided process planning	CAPP	The use of computers in production planning of garment manufacturing	Production planning, linkage between CAD and CAM
Computer-aided quality control	CAQC	Application of computers to inspect the garment quality	Garment inspection, statistical process control
Computer-aided testing	CAT	Testing the components by the use of computers	Intermediate testing of semifinished garments, final inspection
Automated inspection	AIN	Presentation of the components and inspection are both done automatically	Fabric, trims inspection
Automated material handling devices	AMHD	Used to automatically handle the fabric and other cut components	Fabric, patterns, semifinished garment handling
Artificial neural network	ANN	Computational model based on the structure and functions of biological neural networks	Fabrics inspection, color solutions, garment inspection, supply chain, retail management
Pick/place robots	PPR	Robots are used to pick products from one location to another	Fabric handling for sewing

Continued

Table 1.1 **Continued**

Technology used in automation	Abbreviation	Description	Areas of application
Other advanced tools			
High-speed sewing machine	HSSM	A modern sewing machine that can run at very high speeds	Used for different types of stitches to make garments
Numerical control	NC	Computers are used to perform preprogrammed sequences of machine-controlled commands	Sewing, button holing, button attaching
Modern fusing and pressing machine	MFPM	Fusing and pressing equipment for automatic temperature control, automatic on-off	Fusing and pressing operations
Manufacturing resources planning	MRP	Effective planning of all resources in a manufacturing facility	Production planning, process planning
Enterprise resource planning	ERP	A software that integrates several operation of a plant relating to technology, human resources, and other services	Fabric storage, spreading, cutting, sewing, pressing packaging, human resources, inspection, supply chain, and retailing
Computer used factory floor	CUFF	Computers are used to monitor various operations in the production floor	Spreading, cutting, sewing, and inspection
Internet	IT	Global connecting system that connects millions of computers worldwide	Production planning, sewing, quality control
Communication	CM	The exchange of information between departments	It can be between any departments during production, distribution and retail.

Garment manufacturing in many countries is a labor-intensive process. Although automation is widely used in many other sectors, garment manufacturing is still considered as a labor-intensive process. The technology of sewing by machine has not changed much since its invention in 1790. The level of adoption of automation or advanced technologies by a specific garment manufacture depends on the following factors:

- **Industry size**: The size of the industry plays a major role on the implementation of automatic and advanced technologies. Although smaller industries have advantages such as operational speed, flexibility, and adoptability, they are not in favor of automation because of low volume of production. Larger industries on the other hand adopt the automation techniques more easily. This can be due to the high volume of production that compensates the additional cost of installing the automated equipment. Larger industries focus on the research and development of newer technologies and more eagerly engaged in utilizing the technology.
- **Export market**: The export potential of an industry influences its level of adoption of advanced technologies, which help them to gain competitive advantage, keep the product price low, and face more readily the risks involved in global volatile fashion market. An industry working for the domestic market can perform well without the advanced tools and automation; however, for export market it is quintessential to adopt the advanced technologies.
- **Garment styles**: In several instances the styles and design of the garments influence the level of adoption of advanced technologies and automation. For example, a garment manufacturer producing men's shirt can adopt automatic equipment for the attachment of cuffs and collars, which are readily available now at competitive price.
- **Profitability**: The profitability of a plant also influences the level of technology adoption. An industry with higher profitability can easily install advanced technologies.
- **Available budget**: An industry's success on adopting the new technologies is also influenced by the quality of its capital stock. The amount of planned budget for investing on technology adoption influences the level of technology. As majority of the advanced technologies are expensive, a limited amount of budget for adopting the technologies makes it difficult to gain technological competitiveness. Furthermore, the available budget for installation, training, and care and maintenance influences the adoption of advanced technology.
- **Management policy**: The top management of an industry manages its external relationship and implements policies for the adoption of advanced technologies. The top management is involved in the strategic decision-making process, planning and execution, research and development policies, and innovation and exporting policies. The commitment of the top management to technology adoption will shape the level of adoption of the advanced technology by the plant. The commitment of top management toward technology adoption is defined as "the degree to which the values and perceptions of the management are in favor of and open to technology adoption." Hence, an industry with a dedicated team for technology adoption will have higher level of advanced technology in its production floor.
- **Technical skills**: As the global demand for high-skilled operators is increasing, the adoption of automatic tools and equipment can help in this matter. Today's manufacturing industries sought the operators to be multiskilled, but the number of skilled operators is dwindling. Hence, these skill requirements can be addressed by the advanced technology-based manufacturing systems. A lack of adoption of newer technologies because of poor understanding of the technical advantages and the potential usage will sought qualified engineers and technicians. The availability of skilled labor in an industry will help the adoption of the new technologies easier as the skilled operators can better manage the new technologies with their technical skills.

- **Competitive advantage**: The globalization of apparel manufacturing has led to stiff competition among various global partners. Hence, in a highly competitive atmosphere, there is a need to adopt newer technologies and automation to gain the competitive advantage. When the industries gain competitive advantage with the new technologies, it is likely that they adopt it. The use of advanced technologies can better satisfy the firm's requirements and fulfill the requirements of the customers. The advanced technologies can help in solving complex problems, produce improved quality, and reduced defects.

1.2 Automation in garment production

A garment industry's competitive advantage in global market depends on the level of advanced technologies and automatic tools and equipment that are used in its designing, production planning, manufacturing, supply chain, and retailing. Clothing manufacturers can meet the global market demand for high quality and reduced cost by constant adoption of newer technologies and automation for quick response (QR) and just-in-time production. Budget limitations in many developing countries prevent the garment manufacturers to adopt the advanced technologies.

1.2.1 Requirements of automation

Skilled labors are used in almost all the operations involved with garment manufacturing. The quality control of final garment is more subjective in nature based on nonnumeric description of quality and understanding of the garment style and design requirements. There is no doubt that automation can increase the efficiency of production, reduce the number of defects, and reduce the overall cost of manufacturing. The global demand for quality garments, low cost of production, and competitive advantage can be achieved by the adoption of automation. However, the adoption of automation in garment manufacturing will take some time before it is fully realized in garment manufacturing.

1.2.2 Fundamentals of automation

Millions of dollars were spent in the developed countries including the Europe and the United States to automate the garment manufacturing process in 1980s (Nilsson, 1983). However, this attempt did not achieve large-scale automation in garment industries, although some processes were automated. Although there have been a good number of research to automate garment manufacturing after 1980s, the progress in achieving fully automation has not been realized yet. This can be attributed to the associated difficulties in fabric handling, which is discussed in Section 1.4. The principles of automation in garment manufacturing can be started from the very beginning stage, i.e., fiber production through yarn manufacturing, fabric manufacturing, and finally the apparel manufacturing as shown in Fig. 1.2.

Majority of the earlier researches on fabric handling are based on using an industrial style robot arm, which can grip the fabric with a custom end effector and rotate

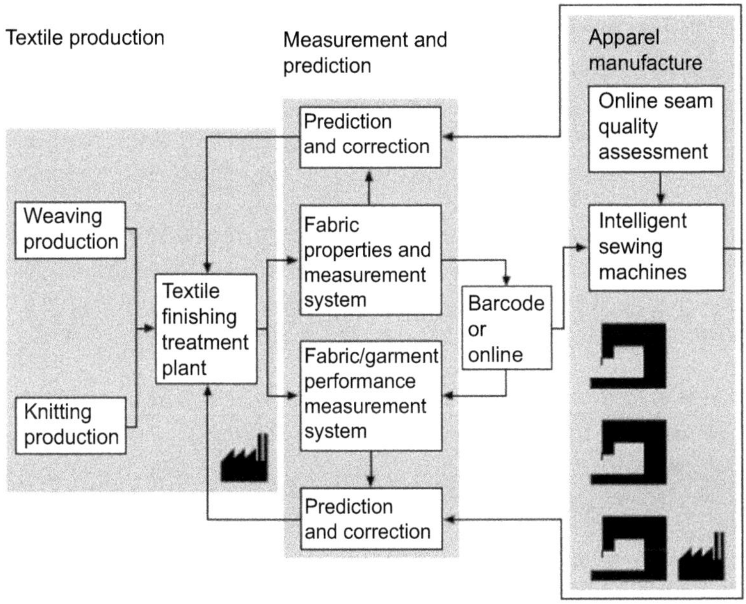

Figure 1.2 The intelligent textile and garment manufacturing environment.

it during the fabric feeding using the feed dog systems. Frank Paul (Paul and Becker, 1983) designed a fabric handling system to detect the edge of a fabric using machine vision. This system can determine the placement of the end effector on the fabric and accordingly plan a seam path at an offset to that edge. However, this system was not much successful as it was lacking robustness because of outgoing filaments and unable to handle inhomogeneous cuts and wrinkles in the fabric. Furthermore, this system was unable to handle multiple pieces of fabric used in a seam and was not very useful for automatic fabric handling.

Programmable logic controllers (PLCs) are used while automation is incorporated in the manufacturing processes (Gungor and Lambert, 2006). Although PLCs are similar to computers, they are optimized for task control during industrial applications compared with computers, which are optimized for calculations (Pinto et al., 2007). Programmable memory is used in PLCs, which store instructions and functions such as logic, counting, sequencing, and timing. The processing system of a PLC uses simple programming to vary the controls of inputs and outputs. The flexibility of the PLCs is their greatest advantage as the same basic controller can operate with a range of control systems. The flexibility also helps in cost saving while designing complex control systems.

The technological advancements in an apparel industry can be classified as: (1) software technology and (2) hardware technology. The software technologies include the CAD, CAM, ERP software, statistical process control, software for production planning and inventory management, and data management; whereas the hardware technologies include automated sewing, automated identification, programmable production controllers, automated material handling, automated inspection systems, and robotics (Kumar et al., 1999).

The use of robotics is also increasing in the garment manufacturing mainly in the sewing floor (Bailey, 1993). Robotics is the branch of electronic technology that deals with the design, construction, operation, and application of robots. Various mechanical, electrical, and electronic components are used including the computer software to make the robotics accurate and fast. The application of industrial robotics started after World War II as there was a need for quicker production of consumer goods (Vogel, 1986). The technological advances has helped to design much advanced robots, which are employed in manufacturing, domestic, commercial, and military applications. Robotics is also applied in areas where there is potential threat or the job is repetitive in nature as in garment manufacturing.

1.3 Areas of automation

There are several areas of automation in garment production, which also includes yarn and fabric production processes. A brief description has been given earlier on the automation of yarn and fabric manufacturing. This section will focus on the automation of processes involved in garment production, which included fabric inspection, CAD and CAM, fabric spreading and cutting, sewing, pressing, material handling, and the role of radio-frequency identification (RFID) in automation.

1.3.1 Automatic fabric inspection

Fabric inspection is performed by the skilled workers on a lighted surface who perform a subjective evaluation of the fabrics. As it is a manual process, many times the faults are not detected accurately. Furthermore, the inspection is also affected by the psychological factors, tiredness, and physical well-being of the inspector. Hence, the inefficiency and inaccuracy of the inspection can be passed into the fabric, which can result in the production of defective garments. The use of automation tools and equipment can help in increasing the efficiency of the inspection process.

Online automated inspection systems can detect the faults during the fabric production as well as during the fabric inspection process. Various techniques such as statistical approach, spectral approach, and model-based approach can be taken for automatic fabric inspection (Ngan et al., 2005, 2011; Park et al., 2000). In all these approaches fabric image is manipulated by a software or modeling tool to extract the information relating to the severity of fabric faults. The faults detected are automatically marked in the fabric and some points are allocated depending on the fault dimension and severity. If the fabric lot exceeds a certain threshold, they are rejected.

1.3.2 Computer-aided design and computer-aided manufacturing

Introduction of computer-aided processes and appropriate information systems to support the area of technological preparation of production started in the clothing industry in the mid-1970s. This was a logical result of rapid development in computer technology and is becoming both a matter of urgency and a decisive factor in the clothing producer's success. The use of modern and capable computer hardware and software can assure high and constant quality of garments, increased productivity, flexibility, and QR to the requirements of the fashion market. Computer equipment is widely used for design and production of garments as well as for the assurance of effective information flows. The producers of such computer equipment, such as graphic workstations, have successfully adopted the characteristics of the engineering area of clothing technology.

The measurement of body dimensions is a manual and time-consuming process. For the production of traditional mass customized garment, different body dimensions are measured and recorded in a paper. These measurements are used by the designer or tailor to produce the customized garment. These practices although inaccurate, inconsistent, and tedious, are still followed in many countries for the production of customized garments. However, for the production of mass customized garment in a retail store, the advanced tools such as 3D body scanning should be used to automatically extract the measurement of the body dimensions. The 3D body scanning devices can capture the three coordinates (X, Y, and Z) for the whole human body. Then appropriate software can convert these data into accurate body dimensions.

3D body scanning is a noncontact technique that captures body dimensions over 360 degrees by the use of white light or laser light (Nayak and Padhye, 2016). The data collected are accurate and represent the three-dimensional shape of the real body, which can be used in the formation of the body shapes and contours to create a 3D virtual model (Nayak and Padhye, 2011). These scanned data can be used to create patterns for different types of garments. For creating patterns, an automatic system need to be developed that can locate the referencing points or landmarks needed for generating body measurements from the scanned data by using a model-based feature recognition algorithm. The scanned data from the 3D scanner have a format of three-dimension point cloud, which indicates many points on the body surface (Kim and Kang, 2003).

These scanned data can also be used for developing the virtual fit model, which are similar to virtual clothing samples. These virtual clothing samples can be presented to the buyers, retailers, or even to the consumers. The virtual fit models eliminate the cost and time involved in the creation of physical samples, and the style is approved in the first attempt. The virtual fit models can help the customers to visualize the mass-customized product before making the purchase (Nayak et al., 2015a). The right type of fabrics can be selected as per the customer's choice and then the virtual fit and appearance of the clothing can be evaluated before making the purchase decision. The virtual fit model is used by many online retail businesses such as eBay.

1.3.3 Fabric spreading and cutting

Fabric spreading can be accomplished by automatic machines on the spreading table. Some machines can work for fabric used in a wide range of applications such as workwear, automotive, container bag, industrial applications, high-performance applications (e.g., Nomex, Kevlar, and carbon), nonwovens, and felts including the apparel fabrics. The fabric parameters such as length, width, and ply counts can be entered into the liquid crystal display touch screen of the machine. The fabric is automatically spread by the machine for the number of plies and stops when the number of plies has been completed. In addition, the machine has the provision to slow down when it approaches both the ends and take care of the alignment of the fabric grain line with the help of sensors.

Similarly automatic cutting machines are available to cut multiple plies of a range of fabric types ranging from lightweight apparel fabric to high-performance industrial fabrics. The marker is fed to a computer using a USB and the cutting head automatically moves to cut the pattern pieces as per the marker. Cutting can be performed by the use of laser, knife, or water-jet. Some of the other features include auto-detection of blade sharpness and indication when the blade is blunt, automatic drilling, and notching. Laser cutters can provide certain degree of advantages than the other cutters in terms of accuracy, no fraying of fabrics, precise and smooth cutting edges, and no change of blades (Nayak et al., 2008). The advantages of automatic cutting over manual cutting are increased efficiency and accuracy; ease of cutting single and multiple plies; and perfect cutting in the first time.

1.3.4 Sewing

As mentioned earlier, majority of the fashion brands and garment retailers have already shifted their production to the ASEAN (Association of Southeast Asian Nations) countries such as Vietnam, Cambodia, and Laos. In these countries, most of the garment manufacturing processes especially the sewing process is still done by skilled labor (Manchin and Pelkmans-Balaoing, 2008; Mirza and Giroud, 2004; Yue, 2005). Substantial progress has not done by the manufacturers on purchasing automated tools and equipment. This has helped them to keep their investments low. On the other hand, there are some manufacturers with automated tools and equipment for sewing and other activities that can produce value-added products more efficiently. The manufacturers not investing on the modern tools and equipment are facing very stiff competition to keep the labor cost low.

For automation of sewing process, industrial robots are recently being developed that can handle the fabric during sewing operation (Lu et al., 2010). The concept of automatic sewing robots was derived from a motorized hand-held medical sewing machine used to close the edge of wounds by spherical seams (Zöll, 2003). Fig. 1.3 shows the image of a compact and light robotic sewing machine (Moll et al., 2009). In this machine the mechanism of seam formation is similar to a traditional sewing machine. The difference lies in the technology the machine operates, the weight, and dimensions. Being robotic, it carries miniaturized components performing specific functions. The machine works with an industrial robot by a coupling unit. Various types of stitches such as overlock stitch, double chain stitch, and double lockstitch can be formed by the machine. The technical challenges with this machine are: (1) the synchronization of the continuous robot movement and discontinuous sewing process;

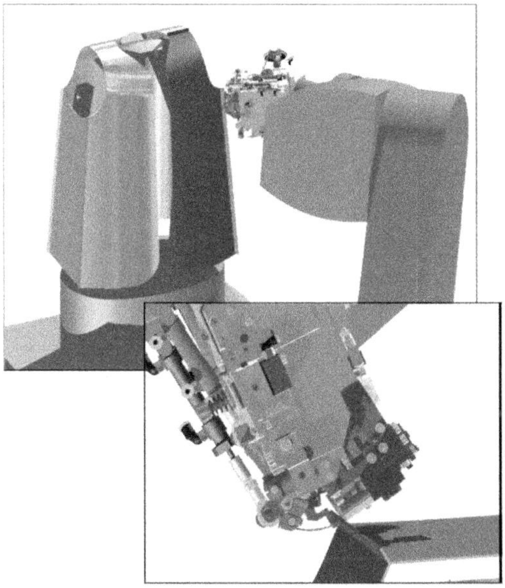

Figure 1.3 The use of robotics in garment manufacturing.
Source: (Top left): Basic principle: Robotic 3D sewing technology/Philipp Moll GmbH & Co KG/ and (bottom right): Spherically positioned fabric is assembled by robot guided sewing machine/Philipp Moll GmbH & Co KG/.

and synchronization of the time sequence of vertical sewing foot movement, horizontal needle movement, and robot speed.

There have been some experimental trials to stitch the whole garment by the use of robots. One such example is Zornow's robot "Sewbo," which can handle the fabric components during automatic sewing (Graham-Rowe, 2011). The fabric need to be stiffened by the application of a water-soluble and nontoxic polymer (polyvinyl alcohol), which makes the handling operation easier. This polymer has been successfully applied to the yarn as a sizing material. The polymer can be removed from the yarn and fabric by the application of hot water. The fabric also retains its original softness after washing.

Invented by Zornow in 2015 (in Sewbo Inc.), the robot "Sewbo" can sew a T-shirt from start to the end. This success was a milestone in achieving 100% automation to manufacture a complete garment. The sequence of operations includes cutting of the panels of the T-shirt by a machine, drenching and stiffening the panels with the polymer, laying them in a flat surface. Then the robotic arm lifts the panels by its suction cups and positions them in a commercial sewing machine. Once stitched, the robot lifts from the sewing machine and the T-shirt is ready. The industrial robot has been successfully applied in the manufacturing of a T-shirt.

The robot can be programmed to grip and position the fabric to the sewing machine repeatedly for a specific size and specific operation. When the size or the style of the garment changes, the robot need to be reprogrammed. The robot is now successful to completely finish all the operations for a T-shirt. This technology can be extended to other garment styles by the program and design modification. However, multiple robots may be needed to perform all the operations relating to a particular garment style.

The use of sewbots such as "Sewbo" will help to achieve high-quality garments at reduced cost (Fig. 1.4). This will also help in solving the labor-related issues, reducing the lead time, reducing defects, and reducing the supply chain, which is a major concern for many global retailers. These robots can work with a wide range of fabrics except the fabric applied with hydrophobic finish or other specialty finish or leather material as it is hard to apply the stiffening polymer. In these cases operations such as attaching the cuffs and collars can be performed automatically.

Although reasonable progress has been made in the sewing machine settings and stitching quality to change with the fabric quality, there are areas of complexity such as needle–fabric interaction while sewing. As the relationship between the physical and mechanical properties on fabric behavior is nonlinear, the interaction between the fabric and needle is nonlinear (Shishoo, 1995). Relating to the nonlinear interaction, the behavior of individual materials is different, which makes the automation process difficult. Furthermore, during the sewing operation, the pulling or slipping of fabric cannot be precisely controlled by the automatic equipment, leading to seam pucker. To resolve this problem the control algorithms should be robust enough to work successfully where no transfer function exists.

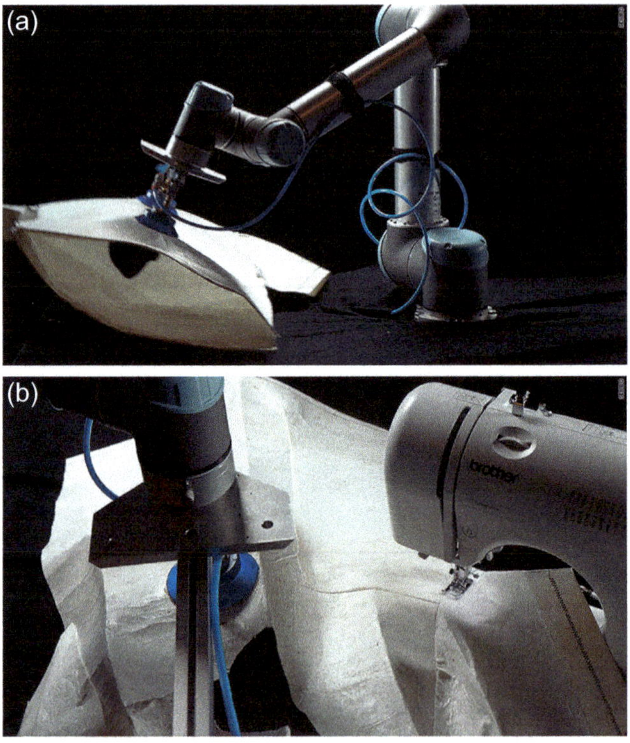

Figure 1.4 Industrial sewing robot "Sewbo": (a) fabric gripping and (b) sewing.

The use of robotic 3D sewing technology can explore new dimensions in sewing as it can produce high-tech garments with high quality. Furthermore, the 3D sewing technology can help in cost reduction and fast response to customer demand. One of the 3D sewing technologies was developed by Philipp Moll GmbH & Co., which can make 3D seams automatically (Moll et al., 2009). The cut components of a garment are placed in a 3D mold and an industrial robot guides a special sewing machine along the spatial seam course. The adjustable mold can adapt to different shapes and sizes of the garment. The 3D sewing technology can be used to manufacture apparels (trousers, jackets, shirts) and car seat covers, airbag fabrics, and many other 3D shapes. This 3D technology can help in achieving better quality of sewn products at higher efficiency. The characteristic features of 3D sewing technology have been discussed below:

- The cut components are placed in the 3D mold in their spatial shape free from creases and tension.
- The cut components are not handled manually during 3D sewing operation.
- The industrial robot guides the sewing machine along the spatial seam path, and the sewing is free from manual interference.
- All the steps involved in sewing such as fabric positioning, transportation, sewing, and off-loading are performed by the robot integrated with the special sewing machine. As multiple operations are handled by the robot, the efficiency of the process increases.
- The 3D sewing technology can be extended to several fabric types, flexible, accurate, and free from skilled worker and their psychological well-being.
- The 3D sewing technology helps to reduce the labor cost, increases productivity, and independent of labor costs and manufacturing location.

1.3.5 Pressing

The other area lacking automation is pressing, which is one of the important steps to enhance the aesthetics of the product before going to the customers. Pressing is done to remove any creases in the garment so that it is attractive when the customer purchases it. A better way of presenting the purchased garment to the customers helps in brand recognition. There have been a wide range of developments in automating the pressing operation, which has been discussed elsewhere in detail (Nayak and Padhye, 2015).

In garment manufacturing, pressing is a taxing task and it is performed in relatively inhospitable environment. Often, it is performed by the workers with less skill of sewing. It is more suitable for male workers as the strenuous work in poor working conditions are counterbalanced by higher pay. There is always a challenge to find and retain skilled workers for pressing operation. The operators when acquire adequate skill, migrate to other industries for higher salary. The labor turnover in pressing is higher than sewing. Hence, getting consistent pressing quality is always a problem. These problems can be addressed by adopting the automation technique in pressing. Furthermore, the problem of labor shortage and human error during pressing are eliminated by automation.

Although there have been several technical advancements in the pressing technology, the number of automation tools is limited. A number of advanced technologies such as pressing robot, jacket finisher (front), shirt finisher, and shirt press have been commercially available. However, the labor application is still prevalent in loading or removal of the garment into the buck, smoothing, and shaping. For successful automation of the whole garment manufacturing process, it is essential to gain significant improvement in pressing in addition to the other areas (cutting, sewing, and material handling).

1.3.6 Use of radio-frequency identification

RFID is an identification system that uses electromagnetic fields to identify and track the movement of objects (Jones et al., 2005; Rekik et al., 2008). The use of RFID system can assist the automation process during garment manufacturing by helping to trace the products during the whole manufacturing process (Nayak et al., 2015c). The identification technique helps to accurately identify and monitor the progress of the semifinished and finished garments from remote places in a real-time environment. The collected information can be used to increase in inventory turnover, improve quality, and automate the manual processes. RFID technology is easily installed and works in the production environment to increase the production efficiency.

For example in cutting, the cut bundles can be attached with an RFID tag, which can be used for the identification of the components, style, color, and other relevant information (Nayak et al., 2015c). Furthermore, the information on the processes that has been completed and the processes that is pending can be obtained from the tags. The use of RFID technology can help in the intelligent garment manufacturing and automation (Nayak et al., 2007). As the RFID tags contain the information on the processes to be performed, once a process is complete, the readers can update the information and transport the components for the next operation. The use of RFID tags can help to generate reports, monitor the progress of each operator and each production style.

RFID technology has many applications in textile and garment manufacturing such as inventory management, product tracking, production monitoring and control, retail management, and brand segregation. RFID technologies may improve the potential benefits of SCM through reduction of inventory losses, increase of the efficiency and speed of processes, and improvement of information accuracy. The basic of success lies in understanding the technology and other features to minimize the potential problems.

1.4 Difficulties in automation

Unlike the large-scale automation in other industries, garment industries are much slower in adopting the technology. The major problem hovers around proper handling of the fabric, due to high flexibility (Paul and Becker, 1983; Brown et al., 1990). One of the major areas of research for several groups is related to the automation in fabric handling (Winck et al., 2009; Ono et al., 1992; Paul, 2004; Koustoumpardis and Aspragathos, 2003). The major component of clothing is the fabric, and in many operations they need to be moved from an operation or placed for a new operation. For moving the fabrics, they need to be held by an appropriate device and transferred to a movable component and then replaced for another operation. The selection and designing of such devices depend on the fabric properties, the operational speed, accuracy required, and the position of the points on the material for which such accuracy is required. The fabric may need to be gripped and transferred as a single component placed on a surface (e.g., table) or from a bulk of other fabrics. For gripping a single component, a number of approaches can be taken as discussed in Table 1.2.

Some of the commercial devices based on the holding methods as described in Table 1.2 are discussed in the following section:

- Clupicker uses the pinching method to hold the fabrics, which is similar to the human fingers picking up the fabric. When a Clupicker is programmed to grip one component, it will be hard to grip the garment assemblies.
- Polytex is based on the method of using pins or needles to pick up single fabric component. As these devices are prepared with high precision, they are slightly expensive.
- Littlewood is also based on the method of using needles, which is a variation of the needle principle used to provide reliability in ply pick up.
- Walton device is based on the combination of air foil, needle, and suction for picking up "oneply only."

Although commercial equipment has been designed using the methods in Table 1.2, they are not very successful in fabric holding and transferring because of the following reasons:

- There is lack of fundamental engineering approach,
- There is lack of quantitative material data, and
- The original equipment manufacturers (OEMs) do not perform dedicated research to solve the problems.

Table 1.2 Various approaches used for fabric gripping and transfer

Holding tool/method	Influencing fabric properties	Comments
Pins or needles	Fabric stiffness	Chances of damaging delicate fabrics
Pinching	Fabric stiffness and surface friction	May not be effective for limp fabrics
Friction	Fabric stiffness and surface friction	Fabric dimensions should be stable
Penetration	The hardness of the fabric	The device used for penetration must not damage the fabric
Electrostatic	Nature of fibers in the fabric, flatness, surface texture	May not be effective in some types of fabrics
Suction	Fabric porosity	Hard to handle porous fabrics such as nets
Adhesive	Nature of fibers, effectiveness in postremoval of adhesive	Additional process of adhesive removal is needed
Freezing	All fabric can be handled. The type of fiber influences the time needed for freezing and heating operations	Additional time needed for freezing and heating operations

The detection of fabric before gripping can be accomplished by the application of different sensing techniques such as:

- Optical: A light source or infrared ray can be used.
- Mechanical sensing: A mechanical sensor can be used.
- Airflow: Measure pressure drop of airflow.

All the fabrics used for apparel purposes are flexible materials, and the handling is influenced by fabric stiffness (Taylor and Koudis, 1987). The other influencing fabric properties during material handling are the friction, which is characterized by coefficient of friction, and longitudinal extension (EM) (Mahar et al., 1990; Behera, 2015). All these three properties (stiffness, friction, and EM) play important role in fabric handling. The low values of fabric stiffness and friction and high EM make the automatic handling rather a difficult task. Because of high variability of these three factors among different fabrics, it is hard to design automated equipment that can handle all types of fabrics.

Furthermore, these fabric properties change depending on the relative humidity and temperature of the working room. As the working room conditions in many garment industries are not precisely controlled, the change in fabric properties will cause difficulty in materials handling. During automatic placement of cut pattern pieces for sewing operation by automatic machines, mismatching of patterns can occur. Positional errors of 5–10 mm generally occur during the operations such as laying, grasping, folding, and sweeping.

Fabrics are flexible material as they undergo significant out-of-plane bending with the application of small forces. The limpness of the fabric due to low stiffness makes it difficult for automatic handling. Automatic handing is very easy in automotive industry where rigid components are handled by robotic arms. Considering the developments relating to material handling in automotive industry, one would find that almost no progress has been achieved in garment manufacturing. The inherent nature of the fabric for automatic handling has made the universal application of automation a difficult task.

The traditional process of manually joining two fabric components by sewing involves: (1) gripping the fabric components, (2) aligning or matching them at the reference point or notch, (3) stitching for the necessary length, and (4) removing the stitched component and placing them in a position to be picked up by the next operator. Hence, while designing automatic robots for fabric handling and sewing of garments, these operations should be kept in mind. The automatic device should be able to grip and feed the fabric component(s) to a sewing machine, match the reference points, if there are two or more components, form the seam, manipulate the components around the needle, stitch them together, and remove them as the stitching has been finished.

While designing the automatic robots, it is important to consider dimensions of the components to be joined and their range, the physical features of the components (such as stiffness, surface roughness, and porosity), and the amount of stitching needed. There are different ways of moving the fabric from one to the other place such as pick and carry, sliding, rolling, conveying, destacking, alignment, and distortion.

The automation of sewing at high speed can lead to excessive needle heating, which can result in improper sewing and faults in the garment. The detection and remedial action is essential to produce quality garments. To resolve this problem and facilitate high-speed automatic sewing, researchers at the Georgia Institute of Technology (GIT) have developed a device that can identify excessive needle heating and indicate to the operator (Silva et al., 2003). The device is based on the use of certain sound frequencies whose amplitudes increase when sewing needles become worn. In the incident of a thread break or when the needle wear exceeds a preset level, the computer alerts the operator by sending a signal that turns on a light. Researchers at GIT are also designing devices to detect sewing problems resulting from needles and thread before they occur. Piezoelectric sensors can be used to monitor the thread movement during sewing, which send the data into a computer and the computer detects the fault.

1.5 Advantages and disadvantages of automation

The use of automation in garment manufacturing provides several benefits and helps the industry to gain competitive advantage and produces good quality product at lower cost. The automation process also suffers from drawbacks, which are discussed in the following section.

1.5.1 Advantages

There are several benefits of using automatic tools and equipment in garment manufacturing, which are mentioned below:

- **Increase in productivity**: Automation increases the productivity by increasing the efficiency of the process. When the job is performed by a labor, there are chances of error, reduced efficiency due to fatigue, and the breaks taken by the worker. However, automated process of performing the job eliminates these and increases the productivity.
- **Increased inventory turnover**: As the productivity of the industry increases, the material turnover also increases. With manual operations the raw materials, cut components, and semifinished components have to wait longer to get converted into the final garment. Hence, with automation increased inventory turnover is achieved.
- **Improvement in quality**: As mentioned above, automation leads to reduced amount of error of garments because of human intervention is eliminated. This leads to the products with less defects, improved quality, and reduced rejection rates.
- **Replacement of repetitive and monotonous work**: Majority of the garment manufacturers use progressive bundle system (PBS) of production. In PBS one worker performs a specific job and passes to the other. Hence, the work becomes repetitive and monotonous for the worker. This can lead to fatigue and reduced efficiency. However, automation can help to avoid these as all these repetitive works are performed by the machine.
- **Reduction of variability among products and product batches**: As the involvement of labor is reduced, the variability of the products produced by different workers is also reduced. Similarly, the variability of the same product manufactured in different batches (manufactured over different times or in different industries) is also reduced.
- **Performing jobs beyond human capability**: Automation can perform some jobs, which needs high skills of the labor. As today's garments are moving toward the integration of electronic devices and other gadgets, high skill is needed many times to perform these operations. Automation can achieve these objectives much easily.

- **Reduction of direct human labor costs and overheads**: Automation helps to achieve increased productivity and efficiency. Automation also helps to perform the task of multiple operators. Furthermore, the need for training of workers for each new style and other quality-related training is reduced. Hence, the cost of human labor and labor overheads are reduced.

1.5.2 Disadvantages

Although there are several advantages of automation in garment manufacturing, there are several disadvantages associated with it, which are discussed below:

- **High initial cost of installation**: The initial cost of installing automated tools and equipment is high compared with the unit cost of garment. The cost of investment may be beneficial when the automation is applicable to many products over a period.
- **High cost of research and development**: The cost involved in the research and development of automatic tools in garment industry is high. Therefore, it may take long time to realize the benefits and cost savings from automation.
- **Security threats**: As the automatic systems lack intelligence, it is common to encounter errors when there is an unexpected change from the normal operation or deviation from the immediate scope. The automated subsystems cannot apply the general principles for simple logic to solve common problems.
- **High cost of maintenance**: Automated equipment need special spare parts too, to repair and skilled people to do the repair and maintenance. Hence, the cost of care and maintenance will be higher compared with the normal machineries.
- **Unexpected production delays**: This situation will arise when the automated equipment malfunctions or ceases to function. As it will take longer time to repair the automatic equipment, there will be production delays. The whole product line will be suffered in case of mal- or nonfunctioning of automatic equipment.
- **Limited scope**: Automation cannot be extended to all the processes involved in garment manufacturing. Some processes are hard or more expensive to automatize. Fabric flexibility, proper alignment of two components joined together, correct tension during sewing, and slippage of fabric during garment manufacturing are some of the factors that limit the scope of automation in garment manufacturing.
- **Lack of flexibility**: Automation is not a flexible and convertible process in production of clothing. Garment manufacturing demands more flexibility as there are many style and size changes in a short time. Automation should allow the switching of production line from one to the other without much hassle. The use of digital electronics is helping to achieve more accuracy and flexibility in the product line.
- **Unemployment**: Many workers may lose their jobs because of automation as automatic equipment can perform the job done by multiple workers. Hence, the workers doing the job will be at the risk of losing their jobs by automation. However, there will be some new jobs emerging because of the automation, which will provide employment. A recent report published by the International Labour Organization revealed that about 88% of workers in Cambodia's textile, clothing, and footwear industry are at high risk of losing their jobs because of automation. Various advanced technologies such as 3D printing, CAD, 3D body scanning, and robotics application are the potential areas that need less people. Similar problems will be faced by the garment manufacturers in many other developing countries. The use of automatic cutting equipment and the increased use of robots in sewing (or sewbots) will be the most influencing areas. Several manufacturers in the Europe, the United States, and China are facing the problem. Although there is no immediate threat in ASEAN region, it will impact more with increased automation.

1.6 Book contents

This chapter discusses briefly about the garment manufacturing process (from concept to consumer) and the global scenario of automation. This chapter also covers the requirements and fundamentals of automation in garment manufacturing. The areas of automation such as design development, body dimension measurement, sewing, pressing, and material handling are also discussed. Automation not only provides several advantages in garment manufacturing but also suffers from some drawbacks (disadvantages), which are discussed in this chapter. The future scope of automation in garment manufacturing has also been discussed.

Chapter 2, Automation versus modeling and simulation compares between the automation techniques, and modeling and simulation tools. It describes the traditional process of garment manufacturing, the application of automation, and compares this with the results obtained by modeling and simulation regardless of their industrial adoptability. The intention of using automation and digitalization is to reduce production time, increase the diversity of products by acceptable cost for mass production as well as for custom production. It also describes the process of pattern construction using CAD.

Chapter 3, Automation in production of yarns, woven, and knitted fabrics deals with the automation in spinning and weaving, which are used for the production of yarns and fabrics, respectively. The first section on automation in spinning includes the developments by various manufacturers of spinning machineries. A section has been included to discuss on the automation in the sewing thread production. The final section discusses on the automation in fabric production by weaving and knitting (especially weft knitted fabrics). The quality monitoring of knitted fabrics has also been described.

Chapter 4, Automation in fabric inspection discusses the advancements in fabric inspection, which is a primary component of the garment. The principles and applications of statistical, spectral, and model-based approaches for the fabric inspection, which are faster and reliable, are also discussed. The researches on fabric inspection done by several researchers are also highlighted. Various commercial fabric inspection systems such as BarcoVision's Cyclops, Elbit Vision System's I-Tex4, Zellweger Uster's Fabriscan, and Shelton webSpector for online or on-loom fabric fault analysis are also discussed in this chapter.

Chapter 5, Artificial intelligence and its application in the apparel industry focuses on the types of AI techniques that are employed in the garment manufacturing process. Various applications of AI in the textile industry such as fiber, yarn, and fabric production are described. The major applications include, predicting the fabric properties, recipe prediction, and fabric fault detection. It also covers the applications of AI in apparel industry such as pattern design, production planning and control, marker making, sewing automation, sales forecasting, fashion recommendation, SCM, and retail. The challenges faced by the AI and the future directions are also discussed.

Chapter 6, Automation in spreading and cutting discusses the role of automation in important garment manufacturing steps such as spreading and cutting influencing the garment quality. This chapter also highlights on the detection of faults by the use of automated machines. In addition, automatic cutting machines based on laser, knife and water-jet cutting are also discussed. The automatic pattern matching process during garment manufacturing has been explained by matching of images. The advancements of fusing technologies are also covered in this chapter. The future directions in the automation in spreading and cutting have also been discussed.

Chapter 7, Automation in material handling discusses the concept of material handling including the influence of material properties on handling. Various types of grippers and gripping technologies for handling textile materials are also described in this chapter. The handling of high-performance textiles is also illustrated. The role of RFID in material handling has made the process much easier compared with the traditional barcode systems, which is also discussed in this chapter.

Chapter 8, Application of robotics in garment manufacturing highlights various applications of robotics in garment manufacturing, which includes material handling, sewing, and composite production. The role of automation is discussed when a 2D fabric has been converted into a 3D fabric. The use of computer numerical control has also been discussed in this chapter. Various sewing automats and robotics are also covered in this chapter. This chapter also gives some examples of commercial automated sewing machines.

Chapter 9, Automation in sewing technology discusses the basic kinematics for automation in sewing such as continuous and cyclic sewing machines. The functions of key hardware and software components in sewing automation have been discussed in the section on building blocks of automation. The evolution of sewing automats, from early to the recent developments of loading, sewing, and unloading of fabrics in sewing automats is also discussed. Automatic features such as under bed trimmer, automatic bobbin changer are discussed in this chapter. Different categories of sewing automats based on the product categories are also discussed. This chapter also includes automats for preparatory operations to sewing before the sections on future trends and sources of further information.

Chapter 10, 3D body scanning highlights the manual measurement techniques that are used to measure the body dimensions needed for clothing design and the relationship between them. This chapter also discusses about 3D scanning devices used for body scanning and the recent developments. The 3D body scanning techniques and virtual fit of garments is also discussed. The role of standardization initiatives related to 3D body scanning has also been discussed in this chapter.

Chapter 11, Computer-aided design—garment designing and patternmaking highlights the application of CAD in various processes that assists in automation. The benefits of CAD, different software used for garment designing, patternmaking, sizing, grading, and construction are also discussed in this chapter. The use of CAD for woven and knitted fabric design is also discussed. The use of 3D modeling and virtual avatars for extraction of body measurements has been discussed. The role of CAD in mass customization, which is the future trend in garment manufacturing, is discussed.

Chapter 12, Advancements in production planning and control discusses about the automation systems in production and manufacturing, the reasons for adopting automation and the strategies followed along with the upcoming current and future trends that are or would be accepted in the near future by the textile and clothing industries. The emergence of fast changes in fashion has given rise to the need to shorten production cycle times in the garment industry. Garment production has become extremely personnel dependent and therefore, it was cost intensive. Automation offers interesting possibilities and potentials for high-tech and better quality garment manufacturing with lower cost and QR to the consumer market in terms of current designs and orders. Automation costs have come down recently, have been more flexible to accommodate varying customer needs, more and more companies have been keen on accepting this technology.

Chapter 13, Use of advanced tools and equipment in industrial engineering focuses on the role of industrial engineering in garment manufacturing and the tools and equipment used in the garment industry for time and motion study. This chapter highlights the methods-time measurement technique and the software applications development for work method design and to establish the operational time. The applications of software for time study, using the Reichsausschuβ für Arbeitsstudium techniques, are also presented in this chapter. This chapter also focuses on the line balancing specific problems including the case studies.

Chapter 14, Automation in quality monitoring of fabrics and garment seams is based on the application of automation in fabric and seam inspection. Fabric faults detection using machine vision has been discussed in this chapter. The role of image processing on fabric quality monitoring has been discussed. The automatic quality monitoring of seams, the traditional approaches, and the factors influencing the same have also been discussed. The detection of seam puckering, 2D process pattern recognition, quality monitoring of welded seams, and photogrammetry have also been discussed in this chapter.

Chapter 15, Recent developments in the garment supply chain covers the developments in logistics and supply chain starting from the fiber processing especially cotton fiber. This chapter covers the developments achieved in the supply chain to facilitate various activities. The role of logistic service providers, ware housing, distribution and retailing and their current trend has been discussed relating to fashion supply chain. The contemporary issues and the recent trends in the garment supply chain have been discussed. The role of lean and agile manufacturing and the use of RFID in fashion supply chain are also included in this chapter.

1.7 Future trends

Garment manufacturing process is labor intensive and now heading toward automation because of several advantages. The application of automation and robotics can transform the labor-intensive garment production into high-tech production centers. The automation can perform small tasks such as bobbin change to the use of sewing robots to produce the entire garment with improved quality, reduced cost, and reduced lead time compared with human work. Although the manufacturing of entire garment is not commercially successful till now, it will be a reality in near future. When automation is adopted in any process during garment manufacturing, the manufacturers should be aware of the pros and cons of installing automatic equipment.

The automation has not gained much success in garment manufacturing because of the flexibility of fabrics (Yue, 2005). For automated fabric handling, the relative changes in the humidity and temperature can lead to difficulty during material handling. Hence, the precise control of fabric and environment is very essential. For reliable results in automation, the fabric need to be manufactured with consistent quality and the environment should not change rapidly in humidity. The fabric handling area is the most challenging field of research for many researchers. In future, the research and development teams, garment manufacturers, and OEMs should look into alternative approaches for effective gripping and transferring of fabrics by using precise engineering principles. If a real solution is achieved, it will be a large success for a number of industries to adopt automation in fabric handling.

In future, the application of automation and robotics will be increasing in garment production. However, complete automation of clothing manufacturing may not be feasible because of complex nature of the production systems and cost factors. As cost is the prime driving factor in garment manufacturing, the company owners in several instances do not want to install expensive automation tools and equipment. Hence, the scope and level of automation in future will be directly influenced by the labor cost in garment manufacturing. If the labor cost increases substantially, the manufacturers will focus on the automation techniques to reduce the cost of production.

It is believed that in future when the automation becomes a reality, several workers will lose their job, which may not be true always. Although automation can perform the jobs done by multiple workers, there will be new jobs emerging because of automation. For example, the need to run the control software of the automation equipment and robotics can create more high-wage jobs compared with the low-wage manufacturing jobs. Hence, the workers can acquire these skills and earn high wages. There will be always some demand for high-skilled people to modify the program, maintain the machineries for automation.

1.8 Conclusion

Apparel manufacturing is a labor-intensive process since it was first mechanized in the 19th century. Although, there have been several technological developments in many other fields, the technology of sewing and its related processes have not done much progress. The modern apparel manufacturing process can be characterized by low fixed capital investment; a wide range of product designs and hence input materials; variable production volumes; high competitiveness; and often high demand on product quality. To achieve the products at competitive price, several fashion brands have moved their manufacturing facilities to developing countries such as Bangladesh, Vietnam, China, Indonesia, India, and Cambodia. Several garment industries in these countries are still performing manual operations because of availability of cheap labor, high cost of automation, and complexity of processes. The increased labor cost and demand for high-quality clothes cannot be fulfilled by the manual operations. The use of automatic tools and equipment is essential to cater these demands at lower production cost.

Fabric inspection, spreading, cutting, sewing, pressing, and material handling are some of the areas where automation can be adopted in garment manufacturing. Automation is achieved by the use of automatic tools and equipment embedded with sophisticated electronic devices or even by the use of robotics. Although not successful commercially, the use of robots with high-speed sewing machines have helped to produce complete garments without the use of labor. In future these processes will be extended to commercial manufacturing of garments fully by the robots. Increase in production efficiency, quality accuracy and reduction in the lead time are some of the benefits achieved by automation. There are several areas where automation reduces human intervention to a minimum resulting in saving of labor and energy and improved precision. Although automation eliminates the human operators from a specific job, they create new jobs to assist the automatic tools and equipment. In future the garment manufacturing will be fully automated that will eliminate the requirement for high-skilled labor. This will help the industries to gain competitive advantage and keep their product cost low.

References

Aitken-Christie, J., Kozai, T., Smith, M.A.L., 2013. Automation and Environmental Control in Plant Tissue Culture. Springer Science & Business Media.

Bailey, T., 1993. Organizational innovation in the apparel industry. Industrial Relations: A Journal of Economy and Society 32, 30–48.

Behera, B., 2015. Role of fabric properties in the clothing-manufacturing process. In: Garment Manufacturing Technology. Woodhead Publishing, pp. 59–80.

Brown III, P., Buchanan, D., Clapp, T., 1990. Large-deflexion bending of woven fabric for automated material-handling. Journal of the Textile Institute 81, 1–14.

Fang, J.-J., Ding, Y., 2008. Expert-based customized pattern-making automation: part I. Basic patterns. International Journal of Clothing Science and Technology 20, 26–40.

Farrant, L., Olsen, S.I., Wangel, A., 2010. Environmental benefits from reusing clothes. The International Journal of Life Cycle Assessment 15, 726–736.

Gereffi, G., Frederick, S., 2010. The Global Apparel Value Chain, Trade and the Crisis: Challenges and Opportunities for Developing Countries.

Graham-Rowe, D., 2011. Robot tailoring: stitched by the sewbot. New Scientist 210, 46–49.

Groover, M.P., 2007. Automation, Production Systems, and Computer-Integrated Manufacturing. Prentice Hall Press.

Gungor, V.C., Lambert, F.C., 2006. A survey on communication networks for electric system automation. Computer Networks 50, 877–897.

Hoos, I., 2000. When the computer takes over the office. Technology, Organizations and Innovation: The Early Debates 1, 179.

Jarvis, P., 2000. Globalisation, the learning society and comparative education. Comparative Education 36, 343–355.

Jones, P., Clarke-Hill, C., Hillier, D., Comfort, D., 2005. The benefits, challenges and impacts of radio frequency identification technology (RFID) for retailers in the UK. Marketing Intelligence and Planning 23, 395–402.

Zöll, K., 2003. Potentials Provided by Robot Supported Sewing. J DNZ Fashion Industry International, Bielefeld.

Kim, S.M., Kang, T.J., 2003. Garment pattern generation from body scan data. Computer-Aided Design 35, 611–618.

Koustoumpardis, P.N., Aspragathos, N.A., 2003. Fuzzy logic decision mechanism combined with a neuro-controller for fabric tension in robotized sewing process. Journal of Intelligent and Robotic Systems 36, 65–88.

Kumar, V., Kumar, U., Persaud, A., 1999. Building technological capability through importing technology: the case of Indonesian manufacturing industry. The Journal of Technology Transfer 24, 81–96.

Lu, J.-M., Wang, M.-J.J., Chen, C.-W., Wu, J.-H., 2010. The development of an intelligent system for customized clothing making. Expert Systems with Applications 37, 799–803.

Mahar, T., WheelWright, P., Dhingra, R., Postle, R., 1990. Measuring and interpreting fabric low stress mechanical and surface properties: part V: fabric handle attributes and quality descriptors. Textile Research Journal 60, 7–17.

Manchin, M., Pelkmans-Balaoing, A.O., 2008. Clothes without an emperor: analysis of the preferential tariffs in ASEAN. Journal of Asian Economics 19, 213–223.

Mani, M., Wheeler, D., 1998. In search of pollution havens? Dirty industry in the world economy, 1960 to 1995. The Journal of Environment and Development 7, 215–247.

Mirza, H., Giroud, A., 2004. Regionalization, foreign direct investment and poverty reduction: lessons from Vietnam in ASEAN. Journal of the Asia Pacific Economy 9, 223–248.

Moll, P., Schütte, U., Zöll, K., Molfino, R., Carca, E., Zoppi, M., Bonsignorio, F., Callegari, M., Gabrielli, A., Principi, M., 2009. Automated garment assembly and manufacturing simulation. In: Transforming Clothing Production into a Demand-Driven, Knowledge-Based, High-Tech Industry.

Nawaz, N., Nayak, R., 2015. Seamless garments. In: Nayak, R., Padhye, R. (Eds.), Garment Manufacturing Technology Cambridge, UK.

Nayak, R., Chatterjee, K., Khurana, G., Khandual, A., 2007. RFID: tagging the new era. Man-Made Textiles in India 50, 174–177.

Nayak, R., Gon, D.P., Khandual, A., 2008. Application of LASER in apparel industry. Man-Made Textiles in India 51, 341–346.

Nayak, R., Kanesalingam, S., Wang, L., Padhye, R., 2016. Artificial intelligence: technology and application in apparel manufacturing. In: TBIS-APCC 2016. Binary Information Press, Textile Bioengineering and Informatics Society, pp. 648–655.

Nayak, R., Padhye, R., 2011. Application of modelling and simulation in smart and technical textiles. In: Patanaik, A. (Ed.), Modeling and Simulation in Fibrous Materials: Techniques and Applications. Nova Science.

Nayak, R., Padhye, R., 2014. Introduction: the apparel industry. In: Nayak, R., Padhye, R. (Eds.), Garment Manufacturing Technology. Elsevier.

Nayak, R., Padhye, R., 2015. Garment Manufacturing Technology. Elsevier.

Nayak, R., Padhye, R., 2016. The use of laser in garment manufacturing: an overview. Fashion and Textiles 3, 1–16.

Nayak, R., Padhye, R., Wang, L., Chatterjee, K., Gupta, S., 2015a. The role of mass customisation in the apparel industry. International Journal of Fashion Design, Technology and Education 8, 162–172.

Nayak, R., Padhye, R., Wang, L., Chatterjee, K., Gupta, S., 2015b. The role of mass customisation in the apparel industry. International Journal of Fashion Design, Technology and Education 1–11.

Nayak, R., Singh, A., Padhye, R., Wang, L., 2015c. RFID in textile and clothing manufacturing: technology and challenges. Fashion and Textiles 2, 1–16.

Ngan, H.Y., Pang, G.K., Yung, N.H., 2011. Automated fabric defect detection—a review. Image and Vision Computing 29, 442–458.

Ngan, H.Y., Pang, G.K., Yung, S., Ng, M.K., 2005. Wavelet based methods on patterned fabric defect detection. Pattern Recognition 38, 559–576.

Nilsson, N., 1983. FIGARMA-Fully integrated garment manufacture: an extension of the concept of flexible manufacturing systems. Computers in Industry 4, 147–163.

Ono, E., Ichijo, H., Aisaka, N., 1992. Flexible robotic hand for handling fabric pieces in garment manufacture. International Journal of Clothing Science and Technology 4, 16–23.

Ostrouh, A., Kuftinova, N., 2012. Automation of planning and management of the transportation of production for food-processing industry enterprises. Automatic Control and Computer Sciences 46, 41–48.

Oxenham, W., 2003. Developments in spinning. Textile World 34–37.

Parasuraman, R., Riley, V., 1997. Humans and automation: use, misuse, disuse, abuse. Human Factors 39, 230–253.

Park, S.-W., Hwang, Y.-G., Kang, B.-C., Yeo, S.-W., 2000. Applying fuzzy logic and neural networks to total hand evaluation of knitted fabrics. Textile Research Journal 70, 675–681.

Paul, F., Becker, R., 1983. Robotic fabric handling for automating garment manufacturing. Journal of Engineering for Industry 105, 21.

Paul, F.W., 2004. Acquisition, placement, and folding of fabric materials. International Journal of Clothing Science and Technology 16, 227–237.

Pinto, V., Rafael, S., Martins, J., 2007. PLC controlled industrial processes on-line simulator. In: IEEE International Symposium on Industrial Electronics, 2007. ISIE 2007. IEEE, pp. 2954–2957.

Pui-Yan Ho, H., Choi, T.-M., 2012. A Five-R analysis for sustainable fashion supply chain management in Hong Kong: a case analysis. Journal of Fashion Marketing and Management: An International Journal 16, 161–175.

Rekik, Y., Sahin, E., Dallery, Y., 2008. Analysis of the impact of the RFID technology on reducing product misplacement errors at retail stores. International Journal of Production Economics 112, 264–278.

Risch, F., Tremel, J., Klier, T., Franke, J., 2014. Flexible automation for the production of stators and rotors of electric vehicles. Advanced Materials Research 321–327 Trans Tech Publ.

Shishoo, R.L., 1995. Importance of mechanical and physical properties of fabrics in the clothing manufacturing process. International Journal of Clothing Science and Technology 7, 35–42.

Silva, L.F., Lima, M., Carvalho, H., Rocha, A.M., Ferreira, F.N., Monteiro, J.L., Couto, C., 2003. Actuation, monitoring and closed-loop control of sewing machine presser foot. Transactions of the Institute of Measurement and Control 25, 419–432.

Stylios, G., 1996. The principles of intelligent textile and garment manufacturing systems. Assembly Automation 16, 40–44.

Stylios, G., Sotomi, O., Zhu, R., Xu, Y., Deacon, R., 1995. The mechatronic principles for intelligent sewing environments. Mechatronics 5, 309–319.

Taylor, P., Koudis, S., 1987. Automated handling of fabrics. Science Progress 351–363.

Viswanadham, N., 2002. The past, present, and future of supply-chain automation. IEEE Robotics and Automation Magazine 9, 48–56.

Vogel, E.F., 1986. Pax Nipponica? Foreign Affairs 64, 752–767.

Wang, C.C., Wang, Y., Yuen, M.M., 2005. Design automation for customized apparel products. Computer-Aided Design 37, 675–691.

Winck, R.C., Dickerson, S., Book, W.J., Huggins, J.D., 2009. A novel approach to fabric control for automated sewing. In: IEEE/ASME International Conference on Advanced Intelligent Mechatronics, 2009. AIM 2009. IEEE, pp. 53–58.

Yan, H., Fiorito, S.S., 2007. CAD/CAM diffusion and infusion in the US apparel industry. Journal of Fashion Marketing and Management: An International Journal 11, 238–245.

Yue, C.S., 2005. ASEAN-China economic competition and free trade area 1. Asian Economic Papers 4, 109–147.

Automation versus modeling and simulation

2

Katerina Machova
Hochschule Reutlingen, Reutlingen, Germany

2.1 Introduction

The world is increasingly becoming digitally connected through networking and thus also production, which of course also affects the textile industry. The German government has named this process of the digital networking of products and production "Industry 4.0," the fourth stage of the development of industrialization, after mechanization, electrification, and digitization. Today, modern companies already connect information and communication technology in production (Dewan and Kraemer, 2000; Eason, 2005). In the next stage of the industry, however, this will be controlled and documented around the entire chain from the idea of a product (partly through the cooperation of customers) through development, production, use, and maintenance to recycling (Roth, 2016).

Automation has already enabled manufacturers to produce large volume of production within a short time. In the clothing industry, the automation of individual process steps for low-cost production has been the main focus (Nayak and Padhye, 2015). Without such automation, countries with high wages and energy costs are not able to offer competitive products (Wang et al., 2005). The price of such mass products is thus significantly reduced, which has contributed significantly to the constantly rising operational costs of today's industries. Through automation, production is enormously accelerated and precise, and production time is calculated effectively (Boer et al., 2005; Forza and Vinelli, 2000). However, the introduction of automation is also cost-intensive and producers have to calculate precisely whether the use of a robot or automation tool is necessary or whether personnel are more cost-effective than high investment in an automated system (Nayak and Padhye, 2014). It is already technically possible to assemble a highly automated production line in the clothing industry. The production costs are reduced enormously owing to high production output and a small number of operating staff, but production becomes more inflexible and is profitable because of high investment costs resulting from high throughput (Joshi and Singh, 2010).

Automation can also be used to make limited product customization affordable for customers' requirements. Today's production chains, for example, are often so flexible in the automotive industry, despite their high degree of automation, that they are able to produce even smaller quantities without a large number of shifts. In this chapter, items of the clothing productions are described where automation and simulation is possible to use. All described processes are possible to apply on industrial or customized production. The reasons for using automation and digitalization are to reduce production time and increase the diversity of products by keeping costs acceptable for mass as well as custom production.

Automation in Garment Manufacturing. http://dx.doi.org/10.1016/B978-0-08-101211-6.00002-1

2.2 The way from idea to technical sheets

Digital modeling and the simulation of clothing are a kind of automation (Kim and Park, 2004; McCartney et al., 2000). With the invention of the sewing machine, embellishments spread rapidly, and also the technique of pattern making. In small factories, patterns were and still are handcrafted, modeled, and edited into cut templates. About 80 years ago, these cuts were performed exclusively by hand.

With the introduction of computer-aided design (CAD), the cutting design and modeling of the cuts has been considerably simplified and, above all, accelerated (Au and Ma, 2010, Nayak and Padhye, 2011). Interfaces for automatic nesting, electronic data processing systems, product life-cycle management (PLM), and automatic cutting have enormously accelerated garment production (Nayak and Padhye, 2015).

In the area of design, however, the creative process of automation is untouched (Hu et al., 2008). Gathering inspiration and keeping pace with the rapidly evolving trends requires good networking in the world of fashion and fast communication between designers and producers. Gathering, sorting, and applying the inspiration depends exclusively on people.

To communicate ideas in a comprehensible way, the first approaches to automation are applied. Sketches and technical drafts are created using drawing software such as Illustrator, CorelDRAW, GNU Image Manipulation Program, or other vector-based drawing programs (Kim and Kang, 2003). Individual elements such as buttons, cuffs, and zippers are created once and applied or copied for several designs. The time savings are enormous. The vector-based software also allows the image size to be changed without loss of quality. The concept of designing a sketch and technical drawing have been explained in Figs. 2.1 and 2.2, respectively.

Figure 2.1 Design sketch.

Figure 2.2 Technical drawing.

The drawings are positioned on a figurine. The figurine has the right proportions and the pose is individually tailored according to the choice of every designer. In this case, artistic representation is already dispensed with and ruffles and shading are reduced to a bare minimum.

The proportions of the figurine match reality. The lengths correspond to the dimensions; it should also be possible to read the lengths and depths of the outfit from the technical drawing. The dimensions do not have to be marked separately, which helps the designer to shorten the processing time significantly. However, this method is possible only if the designer and the production agree. The widths must be viewed as mirror widths. This means that all sizes are not readable from the technical sketch and must be defined by the final size table.

With the introduction of technical drawing, communication between the design process and realization was significantly shortened and more precisely defined. Good knowledge of vector-based drawing programs is required, which is taught in standard fashion or textile schools.

Technical documentation of designs has not yet been standardized. For fast orientation and specification of all necessary details, it is best to use software that contains tables and offers designers or technologists a structure tree with detailed specifications for individual designs (Eberle et al., 2014, Fontaine, 2011).

The table of end measurements belongs to technical documentation, as do accessories and pricing. For pricing, knowledge of production times and material costs is necessary. Calculation systems integrated into CAD systems or added to PLM help to organize production. The information can be saved into a cloud to give access to data to partners that are involved.

Industry 4.0 means digitalization of all processes involved in product creation and manufacturing, combining all technologies in real-time information systems. The question arises, however, how this concept may influence the development departments of the fashion and apparel industry in the near future? At first, there may be only limited work using pen and paper. The ideas of the designers will be created and communicated electronically from the start in three dimensions (3D).

For this kind of work, the content of education has to be adapted because many schools, colleges, and universities that teach fashion design, still work in the classical way with paper and pen. Many fashion designers see themselves as artists and misunderstand supporting computer software as an enemy of creativity rather than something that facilitates and supports work. Therefore, many in this area still see computers or tablets predominantly for use in communication in social networks. At this point, it is necessary for conversion to digital technologies to be carried out in the creative field, as was largely achieved in technical education, to ensure integration into the digitized economy in the future.

A digitalized textile industry with fashion and clothing could look like this from a customer's perspective. The customer no longer buys according to measurement tables, which are based on various standards in online shops and create problems, as it is not using personal measurements. Instead, the customer will use his own avatar in the future, which digitally simulates his body (Nayak et al., 2015; Jo Anderson-Connell et al., 2002). The creation of such an avatar is already possible, because almost every smartphone can be used as a large-format scanner. With this avatar, a customer can choose suitable clothing in every online shop, since it is based on real body measurements. The customer's personal preference determines the fit, but modern algorithms can learn exactly according to recommendations based on personal taste. The decision to buy can be made by the customer or the software can take over completely, depending on the season. This enables new business models, such as in the form of clothing subscriptions.

There are already online shops with tailor-made as well as ready-made clothing. However, there is increasingly less of the second types because automated production makes individual adjustments easy without the product price becoming significantly different from the finished product. The great advantage of prefabrication is that the customer receives the order about 2 days faster. However, it is foreseeable that in the course of the development of Industry 4.0, measurement assemblies will adapt very quickly.

From the point of view of the product development of clothing, there will also be considerable changes (Senanayake et al., 2015). A designer works completely electronically using software that translates a sketch into a technical drawing. This drawing includes correct fit, seams, labels, or even accessories. Based on the technical drawing, a 3D model is created and put together in numerous color combinations. An algorithm, based on sales figures and past data, supports which designs and colors are promising for the next season. Creative human intervention in the design phase and in the compilation of the collection is also possible at any time.

The production of clothing will be completely paperless. Data for the 3D model include not only the cut parts themselves, but also the parts lists for the accessories, the manufacturer or the purchasing point, the material and the storage position, and further production information. For the production of ready-made goods, the order is forwarded electronically to the production site, which has the appropriate technology and material. The selection of machines for the realization of the garment is made automatically on the basis of the technology and current use (Boer et al., 2005). The disposition of materials and accessories in the production facility is also carried out without human intervention. The fabric is placed and cut with the help of robots at the cutting table, also as the multi-ply. The arrangement on transport trolleys and the transport of individual piles with seams and accessories for sewing robots are also fully automated. Even in the sewing process, individual steps are increasingly being implemented automatically, but it can be assumed that the sewing process will not reach a high degree of automation in the foreseeable future such as that seen in the automotive industry.

The second production route involves the production of tailor-made assembly lines. There are only a few differences from standard clothing. However, the cut is only one-ply here and production will take place even closer to the customer. This is to reduce transport distances and transport times, but also because the share of labor costs owing to automation of the total product will be even lower in the future. In contrast to production, logistics already has a high level of automation (Echelmeyer et al., 2008). In the future, environmental aspects, which are mainly influenced by the energy supply of the locations for production and logistics, will become even more important. Because the production networks will become more limited, it is to be expected that the selection of materials will tend to become smaller.

2.3 Pattern development

In 1979, before CAD systems were introduced, the production of garments involved making cuts exclusively by hand, tracing the paper, and making a template. The templates were stored in a room and used again and again. Damaged templates were replaced or repaired.

With the development of computer technology and cutting software, the development of cuts has accelerated considerably. It is no longer necessary to carry out the modeling of basic cuts with the help of paper (Mori and Igarashi, 2007). The consumption of paper has been minimized, and only the final templates are printed. With the introduction of the automatic cutter, paper can be completely omitted. The development process was also accelerated with the development of 3D simulation. Evaluation of the modeled cuts and the correct fit no longer has to be carried out on real patterns, but the simulated and animated models are judged and faults are corrected (Figs. 2.3 and 2.4).

A major step in development was achieved by the possibility of filling the cutting templates with their own printed pattern. With a digital printer, cut templates are printed directly onto the fabric, including patterns as a fill of the cut templates. The full or dyed printed areas of the fabrics fall out. The color that would otherwise be used in the areas where there is waste is saved. A disadvantage associated with this is the pressure capability of the fabrics. Not every fabric can be printed with a digital printer. The fabric to be printed must also have special equipment so that the authentic

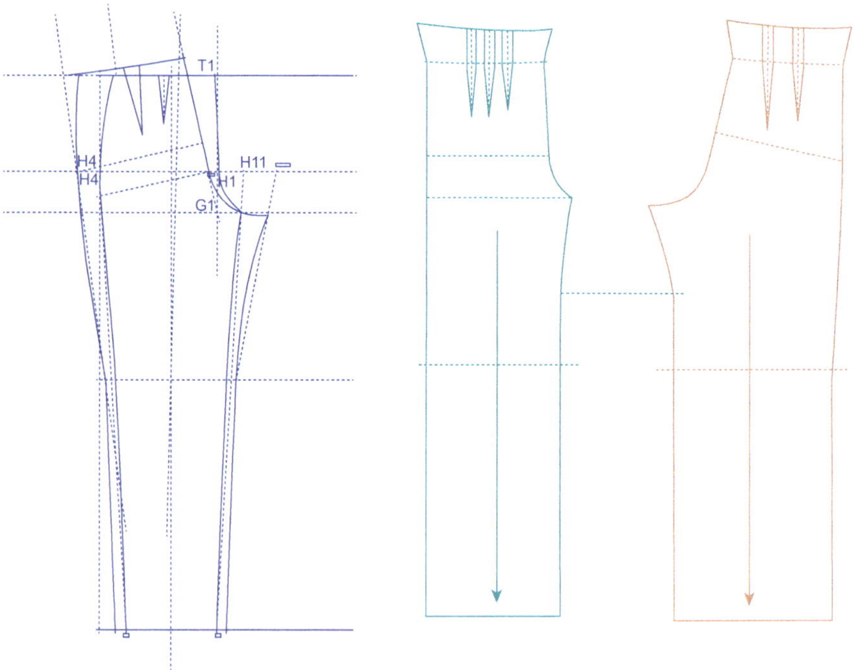

Figure 2.3 Basic pattern made in computer-aided design and its modeling.

color and shades are displayed correctly. The contour lines are even recognized by a camera and are cut out at an exact point. Thus errors that could occur as a result of the materials being warped are avoided.

The possibility of evaluating the models in 3D also makes it easier to decide which colors and fabric structures are to be used for the respective products. Innumerable variations in color and substance can be presented internally or to the customer (Ashdown et al., 2004; Volino et al., 2005). Larger companies use 3D simulation for internal matching and to select collections. In the future, the virtual presentation of an outfit may be considered normal for the customers. The simulations of an outfit are realistic, so it is difficult to distinguish among a simulation, a real image, and a real outfit.

Material properties will also be configured digitally. Existing materials will be examined; elasticity, bending stiffness, and shear behavior can be measured in a textile laboratory. Digital materials will have defined properties in the digital world similar to real behavior. In addition, it will be possible to see the specific influences of material on the avatar. Thus the clothing's comfort can be evaluated before it is produced (Fig. 2.5).

After 3D simulation, the next steps in production are possible (Metaaphanon and Kanongchaiyos, 2005). The cuts with a filled color pattern can be printed directly onto the fabric. Digital printing machines can work with pigments and binders, so it is possible to print onto materials such as cotton and viscose or synthetic materials such

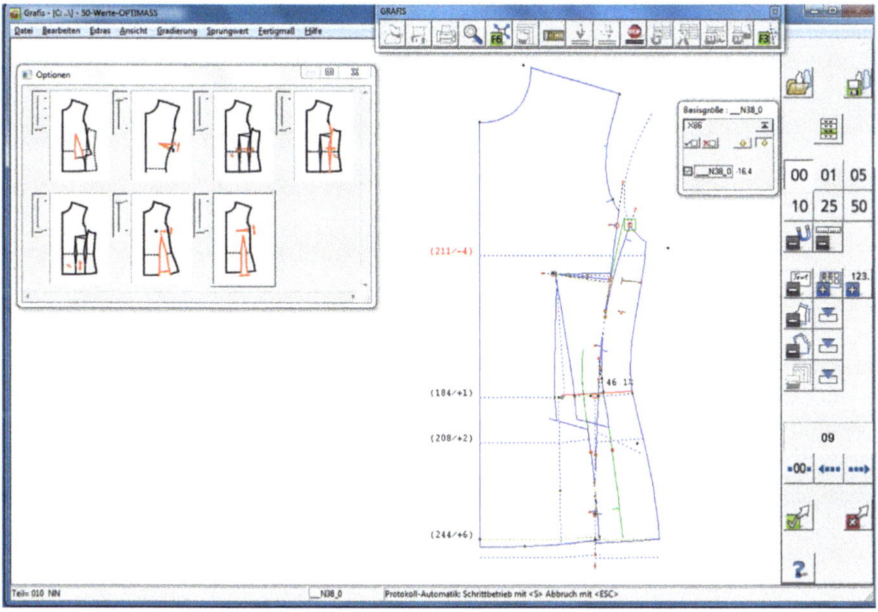

Figure 2.4 Basic pattern by GRAFIS with interactive features.

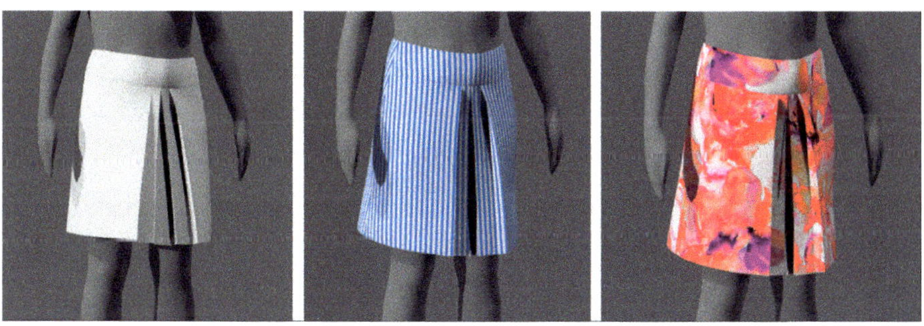

Figure 2.5 Three-dimensional simulation by Assyst Vidya.

as polyester and polyamide. The print has to be thermally fixed. The major advantage of this technology is that the pigments will be used only on areas where the cut is actually placed. The space between the cut pieces remains untouched by color or any other treatment. This area is waste but the material can be recycled as unpolluted material with few problems. The cut with the color filling will be nested and printed as a marker. The printed fabric material is then placed on the roll to be transported to the cutter. There, problems might occur because the nesting picture on the fabric material may no longer be the same as the marker on the screen. Special devices such as Gemini VisionCUT are placed on the cutter. This device can detect the new position

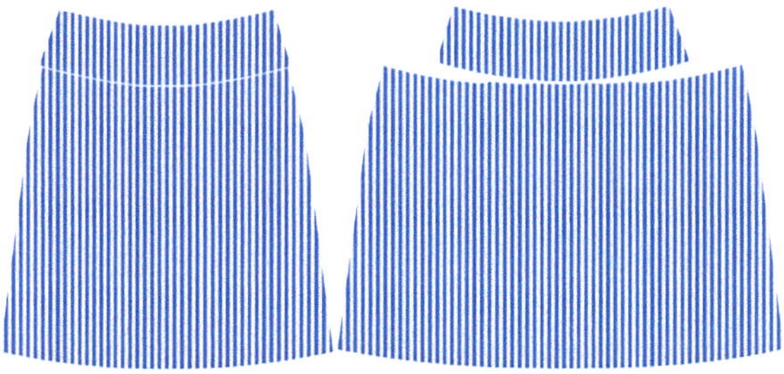

Figure 2.6 Marker prepared for print.

of the cut pieces and compare it with the original marker. The cut piece will then be cut at the correct position.

Other systems such as those used by Bullmer or Zünd employ a scanner or camera to recognize edge contours or special notches as cutting lines. This connection between printing and the cutting process is a remarkable step forward for small and middle enterprises, especially in Europe, because it enables the production of small collections involving high quality and individual designs (Fig. 2.6).

2.4 Basic pattern

2.4.1 System pattern construction

In contrast to the way in which clothing fits, the system pattern is based on mathematical principles. Anthropometric values of the body are calculated from a few measured values. Geometrical connections among lengths, widths, and circumferences are systematized. The construction process is precisely defined. There are countless systems developed by various institutes and authors.

The decision regarding which system should be used is made individually. It is not possible to say which system will work well. The general advantage is that it is applicable regardless of the producer's origin. The use of a systematic pattern makes it possible to model and trace individual patterns from a basic pattern. The basic design is designed once and stored on paper or in a CAD system. In all basic patterns can be applied producer-related added values. Thus each manufacturer develops its own fit form; also, the clothing sizes of various manufacturers can differ.

The use of a systematic pattern has made it possible for the company GRAFIS to develop a CAD system in which the basic cuts are preprogrammed. With the aid of tools, it is possible to adapt individual dimensions to the customer's dimension tables; other measures that represent a mathematical relationship are automatically adapted. The lengths of seams are controlled; thus, errors in patterns are avoided. In other software it is necessary to draw the basic pattern and trace the cut pieces of the model. All software has advantages but none is usable for all demands of current producers (Fig. 2.7).

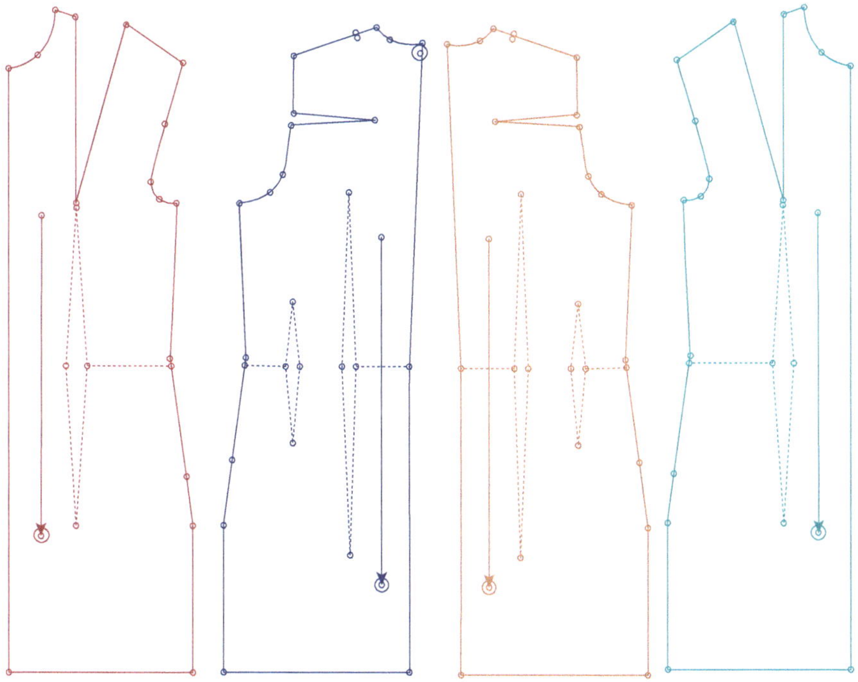

Figure 2.7 Basic pattern for dress as a system pattern.

2.4.2 Basic pattern as a modification of three-dimensional forms

The next method is not usually employed. It is an alternative method suitable for customized pattern making that creates a 3D avatar by changing predefined parameters on an existing shape (Yunchu and Weiyuan, 2007). Usual 3D software is able to roll up the surface to transfer a 3D form into a 2D flat surface with the measurements of a 3D form (Luo and Yuen, 2005). This method can be used when a scan of the body or its avatar exists. The seams of the basic pattern are drawn on the surface as virtual seams. Areas between the seams are then rolled up. The surface of the 3D form has to be reworked so that it is possible to use it as a basic pattern. This method can be implemented using different types of 3D software, e.g., Lectra (Fig. 2.8).

Currently there is no available way to obtain a body scan at a low cost (Bye and McKinney, 2010). With the quickly developing technology, smartphones and tablets might not only take great photos, but in the near future 3D scans may create avatars. Therefore, as an iterative step, different software solutions will create an avatar using only limited and easily measured data for a person, such as age, height, and the circumferences of the breast, waist, and hip. Remaining and not easily measured information will be assumed on the basis of statistical data. These will be evaluated by using the full measurements of thousands of individuals.

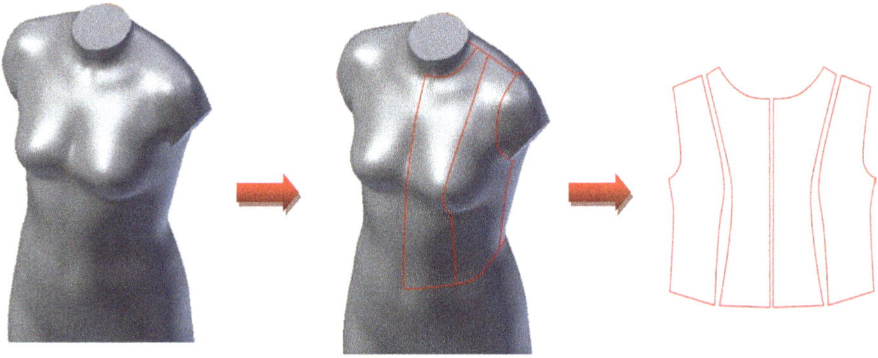

Figure 2.8 Two-dimensional form modeling: avatar-cut lines, basic pattern.

Data from the avatar can be used in online shops or stores, and clothing may be chosen or recommended according to the customer's taste and size. If problems remain with different sizes from different producers, more information about the exact dimensions of their products will have to be made available.

2.4.3 Basic pattern as drapery

Through the drapery method, cut templates are developed directly on a dressmaker dummy or model. Drapery is a kind of art. It is used for individual models as well as complicated cut shapes. A fabric web is built; and the darts, folds, or rafts are pinned together, marked on the fabric, designed as a 2D template, and marked as a paper template (Wang et al., 2003). In this area the experience and skill of the cut maker are extremely important. It is also difficult to find some kind of automation in this filed. This is a handmade method. The computer is used exclusively to rework and grade cut templates.

The form of clothing, also called the silhouette, is made from woven cotton or other fabric. The fabric is copied onto a paper sheet as a cut template. This cut template is made in one size, equal to the dressmaker dummy. In the area of custom tailoring it is not necessary to rework the cut pattern in any other form. The cut template will be used only once for one single person.

Of course, there are companies that use drapery to develop cuts for whole collections. There it is necessary to drape the clothing onto dressmaker dummies of specific sizes. After draping, the fabric is copied onto the paper. The draped cut templates are digitized and reworked into CAD software. The seam allowances, notches, grain lines, and other cut information are then added to the cut template. Grading, which means enlarging or reducing different sizes, is also carried out in CAD (Fig. 2.9).

This method may seem old-fashioned or used only for custom tailoring. However, it is still the method of choice for developing complicated pattern forms and presumably will never be completely eliminated.

Figure 2.9 Drapery: handwork on dressmaker dummy.

2.5 Cutting and printing systems

Pattern that are developed or digitized in a CAD system are simultaneously prepared in a suitable format for cutout by an automatic cutter (Vilumsone-Nemes, 2015). Lines, notches, drill holes, and other marks are detected owing to the cutter software. Cutting through an automatic cutter is not only fast but also extremely precise. Further features can also be introduced automatically, e.g., printed labels.

Automatic cutting systems connected to CAD software enormously optimize the use of materials (Vilumsone-Nemes, 2012). In a short time it is possible to compare the use of materials for types of markers: one or more size markers as separated or combined sizes. Information about results is integrated into PLM and production costs for materials can then be calculated (Fig. 2.10).

Another advantage, which is not negligible, is that the operation of a system can be carried out by one person. For cutting, fabric is placed on a worktable. The laying machine is part of the cutting table; optionally it is selected by the cutter manufacturer. The laying system facilitates manipulating fabric rolls and the placement of individual fabrics into position is much more precise than laying them by hand. Edges lie exactly one against the other, and thus waste is substantially reduced.

In the clothing industry, different types of cutting systems are used. The most commonly employed cutting system is the oscillating or round knife. The advantages of the oscillating knife are that it is possible to cut several layers at once and the precut is small, so there is no size difference between the top and bottom layers. The cutting speed depends on the material cut; however, it is generally faster to cut with a round knife.

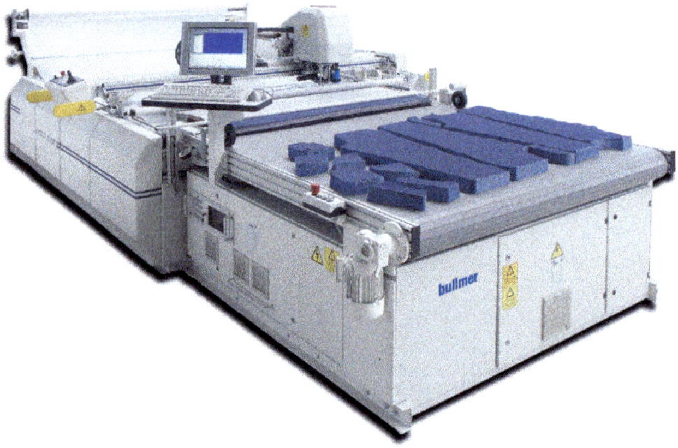

Figure 2.10 Automatic cutter PROCUT from Bullmer.

The automatic cutting method generally used in the apparel industry is laser cutting; less commonly, ultrasonic cutting is used. Laser cutting is extremely precise and fast (Nayak et al., 2008; Nayak and Padhye, 2016; Vilumsone-Nemes, 2012). It is possible to cut almost all kinds of material, but it is necessary to think about combustion gases when cutting materials with a fluorine or chlorine content. When cutting several layers of synthetic materials, the layers can become connected at the edges owing to the melted mass.

Hand cutter devices are still used. Beside the classical scissor band saw, jigsaws and round knifes are often used (Nayak and Padhye, 2015). Cutting by hand is not controlled by software, and the person who uses the saw or knife cuts according to a pattern that is usually printed on paper. The decision to use this slow kind of cutting depends on the finances of the company but also the amount of production and the diversity of the collection. Within the context of Industry 4.0, the interaction between the pattern marker and the handling of the cut pieces is of primary importance.

The process of cutting is closely connected to printing. Modern digital printers are able to print directly onto fabrics or as impressions on paper so that the print can be transferred onto fabric. The biggest challenge is to keep the fabric flat in position without warping as contractions, skews, or bows (Figs. 2.11 and 2.12).

The technologies of printing and cutting both aim for the same goal: the fabric is printed and pattern pieces need to be cut. Because of the elastic properties of the fabric, it is likely that the position of the marker will change. Therefore, the position of the pattern pieces has to be detected automatically to compensate for the position of the new cutting contour, e.g., by Gemini VisionCUT. Such technology is a big step in the processing of individual apparel or small collections, because pattern pieces can be cut without losing of their original shape (Figs. 2.13 and 2.14).

Furthermore, pattern pieces have to be identified after cutting. Nearly all cutter machines include a labeling system that mark pattern pieces with words or codes used by the company and its quality management system. The cutting bridge outfits the equipment for labeling and ultrasonic sealing, an inkjet printer, and other optional tools (Fig. 2.15).

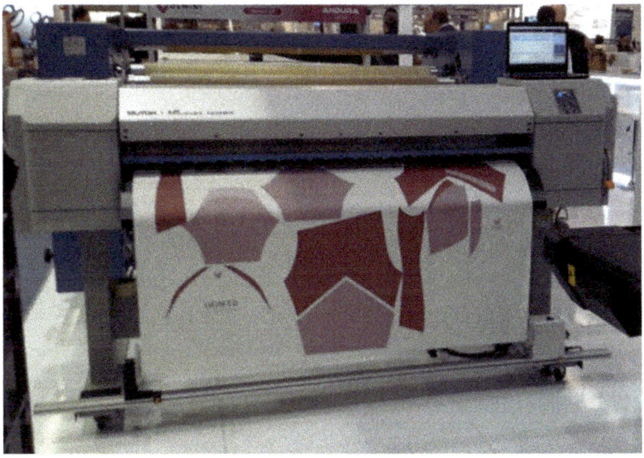

Figure 2.11 Digital print on paper.

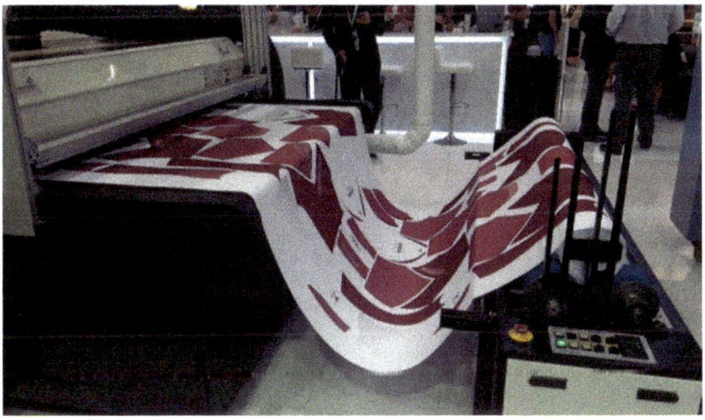

Figure 2.12 Transfer print on fabric.

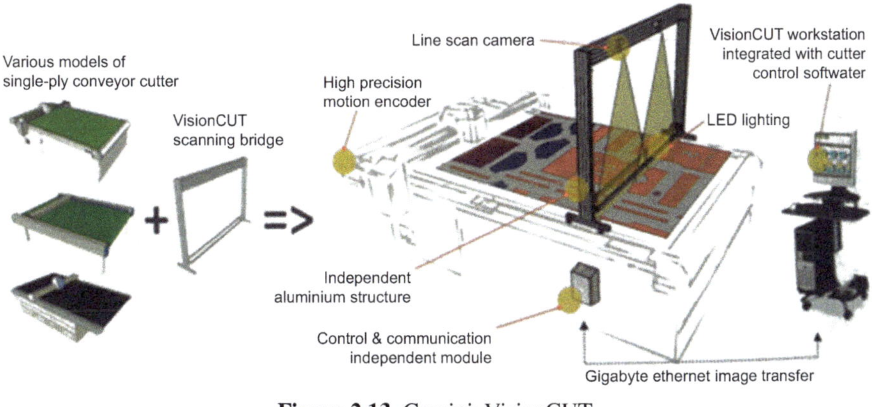

Figure 2.13 Gemini: VisionCUT.

Figure 2.14 Perfectly cut pattern pieces.

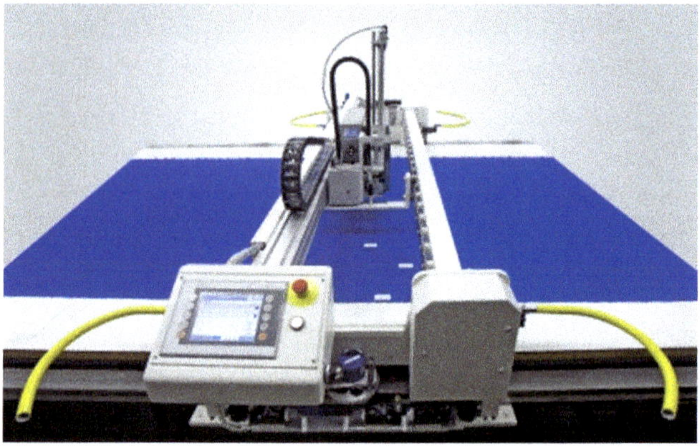

Figure 2.15 Bullmer: labeling device.

Between cutting and joining, it is still necessary to use physical labor because no technical device exists that transports cut pieces safely to the sewing or joining station. The first reason is that the fabric is usually flexible; the second (Ono et al., 2001) is that there is a chance of mixing of different sizes.

Standard grippers to hold and transfer the materials work on the basis of a vacuum or needles and therefore are not suitable for clothing fabric (Koustoumpardis et al., 2006). In the area of technical textiles, it is far more easy because the fabrics are less sensitive. Because of this manipulation, using gripper has become more of a standard (Fig. 2.16).

2.6 Joining systems

Complete automation of the joining process in clothing technology will never be possible (Tyler et al., 2012; Shi and Little, 2000). The human hand still needs to control the sewing process, which directs, positions, and composes substances. Despite this,

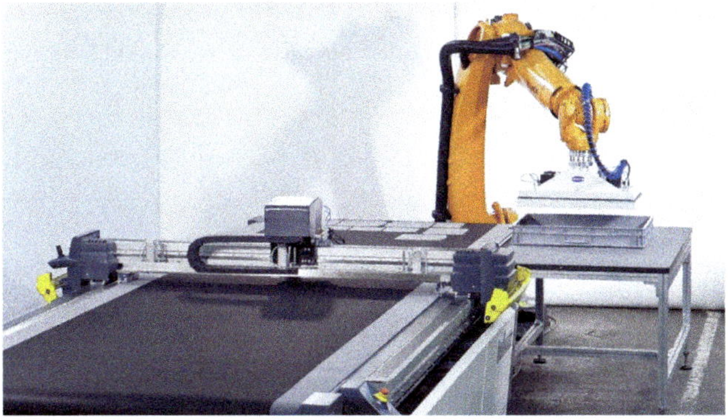

Figure 2.16 Bullmer and KUKA robots with vacuum gripper manipulate pattern pieces.

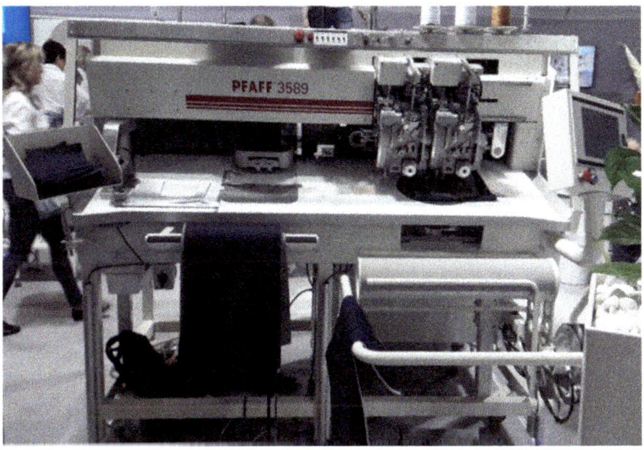

Figure 2.17 PFAFF industrial automatic sewing system.

more and more individual parts of production are being automated. For example, items such as neck openings for polo shirts, piped pockets, attached pockets, and zippers are manufactured using computerized numerical control–programmed sewing machines. Juki, PFAFF Industrial, and Dürkopp Adler developed sewing machines and suitable accessories for numerous sewing processes. Sewing with automatic sewing machines delivers highly professional results in the shortest time that a person would never be able to achieve (Fig. 2.17).

Many experts say that sewing processes will be increasingly robotized, but complete automation will not be possible in the near future. Automation levels of more than 95% as in the automotive industry cannot be reached. Clothing styles change too quickly and there are nearly countless cuts of forms. With the development of robots, it is within reason, however, that more and more individual processes of the production will become automated. In this way, smaller companies will be able to keep the prices of the outfits at an affordable level and still offer individualized clothing.

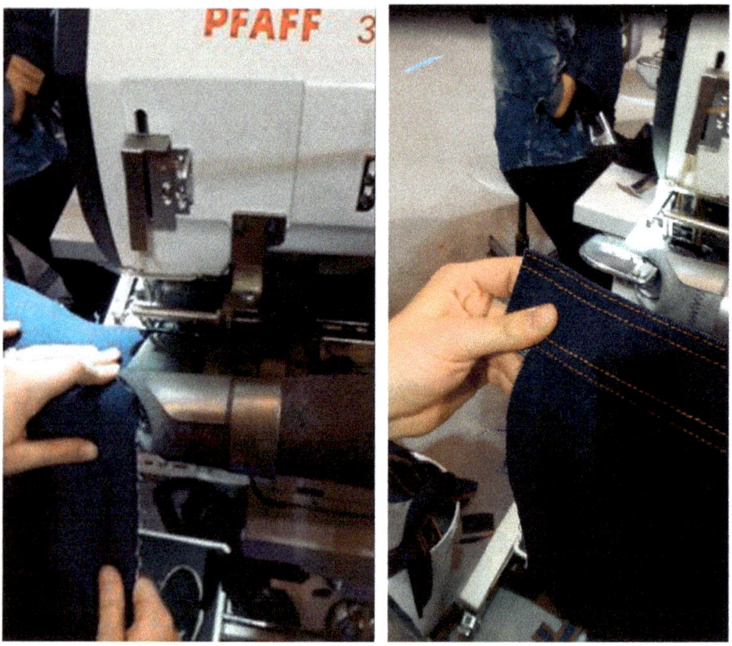

Figure 2.18 PFAFF industrial automatic sewing machine for waist belt.

Despite difficult conditions because of their soft material properties, the automated processing of fabrics is the focus of several ongoing projects. For example, Adidas successfully developed the fully automated production of shoes in the so-called "Speedfactory." The shoe sole is printed using a 3D printer and the shaft is knitted as one piece. The connection of the sole and the shaft is made by a robot that manipulates all necessary components. The customer can design his or her own pattern and obtain a shoe in the exact size.

Especially for the production of jeans, automation is already being used. Pockets, zippers, side seams, and waist belts can be sewn automatically (Caputo and Palumbo, 2005). This automation level significantly reduces time and is suitable for large production. When the collection changes the form of patterns, it is possible to set up the pocket form, the width of the waist belt, or the length of stitches in the automatic sewing system (Fig. 2.18).

2.7 Fitting systems

After production, the products are ironed, folded, and packaged. Various aids such as finishing equipment and pressing machines are used to automate the process. Here, humans take over the task of the machine. The garments are brought to the machines and installed on the figurines, and the system is activated (Fig. 2.19).

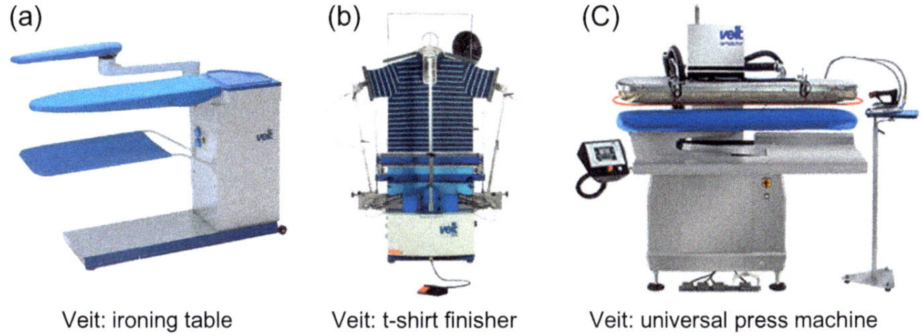

Figure 2.19 VEIT finishing devices: (a) ironing table, (b) t-shirt finisher, and (c) universal press machine.

2.8 Conclusion

In the current age of increasing product customization in all areas of life, this will become the standard, which is why individually designed, modeled, and simulated products has to be automated even more. In the course of Industry 4.0, this is intended to enable the networking of products and production to the extent that without additional outside manual intervention, customers will be able to create their own product design using simple tools and their own measurements, to then produce materials. However, the data required for production require comprehensive modeling and simulation of products and processes to demonstrate the necessary flexibility.

For the clothing industry, the greatest potential is in the networking of digital systems from cutting development to automatic sewing machines to achieve the highest degree of flexibility and individualization of products along with reasonable costs. The particular advantage of the digitized world is that modeling and simulation or production do not need to take place in one location, because the transmission of data for the entire production chain is possible from anywhere in the world. This means that the design can be carried out directly with the customer selecting the fabric at a different location. The cuts are automatically done from the correct fabric and an automatic sewing machine makes up all of the elements that are planned for the design, such as bags, zippers, and quilting seams. Completion of the product remains in the hand of the seamstress, who may be in another place.

By modeling and simulating it will be avoided to buy wrong size or bad fit. The customer is scanned and his or her clothing size is precisely adapted or modeled and simulated without further detail. In doing so, new possibilities open up for online shops to reduce the rate of complaints or to sell tailor-made products directly.

Modeling and simulation create new prospects for such tailor-made products. Customers who are willing to wait for their own clothing or design for a short time are rewarded by a perfect fit and individual look through their own composition of fabrics. The entirety of all of the different steps and the automation of production enable digital designs to be realized within a short time at a reasonable price. However, the networking of modeling, simulation, and production has to be done from a single source, or future standards have to be established that will allow the specialization of individual aspects of the overall process.

One of the biggest challenges of changes resulting from digital production is in training employees. In the future, not only will technical skills be required, the ability to use simulation software and machines will be critical. For this purpose, the training center must adapt the content of its training to the new requirements of a networked automation system. New courses of study must be developed that are directly connected to the industry.

Furthermore, sociocultural questions arise as to the influence that networking will have on our environment and culture. There is already an overproduction of clothing and thus an enormous amount of old clothing that is not always properly disposed of or is blamed in developing countries for negative effects on local industry. On the other hand, such countries still benefit from the high demand for human labor, which will not be needed as much in this form.

The short life span of today's clothing is mainly attributable to low prices. According to a study by Greenpeace, 19% of clothing that is bought is never even worn. Another reason may be that the fit is not correct. By simulating the clothing and modeling its dimensions, in the future garments might correspond better to the fit of the wearer, and thus the piece may actually be worn. This shows an example in which changes in networking in clothing production can have positive as well as negative aspects.

Sources of further information

- www.grafis.de; Retrieved 09/2016
- www.assyst.de; Retrieved 09/2016
- www.optitex.com; Retrieved 09/2016
- www.hugoboss.com; Retrieved 09/2016
- www.optitex.com; Retrieved 05/2017

References

Ashdown, S.P., Loker, S., Schoenfelder, K., Lyman-Clarke, L., 2004. Using 3D scans for fit analysis. Journal of Textile and Apparel, Technology and Management 4, 1–12.

Au, C.K., Ma, Y.-S., 2010. Garment pattern definition, development and application with associative feature approach. Computers in Industry 61, 524–531.

Boer, H., Drejer, A., Buxey, G., 2005. Globalisation and manufacturing strategy in the TCF industry. International Journal of Operations and Production Management 25, 100–113.

Bye, E., Mckinney, E., 2010. Fit analysis using live and 3D scan models. International Journal of Clothing Science and Technology 22, 88–100.

Caputo, A.C., Palumbo, M., 2005. Manufacturing re-insourcing in the textile industry: a case study. Industrial Management and Data Systems 105, 193–207.

Dewan, S., Kraemer, K.L., 2000. Information technology and productivity: evidence from country-level data. Management Science 46, 548–562.
Eason, K.D., 2005. Information Technology and Organisational Change. CRC Press.
Eberle, H., et al., 2014. Clothing Technology. Europa-Lehrmittel.
Echelmeyer, W., Kirchheim, A., Wellbrock, E., 2008. Robotics-logistics: challenges for automation of logistic processes. In: ICAL 2008. IEEE International Conference on Automation and Logistics, 2008. IEEE, pp. 2099–2103.
Fontaine, A., 2011. Technologie für Bekleidungsberufe. Bildungsverlag EINS.
Forza, C., Vinelli, A., 2000. Time compression in production and distribution within the textile-apparel chain. Integrated Manufacturing Systems 11, 138–146.
Hu, Z.-H., Ding, Y.-S., Zhang, W.-B., Yan, Q., 2008. An interactive co-evolutionary CAD system for garment pattern design. Computer-Aided Design 40, 1094–1104.
Jo Anderson-Connell, L., Ulrich, P.V., Brannon, E.L., 2002. A consumer-driven model for mass customization in the apparel market. Journal of Fashion Marketing and Management: An International Journal 6, 240–258.
Joshi, R., Singh, S., 2010. Estimation of total factor productivity in the Indian garment industry. Journal of Fashion Marketing and Management: An International Journal 14, 145–160.
Kim, S., Park, C.K., 2004. Parametric body model generation for garment drape simulation. Fibers and Polymers 5, 12–18.
Kim, S.M., Kang, T.J., 2003. Garment pattern generation from body scan data. Computer-Aided Design 35, 611–618.
Koustoumpardis, P., Zacharia, P., Aspragathos, N., 2006. Intelligent robotic handling of fabrics towards sewing. In: Industrial Robotics: Programming, Simulation and Applications. InTech.
Luo, Z.G., Yuen, M.M.-F., 2005. Reactive 2D/3D garment pattern design modification. Computer-Aided Design 37, 623–630.
Mccartney, J., Hinds, B., Seow, B., Gong, D., 2000. Dedicated 3D CAD for garment modelling. Journal of Materials Processing Technology 107, 31–36.
Metaaphanon, N., Kanongchaiyos, P., 2005. Real-time cloth simulation for garment CAD. In: Proceedings of the 3rd International Conference on Computer Graphics and Interactive Techniques in Australasia and South East Asia. ACM, pp. 83–89.
Mori, Y., Igarashi, T., 2007. Plushie: an interactive design system for plush toys. ACM Transactions on Graphics (TOG) 45 ACM.
Nayak, R., Gon, D.P., Khandual, A., 2008. Application of LASER in apparel industry. Man-Made Textiles in India 51, 341–346.
Nayak, R., Padhye, R., 2011. Application of modelling and simulation in smart and technical textiles. In: Patanaik, A. (Ed.), Modeling and Simulation in Fibrous Materials: Techniques and Applications. Nova Science.
Nayak, R., Padhye, R., 2014. Introduction: the apparel industry. In: Nayak, R., Padhye, R. (Eds.), Garment Manufacturing Technology. Elsevier.
Nayak, R., Padhye, R., 2015. Garment Manufacturing Technology. Elsevier.
Nayak, R., Padhye, R., 2016. The use of laser in garment manufacturing: an overview. Fashion and Textiles 3, 1–16.
Nayak, R., Padhye, R., Wang, L., Chatterjee, K., Gupta, S., 2015. The role of mass customisation in the apparel industry. International Journal of Fashion Design, Technology and Education 1–11.
Ono, E., Kitagaki, K., Kakikura, M., 2001. Picking up a piece of fabric from layers by a hand with 3 fingers and a palm. In: Proceedings. 2001 IEEE/RSJ International Conference on Intelligent Robots and Systems, 2001. IEEE, pp. 931–936.

Roth, A., 2016. Einführung und Umsetzung von Industrie 4.0. Springerverlag.

Senanayake, M., Nayak, R., Padhye, R., 2015. Product development in the apparel industry. Garment Manufacturing Technology 21–57.

Shi, W., Little, T., 2000. Mechanisms of ultrasonic joining of textile materials. International Journal of Clothing Science and Technology 12, 331–350.

Tyler, D., Mitchell, A., Gill, S., Shishoo, R., 2012. Recent Advances in Garment Manufacturing Technology: Joining Techniques, 3D Body Scanning and Garment Design. Woodhead, Cambridge.

Vilumsone-Nemes, I., 2012. Industrial Cutting of Textile Materials. Elsevier.

Vilumsone-Nemes, I., 2015. Fabric spreading and cutting. In: Garment Manufacturing Technology.

Volino, P., Cordier, F., Magnenat-Thalmann, N., 2005. From early virtual garment simulation to interactive fashion design. Computer-Aided Design 37, 593–608.

Wang, C.C., Wang, Y., Yuen, M.M., 2003. Feature based 3D garment design through 2D sketches. Computer-Aided Design 35, 659–672.

Wang, C.C., Wang, Y., Yuen, M.M., 2005. Design automation for customized apparel products. Computer-Aided Design 37, 675–691.

Yunchu, Y., Weiyuan, Z., 2007. Prototype garment pattern flattening based on individual 3D virtual dummy. International Journal of Clothing Science and Technology 19, 334–348.

Automation in production of yarns, woven, and knitted fabrics*

3

Marco Saggiomo, Marko Wischnowski, Kristina Simonis, Thomas Gries
Institut für Textiltechnik der RWTH Aachen University, Aachen, Germany

3.1 Introduction

In the global competitive textile and clothing market, the cost and quality are playing major roles in the survival of the industries (Nayak and Padhye, 2015b). This in turn is influenced by the adoption of the automated tools and equipment by the industries. The industries that are unable to adopt these technologies are thriving to survive. On the other hand, the industries that adopt automation are gaining competitive advantage. In a global market where the cost of labor is increasing, automation can help to keep the labor cost low (Nayak and Padhye, 2015a).

Spinning industries provide yarn to the weaving industries, which provide fabric to the garment industries. The quality of yarn and fabrics greatly influences the quality of the garments. By the incorporation of automation from the preliminary stages can result in the production of good-quality raw materials for the subsequent stages. The automation in spinning mainly focuses on auto-doffing, automatic material handling, auto fault detection, and automatic bobbin replacement. Similarly weaving automation includes the automatic fault detection and mending, automatic splicing, automatic temperature and moisture control in sizing, automatic drawing, and automatic pick repair to name a few.

This chapter deals with the automation in spinning and weaving. Automation in spinning includes the developments by various manufacturers of spinning machineries. A section has been included to discuss on the automation in the sewing thread production. Separate sections describe the automation in weaving and knitting.

3.2 Types of industries

In several industries the spinning and weaving operations are performed in tandem. The yarn prepared from the spinning is directly used as a raw material in weaving. These industries can better monitor the quality of the products (yarn and fabric). These types of industries are shown in Fig. 3.1. It shows the combination of spinning and weaving process in one firm. In this case the automation of spinning and also the quality of spinning processes affect the effectiveness of weaving, which is easier to control.

*Edited by Dr. R. Nayak and Dr. R. Padhye, RMIT University, Australia.

Automation in Garment Manufacturing. https://doi.org/10.1016/B978-0-08-101211-6.00003-3

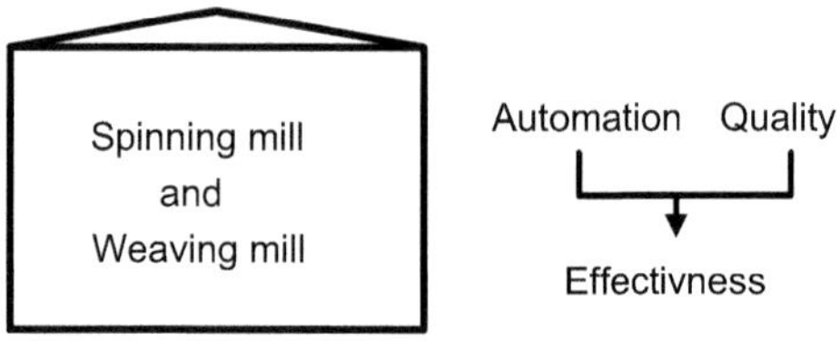

Figure 3.1 Spinning and weaving in one firm.

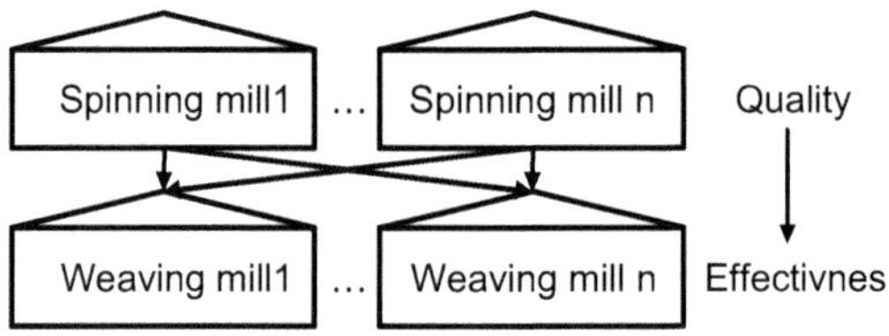

Figure 3.2 Several spinners and weavers.

In Fig. 3.2, the industries producing yarn and fabric as independent organization have been explained. It is the case of decentralized situation, where multiple organizations are involved. In this case there are several spinners and weavers. The weaver buys the yarn in one quality from different spinners. Therefore, the yarn quality is the determining parameter, which may be hard to control. There can be wide variability of the yarn quality received from various industries. The automation in spinning can help to improve the yarn quality and reduce the variability of the yarns. Improper yarn quality can result in yarn breakages during weaving and reduce the weaving efficiency and quality of the fabrics.

As it shown above for isolated spinning and weaving mills and for fully integrated companies the automation in spinning is crucial to ensure sufficient availability of yarn in a certain quality. The following section provides an overview of the automation and quality-control systems of renowned machine manufacturers for spinning systems.

3.3 Global spinning machine manufacturers

The most relevant spinning systems are ring spinning (c.65% market share), followed by rotor spinning (c.30% market share), and air-jet spinning (c.2%–5% market share) (Gherzi). Exotic spinning principles such as friction spinning are not taken into account here as the utility is not very high. Suppliers of the above indicated spinning technologies include

- Saurer Schlafhorst GmbH & CO. KG, Übach Palenberg, Germany: ring-, rotor spinning, and winding.
- Rieter AG. Winterthur, Switzerland: ring-, rotor-, and air-jet spinning.
- Murata Machinery, Ltd., Fushimi-ku, Kyoto, Japan: air-jet spinning and winding.
- Savio Macchine Tessili S.p.A., Pordenone, Italy: rotor spinning, twisting, and winding.
- Lakshmi Machine Works Limited, Coimbatore, India: ring spinning and winding.

All premium models of ring spinning machines provide automatic doffing systems (http://schlafhorst.saurer.com/de/zinser-351/; http://www.rieter.com/en/machines-systems/products/; https://www.lakshmimach.com/). During the mixing stage in blow room, fibers from several bales are drawn and mixed automatically, which results a homogeneous blending. Chute feed systems in carding and auto-levelers help in achieving improved productivity and quality, respectively. Automatic doffing and empty package replacement in ring spinning has led to increased machine efficiency. Automatic splicing and yarn fault detection in autoconers have helped to achieve yarns with higher uniformity and improved quality. The following section describes various automation by different spinning machine manufacturers.

3.3.1 Saurer Schlafhorst GmbH & CO. KG, Übach Palenberg, Germany

Saurer Schlafhorst GmbH & CO. KG, Übach Palenberg, Germany produces machinery for ring- and rotor spinning and winding. The "Zinser 351" is the premium model for ring spinning at Saurer AG, Wattwil, Switzerland. The Zinser 351 model promises increased productivity and can be individually tailored to a variety of production requirement through the installation of additional modules and automation components. The most important spinning parameter and machine component, e.g., spindle drive, FancyDraft, and CoWeMat, is centrally controlled. Central settings minimize adjustment errors and eliminate inaccuracies (http://schlafhorst.saurer.com/de/zinser-351/; http://schlafhorst.saurer.com/de/produkte/).

The winding machine of Saurer AG, Wattwil, Switzerland is called "Autoconer 6." The highlights of this new development are

- Self-adjusting winding unit with Eco-Drum-Drive System and SmartCycle,
- Power on demand with intelligent vacuum control,
- Intelligent doffer functions,
- New feed principle and up to 80 winding units on the type RM machine,
- State-of-the-art automation with VarioReserve, Intelligent Bobbin Sharing, and High-Speed Feeding,
- Intelligence inside for maximum reliability,
- Schlafhorst quality packages with standard tension control and optimized splicing technology (http://schlafhorst.saurer.com/de/zinser-351/; http://schlafhorst.saurer.com/de/produkte/).

The rotor spinning machine of Saurer AG, Wattwil, Switzerland is called "Autocoro 9". The highlights of this new development are

- Reduce energy costs by up to 25%,
- Maximize efficiency with the unique individual spinning position technology,
- High rotor speeds of up to 180,000 rotations per minute (RPM) and take-off speeds of up to 300 m/min—on all lengths of machine and also for very large packages up to 350 mm in diameter,
- Individually customizable machine length, perfectly automated spinning positions with up to 6 doffers and SynchroPiecing 24,
- Maximum productivity/m^2, which leads to optimal space utilization,

- Reduced maintenance time and personnel requirements,
- Maximum economic efficiency in all markets (http://schlafhorst.saurer.com/de/zinser-351/; http://schlafhorst.saurer.com/de/produkte/).

3.3.2 Rieter AG. Winterthur, Switzerland

Rieter, Winterthur, Switzerland produces machinery for ring-, rotor- and air-jet spinning. The main ring spinning machine of Rieter, Winterthur, Switzerland is called "G 36 ring spinning machine." It contains an electronically controlled drafting arrangement drive Flexidraft, which permits highest flexibility. Customers profit from a high-capacity, dependable, and high-quality machine with multiple technical spinning solutions. Quality is ensured by SERVOgrip, the Rieter innovation for real, under winding-free doffing, eliminates thread under winding, distinctly reduces fiber fly and thereby ensures dependable and rapid doffing. Besides, the proven Ri-Q-Draft drafting arrangement with pneumatically loaded guide arm and the Ri-Q-Bridge for superior spinning conditions form a solid basis for high and consistent yarn quality. Finally, the direct motor-driven intermediate drive ensures consistent quality with long machines (http://www.rieter.com/en/machines-systems/products/).

The fully automatic rotor spinning machine of Rieter, Winterthur, Switzerland is called "R 66." It sets the standard for quality and productivity. Quality is ensured by improved breaking tenacity and yarn consistency with the advanced spin box S 66 with individual centering and easily replaceable spinning elements. Additionally, good properties in further processing through uniform yarn quality are reached. Yarnlike piecing technology "AEROpiecing" and rotor cleaning "VARIOclean" on each piecing cycle helps to achieve better yarn quality. Furthermore, better yarn quality is ensured through cooler nozzle surface with COOLnozzle technology (prevents damage to sensitive fibers) (http://www.rieter.com/en/machines-systems/products/).

The fully automated air-jet spinning machine of Rieter, Winterthur, Switzerland is called "J 26." Quality is ensured by the unique Com4jet yarn structure that provides benefits in downstream processing and in the fabric. Technological innovations enable fabric softness to be adjusted. The well-established Rieter Q 10A yarn clearer is integrated in the J 26 machine. All standard clearer functions and special monitoring channels for polyester spinning are available. With the new edge displacement, packages have homogenous density, heavier weight, and soft edges for best performance in downstream processing. The automated piecing system utilizes the proven principle of progressive fiber feeding that helps to create strong, yarnlike piecings, which are undetectable in fabrics (http://www.rieter.com/en/machines-systems/products/).

3.3.3 Murata Machinery, Ltd., Fushimi-ku, Kyoto, Japan

Murata Machinery, Ltd., Fushimi-ku, Kyoto, Japan produces machines for winding and air spinning. Their air-spinning machine is called VORTEX. The first generation of VORTEX spinning machine contains an automatic yarn piecer for continuous spinning. The second generation of VORTEX implies self-spinning technologies and splicers were added. These lead to more stable and improved joint for all types of yarn. The third generation of VORTEX will answer the expectations in the three concepts of value, vantage, and versatility (http://www.muratec.net/tm/index.html).

Another product of Murata Machinery, Ltd., Fushimi-ku, Kyoto, Japan is the Spinning Tension Stabilizing System (STS). This STS System is the key technology for VORTEX. This mechanism draws the yarn from the spinning nozzle where a friction roller is adopted instead of a nip roller. This system and newly developed spinning sensor realize stable yarn structures as well as consistent and reliable yarn qualities for the third generation of VORTEX (http://www.muratec.net/tm/index.html).

Furthermore, Murata Machinery, Ltd., Fushimi-ku, Kyoto, Japan produces the automatic winder, called PROCESS CONER II QPRO Plus. This automatic winder ensures high quality through auto setting. This keeps uniformity of tension, consistency of package density, and optimized positioning of bobbin (http://www.muratec.net/tm/index.html).

3.3.4 Savio Macchine Tessili S.p.A., Pordenone, Italy

Savio Macchine Tessili S.p.A., Pordenone, Italy produces rotor spinning frames. Their rotor spinning frame is called "FlexirotorS 3000." For Savio Macchine Tessili S.p.A., Pordenone, Italy production flexibility is the main feature they want to offer their clients. Therefore, FlexirotorS 3000 is designed to meet the most different needs of flexibility in open end processing line.

The use of a Suessen spinning unit, together with two separate fronts, associated with a completely electronic machine, offer the customers the ideal solution for production control. Furthermore, FlexirotorS 3000 guarantees the highest speed and take-up performances. It ensures best production planning, high quality yarn and packages, and it minimizes unproductive time (https://www.saviotechnologies.com/savio/en/Products/Pages/default.aspx).

3.3.5 Lakshmi Machine Works Limited, Coimbatore, India

The Lakshmi Machine Works Limited, Coimbatore, India produces the "LR60" line for ring spinning frames. The variants in "LR60" Ring frames are

- LR60/A: spinning frame without doffer,
- LRJ 60/A: spinning frame without doffer and with suction compact,
- LR60/AX: spinning frame with doffer, and
- LR60/AXL: spinning frame with link coner.

All LR60 ring frames contain an inverter drive. This inverter drive controls the speed through the inverter system and leads to less speed variation during the spinning process. The optional features of ring frame "LR60" series are auto doffer, link coner, and compact spinning. The series is an automated and user-friendly machine. The auto doffer leads to rationalization of labor. With it the doff time is less than 3 min. Because of the gentle removal of cops, the life expectancy of spindles is increased (https://www.lakshmimach.com/).

Another product of Lakshmi Machine Works Limited, Coimbatore, India is the Lakshmi Winder "LW72." This winder implies the following features concerning automation and quality assurance of the process:

- Digital signal control technology,
- Square type unwinding accelerator with variable speed control system,
- Variable timing antikink system,
- Electronic antipatterning,
- Dual drive cycle shaft to increase splicing cycle,
- Twin-disc type pneumatic yarn tensioner, and
- Inverter controlled suction motor (https://www.lakshmimach.com/).

3.4 Automation in production of sewing threads

Sewing is still one of the most important joining technologies used to manufacture garments. For sewing the sewing thread is crucial. The sewing thread determines the joining process and the seam quality. Parameters of sewing threads influencing the joining process are (Gries and Klopp, 2007): (1) bending stiffness, (2) tensile strength, (3) shear resistance, and (4) friction. The listed parameters of the sewing thread can be determined by: (1) material and material blends, (2) yarn spinning process (Basu, 2009), and (3) yarn postprocessing (twisting and finishing). Usual materials for sewing threads include (Amann & Söhne: Amann product range, 2016):

- Polyester,
- Cotton,
- Viscose,
- Polyamide.

Materials used for sewing threads can be of pure material or blended material. Blending is possible in fiber stage or in the thread stage. An example for thread blending is the twisting process. Man-made fibers such as polyester and polyamide (nylon) are produced continuously in infinite (very long) filament form for sewing threads. Natural fibers are of finite length, such as cotton. Hence if a fiber blending is needed, the filament yarns need to be cut. After cutting most man-made fibers are crimped to improve the fiber to fiber cohesion. The fibers are blended in the blow room or while drawing in short-staple fiber processing in a draw-frame before yarn spinning. Filaments and spun yarns can be blended in additional process steps such as twisting, doubling, and core spinning (Lawrence, 2003).

Sewing threads of natural fibers such as cotton demand further finishing steps after spinning. A usual finishing process for cotton sewing threads is the Mercerization. Mercerization increases the strength, dyeing, and the brilliance of cotton yarn. The yarns are processed with sodium hydroxide (NaOH) solution and the molecular chains get a parallel orientation (Haller, 1928; Amann & Söhne: Amann product range, 2016). Furthermore, it is possible to reduce the hairiness of a sewing thread by singeing. The singeing is a textile finishing process, which burns protruding fibers of a yarn with a gas flame (Kiessling and Matthes, 1993).

The automation in sewing thread manufacturing is similar to the automation in spinning. The yarns prepared by ring spinning are mainly used for sewing thread production because of the highest strength of these yarns. Hence, the automation process of sewing threads is the same as the yarn production as discussed in the previous section. The sewing threads are doubled yarns prepared by winding and doubling machines. The automation in these processes include automatic splicing, automatic bobbin change, fault detection, and auto-doffing to name a few. The automation in sewing thread production can help to produce threads free from faults that can lead to increased efficiency in garment production.

3.5 Automation in production of woven fabrics

Weaving is the most widely used method for fabric production. Weaving industries have also adopted several automation technologies that include automatic shuttle and shuttleless looms (rapier, air-jet, projectile, etc.). The use of computer technology has helped in the automatic fault detection, design, and artificial intelligence. Various steps involved in woven fabric production such as warping, sizing, and weaving have seen several automation technologies. These automation technologies can be related to the machine operations or material handling.

The automation in warping includes automatic section positioning, pneumatic stop brakes, automatic warp beam loading, doffing and chucking, sensors for machine stop, and automatic creel movement. Automation in sizing includes automatic yarn tension control, automatic humidity and temperature control during sizing, and automatic machine stop in case of fault detection. Automatic detection of warp breakage, automatic pick repair, automatic color selection, and automatic fabric fault detection are some of the automation in weaving. The warp beam change and mending of warp threads is still done by manual operations.

Real (physical) objects melt together with information-processing (virtual) objects to create cyber-physical production systems (CPPS). Through embedding of intelligent, self-optimizing CPPS in process chains, productivity of manufacturing companies and quality of goods can be increased. Textile producers especially in high-wage countries have to cope with the trend toward smaller lot sizes in combination with the demand for increasing product variations. One possibility to cope with these changing market trends consists of manufacturing with CPPS and cognitive machinery. Current approaches for CPPS focus on implementing intelligent algorithms in machines' control systems to self-optimize the production process (Saggiomo et al., 2016). This presents a method for multiobjective self-optimization (MOSO) of the weaving process.

3.5.1 Multiobjective self-optimization of weaving process

Weaving is the most common as well as the oldest process for fabric manufacturing. Until today a fabric is created by crossing warp and weft threads in a right angle, like it was done since ~4000 BC. Today's applications of woven fabrics are, for example

- apparel (jeans, lining fabric, etc.),
- geotextiles (erosion protection, soil reinforcement, etc.), and
- technical textiles (fireproof fabric, airbag fabric, reinforcements for fiber composites, etc.) (Gries et al., 2015).

Because of the low production costs, the textile production has been relocated to the Asian countries, whereas the production of high quality and technical textiles is progressively shifted to Europe and other developed countries including the developing countries. The textile industry in high-wage countries such as Germany is facing numerous challenges today. For example, the tendency to produce small lot sizes requires shorter cycle times and aggravates the economical production of goods (Brecher, 2011; Osthus, 1996).

Small lot sizes in the fabric production often involves a change of the fabric. A weaving machine with about 200 parameters has to be reconfigured after each change of the fabric to fulfill the expectations of the customer. To find the optimal configuration for the machine, the operator of the weaving machine has to conduct weaving trials. These time-consuming and wasteful trials require—depending on the experience of the operator—the weaving of up to 120 m of fabric until the optimal parameters are found (Chen, 1996).

Self-optimization systems apply adoptions of their inner state or structure in case of changes in input conditions or disturbances. Target values for self-optimization can be e.g., capacity, lot size, quality, or energy consumption. Self-optimization systems are characterized by the following continuous steps (Gausemeier et al., 2009):

- analysis of actual situation
- determination of targets
- adaption of system behavior to reach the targets

The presented concept of self-optimizing production systems will now be applied to the weaving process. The following objective functions are considered by the MOSO of the weaving process:

- warp tension
- energy consumption of the weaving machine (active power consumption)
- quality of the fabric

The objective functions are optimized according to the following parameters:

- basic warp tension (bwt)
- revolutions per minute (n)
- vertical warp stop motion position (wsmy)

With the MOSO of the weaving process, a weaving machine is enabled to automatically find an optimal configuration. A program for self-optimization is implemented in a programmable logic controller (PLC). Fig. 3.3 provides an overview of the required hardware and software infrastructure.

For the signal processing and the execution of the self-optimization routine, the ibaPADU-S Module system by iba AG, Fürth, Germany, is used. The system consists of the following modules:

- ibaMS16xAI-20 mA: Analog input module for current signals in the range of [0–20] mA
- ibaMS16xAI-10V: Analog input module for voltage signals in the range of [−10–10] V
- ibaPADU-S-IT-16: Central processing unit (CPU) for the modular system

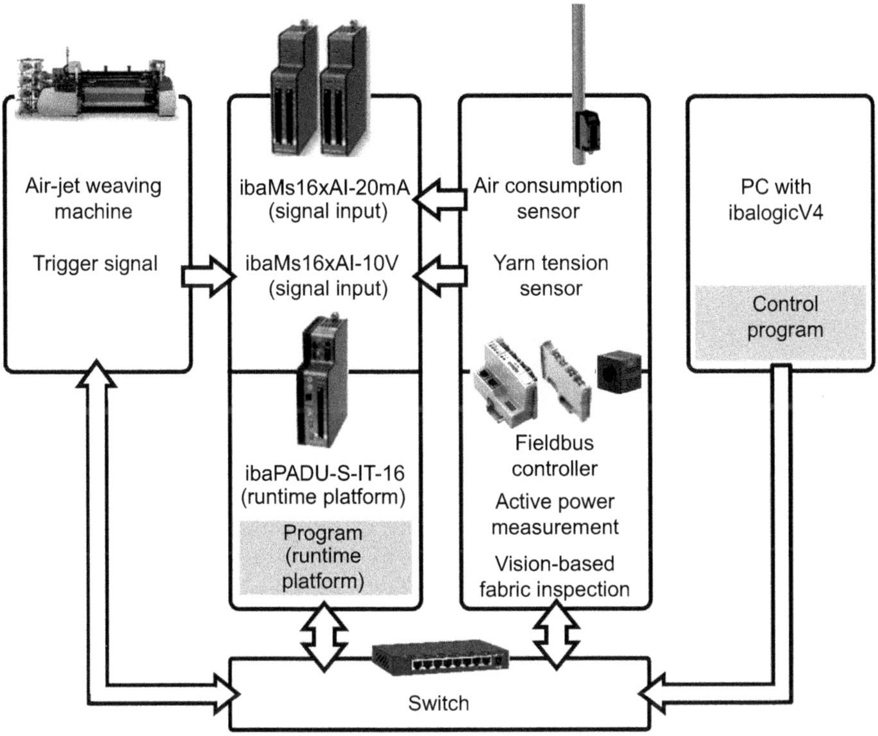

Figure 3.3 Infrastructure (hardware and software) of multiobjective self-optimization of the weaving process.

The analog input modules ibaMS16xAI-20mA and ibaMS16xAI-10V collect and process the signals from the sensor system. Both analog input modules are connected to the base unit ibaPADU-S-IT-16 using a back panel bus. The base unit receives the data from the analog input modules through the back panel bus.

The central unit ibaPADU-S-IT-16 is connected to a computer using the transmission control protocol/internet protocol (TCP/IP) interface. On the computer the software ibaLogicV4 from iba AG is installed. IbaLogicV4 is a programming environment, which forms a software-based programmable logic controller (soft-PLC) together with the introduced modular system. An ibaLogicV4 program is created on the computer and transmitted to the central unit using TCP/IP. The central unit provides a runtime platform for the ibaLogicV4 program (runtime system). In the ibaLogicV4 program the data from the analog input modules are collected and processed. The program for MOSO is developed within the environment of ibaLogicV4 and uses the ibaPADU-S-IT-16 module as runtime system.

For measuring the warp tension, the yarn tension sensor TS44/A250 by BTSR International S.p.A. Partita, Olgiate Olona, Italy is used. The yarn tension sensor generates a voltage signal in the range of 0–10V, which is proportional to the present yarn tension. The yarn tension sensor is placed in the middle of the weaving machine, between the back rest and the warp stop motion. The data connector of the yarn tension sensor is connected to the soft-PLC using an analog/digital converter. For additional information on the yarn tension sensor see Gloy et al. (2015).

The air consumption of the weaving machine is measured using the flow sensor SD8000 by ifm electronic GmbH, Essen, Germany. The flow sensor generates a signal, which is proportional to the compressed air consumption in the range of 4–20 mA. The output data from the flow senor are wirelessly transferred to the soft-PLC.

The power measurement module collects characteristic values of the three-phase supply and saves the values into the process image. To access the measurement values from the power measurement module, the power measurement module is connected to a fieldbus controller using a terminal bus.

The process image of the power measurement module is provided to the fieldbus controller via the terminal bus. The fieldbus controller is connected to the soft-PLC using the TCP/IP Interface. The communication between fieldbus controller and soft-PLC is carried out in the Modbus protocol format. The soft-PLC sends out a specific request (Request) in the Modbus protocol formal to the fieldbus controller and receives the requested value from the process image (Response).

As soon as the response has been received by the soft-PLC, the requested data are available for the signal processing. Both, Modbus-Request and Modbus-Response, consist of binary codes and are organized as bytes. The runtime platform with soft-PLC is the Modbus-Master that sends the request to the fieldbus controller, which is the Modbus-Slave. The request contains information regarding the requested value from the process image of the power measurement module. In response to the request, the fieldbus controller identifies the requested value and stores it into the response. The response is send via TCP/IP interface to the Modbus-Master. Then the active power is requested from the process image of the power measurement module.

At Institut für Textiltechnik der RWTH Aachen University (ITA), Aachen, Germany, a measuring system for online error detection during fabric production was developed (Schneider et al., 2015). A camera takes pictures of the fabric. Subsequently the pictures are checked for defects in the fabric, using digital image processing. The software for digital image processing runs on a separate computer. The camera system is installed over the section of the weaving machine where the fabric is produced.

The camera system is able to detect defects immediately after the fabric is produced. The digital image processing software is calibrated using a flawless piece of fabric. The digital image processing classifies deviations from the calibrated condition as a defect. Depending on the share of incorrect pixels in the pictures, the fabric is assigned to a quality category.

The examination for defects is carried out in real time during the weaving process. The computer running the digital image processing is connected to the soft-PLC via TCP/IP interface. Depending on the status of the fabric, the number of the quality category (0–4) is continuously transmitted via TCP/IP. A quality category of 0 is achieved in case the fabric quality is accurate. A quality category of 4 stands for a destroyed fabric.

The program for MOSO consists of the steps shown in Fig. 3.4. The communication between the weaving machine and the soft-PLC is implemented continuously. This way, the weaving machine is enabled to run the entire program for MOSO autonomously.

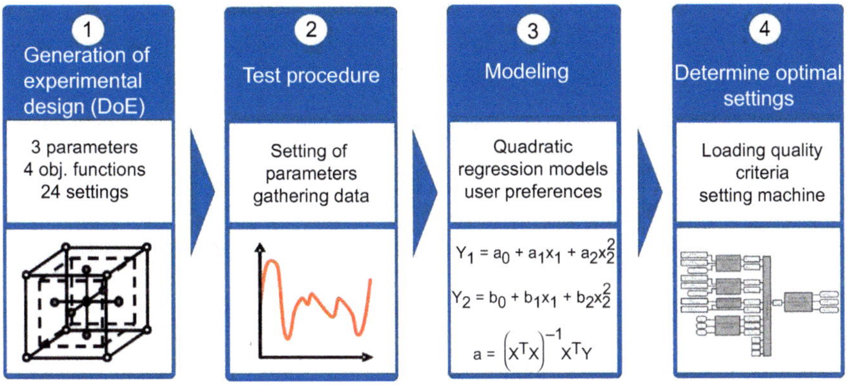

Figure 3.4 Program flow of multiobjective self-optimization of the weaving process.

In the first step, an experimental design is calculated automatically. Within this design, the three setting parameters: static warp tension, vertical position of warp stop motion, and revolutions per minute are varied. The user sets the parameter spaces to ensure that the algorithm acts within a feasible range.

During the second step, the test procedure, the weaving machine sets up every test point. Sensor data describing the objective functions are recorded for the respective parameter setting. In the third step, the obtained data are used to calculate three regression models (one model per each objective function), which describe the objective functions in dependence of the setting parameters.

In the last step, an optimized setup of the weaving machine based on predefined quality criteria is calculated by the application of desirability functions and a numerical optimization algorithm. Before execution of the optimization procedure, user-defined preferences regarding the objective functions (warp tension, energy consumption, and fabric quality) can be integrated through target weights. The preference scale for each objective function is divided into three sections (low, middle, high).

The program for MOSO is implemented within the ibaLogicV4 programming environment and runs on a CPU. The next section illuminates desirability functions and the optimization algorithm used for MOSO.

The origin of the application of desirability functions in the multidimensional optimization goes back to Derringer and Suich (Derringer and Suich, 1980). The aim of using desirability functions is to summarize the objective functions, which need to be optimized into one common function. The aggregation of the objective functions is conducted using the so-called desirability. For each objective function, one desirability function is developed. The desirability function assigns a desirability to each value of the objective functions. The desirability function has a value range of 0–1. If the value of an objective function reaches a desirability of zero, the result is invalid within the optimization routine. In case the desirability reaches the value one, the value of the objective function is optimal. The desirability w_Z is plotted over the normalized objective function Z(X). Desirability functions can be constructed in three different ways. If the goal is to achieve the highest possible value for one objective function, the desirability function for maximizing has to be used. The desirability increases when the objective function value increases, etc.

The aim of the utilization of desirability functions is to aggregate the target functions into one common function, the so-called total desirability d_{tot}. The d_{tot} is calculated by using the geometric mean of the individual desirabilities:

$$d_{tot} = (w_1 \cdot w_2 \cdot \ldots \cdot w_n)^{\frac{1}{n}} \tag{3.1}$$

Whereas w_1, w_2, …, w_n are the desirabilities of n objective functions.

The total desirability reveals how close the individual desirabilities are to the optimal range. Because of the multiplication of the individual desirabilities, d_{tot} is in the range of 0–1. A total desirability of one is reached, when all target functions are in the optimal range. In case only one target function has an invalid value, the total desirability equals to zero.

The combination of process parameters, which maximizes d_{tot}, represents the optimal operating point for the weaving process. Numeric algorithms are suitable for maximizing the total desirability. The application of numeric algorithms is more efficient than, for example, grid search methods (Blobel and Lohrmann, 1998). It is advised in several references to utilize Nelder/Mead algorithm to maximize the total desirability (Bera and Mukherjee, 2010).

The Nelder/Mead algorithm is a numeric optimization procedure (Nelder and Mead, 1965). To find a subjective optimal operating point of the weaving machine, d_{tot} is maximized. The Nelder/Mead algorithm searches for a combination of the three parameters: basic warp tension, revolutions per minute, and vertical warp stop motion position that maximizes d_{tot}.

Setting the start values for the considered parameters leads to the starting point for the algorithm. The start values are set before the first iteration and are moved toward the optimal values during the utilization of the algorithm. Starting from a minimization problem with m parameters, the algorithm considers m + 1 parameter combinations (P_1, P_2, …, P_{m+1}). The values of the objective functions are calculated in the m + 1 points and sorted in ascending order. The next step is the examination of the three points P_1, P_2, and P_3 in the parameter space. At each of the three points the algorithm calculates the value of the objective functions $F(x_1, x_2)$:

$$F_i = F(P_i), \ i = 1, 2, 3 \tag{3.2}$$

Afterward the function values are sorted:

$$F_1 \le F_2 \le F_3 \tag{3.3}$$

Considering this example, F_3 is the worst (highest) and F_1 is the best (lowest) value in the context of the optimization. The minimization of the target function is achieved by applying several iterations of the algorithm. In each iteration one new point in the parameter space is created, which replaces the point P_{m+1} with the largest value of the function to be minimized. In the present case the point P_3 results in the worst value of the target function and is therefore replaced in the next iteration. The replacement of the worst point is achieved through the basic operations of the Nelder/Mead algorithm.

The ibaLogicV4 program for MOSO of the weaving process is validated during a long-term test in the laboratory of ITA. To establish industrial conditions, the duration of the long-term test is 8 h, like usual shift duration. A long-term test is carried out using the MOSO against not using the optimization procedure, respectively, to examine the influence of MOSO on production figures. The long-term test is conducted with an air-jet weaving machine "OmniPlus 800 by Picanol n.v., Ieper, Belgium." During the long-term test, polyester filament yarn with 330 dtex for warp and weft was used (pattern: twill 3/1). The configuration of MOSO used for the long-term test is listed in Table 3.1.

After program execution of MOSO, the algorithm calculates the following optimal parameter settings: bwt = 3.71 kN; n = 522 RPM; wsmy = 20 mm.

The following settings are used as reference that was derived from an industrial weaving mill processing the same material as mentioned above: bwt = 4 kN; n = 900 RPM; wsmy = 0 mm. During the long-term test, the following parameters of the weaving process are recorded:

- efficiency of the weaving machine (i.e., the amount of fabric or the amount of weft insertions)
- warp/weft defects and breakages

Additionally, data of the objective functions are recorded. The results of the long-term test using MOSO and reference settings are shown in Table 3.2.

During the long-term test, sensor data regarding the objective functions are recorded using the software ibaPDA from iba AG, Fürth, Germany. The measured data are illustrated in Figs. 3.5–3.7. Data are plotted over the main shaft position of the weaving machine, which is the rotating angle of the machine's main drive.

Table 3.1 Configuration of self-optimization used for long-term test

Setting	Value
Lower limit BWT	2 kN
Upper limit BWT	4 kN
Lower limit n	400-RPM
Upper limit n	900-RPM
Lower limit—WSMy	0 mm
Upper limit—WSMy	20 mm
Target weight warp tension [1, 2,..., 10]	1
Target weight energy consumption [1, 2, ..., 10]	1
Target weight quality [1, 2, 3, 4]	4
Algorithm start point BWT	3.5 kN
Algorithm start point n	750-RPM
Algorithm start point—WSMy	15 mm

Table 3.2 **Results of long-term test**

	With self-optimization	Without self-optimization
Efficiency	98.6%	97.2%
Produced fabric	8.16 m	15.41 m
Weft insertions	125,157	215,982
Weft defects	2	6
Warp breakages	0	0
Average warp tension	1.27 N	1.49 N
Average air consumption	134.23 m³/h i N	155.26 m³/h i N
Average active power usage	2.49 kW	4.62 kW
Average quality category	0.93	1.55
Setup time	30 min	120 min

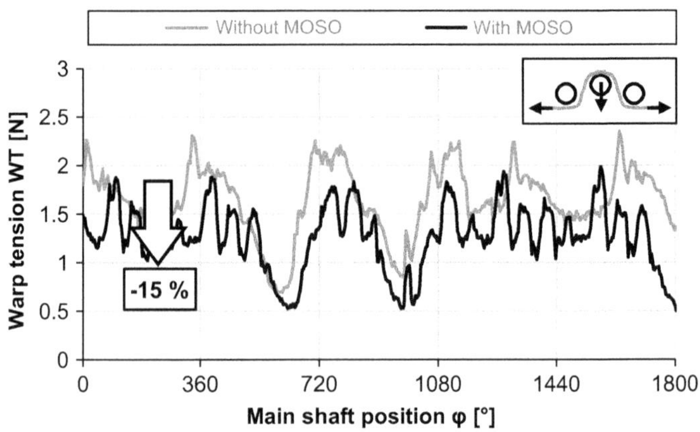

Figure 3.5 Comparison of warp tension with and without self-optimization.

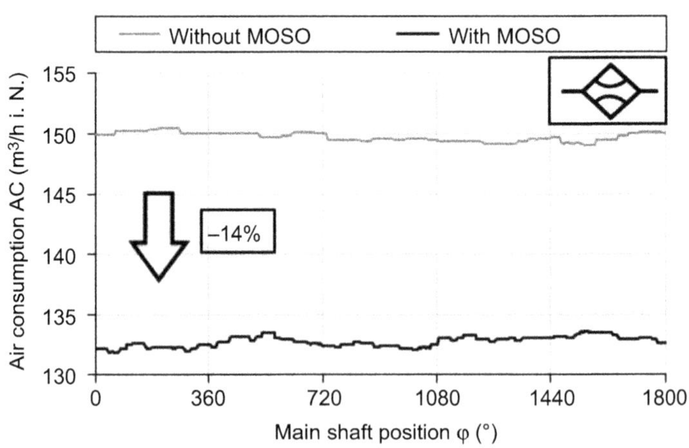

Figure 3.6 Comparison of air consumption with and without self-optimization.

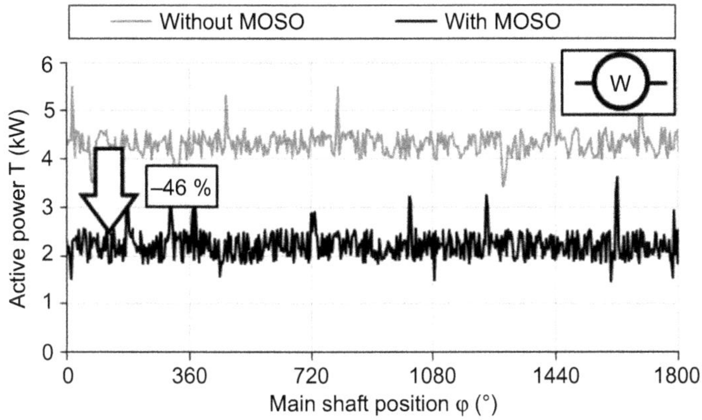

Figure 3.7 Comparison of active power usage with and without self-optimization.

The program for self-optimization enables the weaving machine to autonomously find an operating point, which improves all objective functions compared to conventional (reference) machine settings.

The efficiency presents the relation of production time of the weaving machine to the total time. The efficiency of the weaving machine is higher using the optimal setting than in case of using the reference settings, see Table 3.2. Higher efficiency is mainly achieved by reduced machine downtime.

Using the optimal machine settings, two weft defects caused be the collision of the weft threads with sagging warp threads occurred. In contrast, using the suboptimal settings, six weft defects occurred. During the long-term test it was observed that the machine runs more stable with less RPM. The higher amount of weft defects can be explained by a disadvantageous machine speed of 900 RPM. Weft defects result from the faulty transport of weft threads across the width of the weaving machine.

Without MOSO, a machine operator needs around 120 min for the configuration of the weaving machine and to find appropriate settings for the process. The program for self-optimization is concluded in 30 min and successfully reduces the setup time by 75%.

3.6 Automation in production of weft-knitted fabrics

Knitted fabrics are defined as “fabrics, made of interlooping of one or several threads or one or several thread systems by stitch formation.” (German Industrial) For further explanations of knitwear production see Au (2011), Ray (2011), and Spencer (2001). The most used form of interlooping is weft knitting followed by warp knitting. Both techniques are separated by movement of loop building yarn. In weft knitting, all needles are sequentially supplied with one weft yarn during one knitting cycle.

In warp knitting the yarn feeding and the loop forming occurs along all the needles in the needle bar level at the same time in one knitting cycle. All needles are lapped by different warp guides (Spencer, 2001).

Knitted fabrics are used in the fields of classic apparel and clothing, such as pullovers, underwear or bathing suit fashions, or home textiles such as curtains or cover for furniture (Schrank et al., 2015). As knitting technology has been implementing possibilities to manufacture technical yarns such as aramid, carbon, glass, polyamide, and polyester, the field of knitted technical textiles is rising. Moreover, by variation of process parameters, the porosity of the structure and therefore the stress and strain behavior of the fabrics can be designed according to the application. Typical applications can be found in geotextiles, automotive interior, sport industry, filter materials, or nets in addition to the apparel textiles (Schrank et al., 2015).

In industry three types of knitting machines are mainly used: flat knitting machines, large circular knitting machines (diameter larger than 165 mm), and small circular knitting/body size machines (diameter smaller than 165 mm). Flat knitting machines provide high flexibility in product design. Because of the special needle-thread arrangement, the range of pattern and color variation is higher in flat knitting machines compared with circular knitting machines. Using flat knitting machines, whole garments such as pullovers, can be produced within one knitting process (Nawaz and Nayak, 2015).

This technology of whole garment production is either called *Stoll-knit and wear* by H. Stoll AG & Co. KG, Reutlingen, Germany, or *WHOLEGARMENT* by SHIMA SEIKI MFG., LTD., Wakayama, Japan. So no additional sewing process is needed afterward. This high flexibility is accompanied with a reduction in knitting speed (Fig. 3.8). The stable realization of complex pattern command, such as loop transfer from one needle to the other, requires more time than the realization of, e.g., standard loop formation (single jersey t-shirt pattern). Moreover the construction of the flat needle bed being fed with yarn by the horizontally moving carriage results in a discontinuous knitting movement.

In general, circular knitting machines are more productive (m^2/min) than flat knitting machines because of the continuous circular movement of the needles. The needle movement speed of circular machines can go up to 2.2 m/s, whereas the carriage speed of flat knitting machines is about 1.3 m/s. Moreover, up to 144 threads can be knitted simultaneously using large circular knitting machines. Using flat knitting machines up to four threads can be knitted at the same time (Weber and Weber, 2004; www.stoll.de).

3.6.1 Properties of pattern elements in large circular weft-knitted fabrics

In knitting pattern such as t-shirt or polo shirt structures are realized by the use of pattern elements (pattern cams), such as tuck, miss, or knit elements (Fig. 3.9). By the combination of these pattern elements not only the haptic and optic properties but also the mechanical properties (stress/strain behavior) of the knitted product can be influenced.

Type	Large circular knitting machine	Body-size/seamless machines	Small circular knitting machines	Flat knitting machines
Productivity	++ • High due to continuous needle movement • Up to 144 feeders/ courses per rev. • Designed for 24/7production	+ • Continuous needle movement • Discontinuous product manufactuing • Less amount of feeders thanlarge circularmachine • 4–12 feeders,	+ • Continuous needle movement • Reduction of speed while oscillating • Discontinuous product manufactuing • Ca. 4 feeders	−− • Slowest technology (ca. 1.2 m/sec) • Horizontal movement, no continuous production • Stop and go movement • Ca.4 feeders
Knit -to-wear production	−− • Not possible • Designed for mass production • Cloth bales	+ • Seperate production of corpus and, e.g., arms • Depending on configuration: transfer possible	+ • Tube production with small diameter, • Product wise (e.g., socks)	++ • Final product • Knit and wear
Necessarity of downstream process	−− • Mandatory • Heat setting process • Cut and sew process	− • Downstream assembly	− • Generally no, at most heat treatment • Possibly closing of toe zone in downstream step	++ • No, at most heat treatment • Fully fashioned • Knit and wear
Flexibility	−− • High limitations regarding pattern and color variety • No 3D structure possible	− • Limited size of cylindar diameter per each clothing size (e.g., S, M, L) • 3D shapes by pattern and material combination	+ • Limited size of cylindar diameter per each clothing size • (e.g., S, M, L) • 3D by oscillation of cylindar (heel production)	++ • Machine width limits the product's width • 3D structures realizable • Online decrease and increase of loops • Selvedge knitting

Figure 3.8 Comparison of common knitting technologies (www.stoll.de, www.santoni.com/circular-electronic-knitting-machines.asp, http://www.merz-maschinenfabrik.de/, www.terrot-g).

Fig. 3.10 shows the results of a Drapetest of five different standard knitting patterns in collaboration with the Faserinstitut Bremen e.V, Bremen, Germany. The diagram shows that the fabric resistance is depending on the combination of the three existing pattern or cam elements in large circular knitting: knit, tuck, and miss (Simonis et al., 2016a).

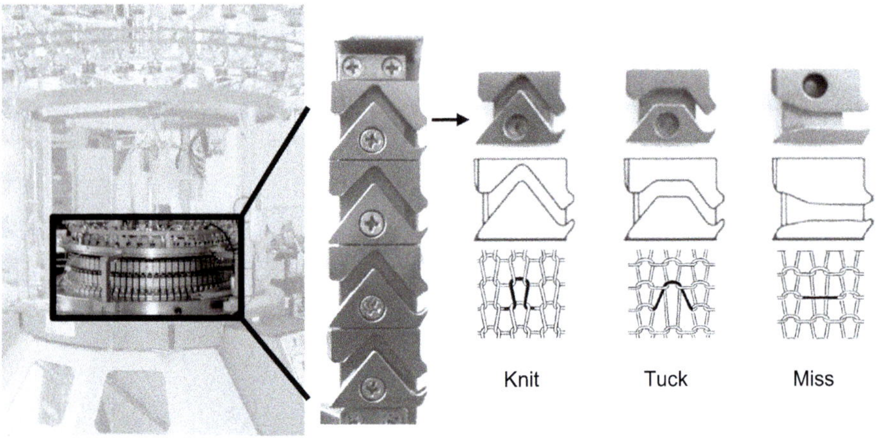

Figure 3.9 Cams and pattern element of single jersey large circular knitting machine.

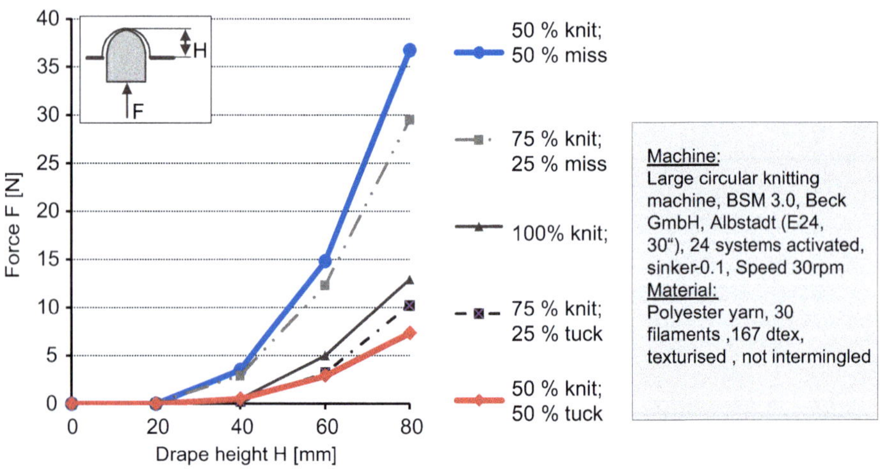

Figure 3.10 Drapetest of five different standard patterns in weft knitting investigated at Institut für Textiltechnik der RWTH Aachen University (Simonis et al., 2016a).

The knitted fabric is produced by defining fabric structure information in pixel images where each pixel refers to one knitted loop. This information is then transferred to the knitting machine. The general input parameters for knitted products are (Simonis et al., 2016b):

Product requirements: Color design and knitted pattern/structure based on desired mechanical and optical/haptic properties.

Machine status: Position of yarn bobbins in the bobbin creel and number/position of used knitting units (systems), diameter of the cylinder of circular knitting machines, respectively, production width of flat knitting machine, machine gauge, electronic or mechanical needle selection, electronic or mechanical sinker height change, type of take-down, type of equipped feeders (positive or storage feeder, elastane feeder).

Process parameters: Yarn tension of each feeder [cN] given by yarn length per revolution of feeder, sinker height, knitting speed [RPM], take-off tension [N], and spreader width [mm].

Material information: material type, filament or staple fiber yarn, yarn fineness, stress and strain behavior of the yarn material, amount of filaments, etc.

Circular knitting machines are initially designed for mass production. Originally this technology is used to manufacture a great batch number of the same fabric being cut and sew afterward to receive the final product (t-shirt). More and more technical products, such as geotextiles, home textiles and automotive interior or mattresses are produced using large circular knitting machines. All of these application markets provide a high amount of sales volume. The market volume combined with the high productivity and high material costs (aramid 20€/kg) makes a production in Europe interesting to knitting companies.

The trend in Germany is to implement automation up to Industrie 4.0 regarding the knitting process. In 2016, a new publicly funded project called "Strick 4.0" has been launched at the research institute DITF-MR, Denkendorf, Germany. The purpose of the project was to analyze new perspectives as well as future approaches regarding Industrie 4.0 in weft knitting. The potential of Industrie 4.0 in textile among others will be the efficient production of individual customized products, notable production of lot size one using smart factories, and the establishment of new dynamic business and engineering processes as well as new business models (Tilebein, 2016; http://www.bmbf.de/pubRD/Umsetzungsempfehlungen_Industrie4_0.pdf).

At the Institut für Textiltechnik of the RWTH Aachen University, Aachen, the use of large circular knitting technology for also lot size one application is investigated. The researchers have implemented a programming process pipeline, which enables to change products quickly. The first step of enabling flexibility and production of lot size one in the textile industry is to make the large circular weft-knitting machine more flexible and to allow subsequent production of different designs. The first step is to digitalize the entire process, which is done by, e.g., setting up a virtual bobbin creel and telling the positions of each yarn bobbin in the bobbin creel to the process pipeline. A process pipeline will process the input data by, e.g., scale knitted fabric in course and wale direction including the influence of the knitting as well as the downstream heat setting parameters. Moreover the process parameters setup, which is suitable or necessary for the production is included to the product programming. All information is sent to the machine via Wi-Fi and the machine immediately starts running.

To increase the possibilities to produce lot size one using large circular knitting machines, at the Institut für Textiltechnik invented a new way to manufacture three-dimensional weft-knitted fabrics according to the desired target geometry by the use of common large circular knitting machines. By the use of this new methodology, a next step toward increase of flexibility of circular knitting machines has been taken on the way to the production of lot size one.

Cross-linking between production steps can save material and human resources and machine capacity. The two great manufacturers of large circular knitting machines in Germany (Mayer & Cie. GmbH & Co. KG, Albstadt and Terrot GmbH, Chemnitz) both have invented knitting processes, which interlink the yarn processing with the process step of weft knitting. Mayer & Cie. GmbH & Co. KG, Albstadt calls its technology Spinit technology or Spinitsystems, whereas Terrot GmbH, Chemnitz, Germany, has launched the cross-linked technology called Corizon (Tuschak, 2016; Mutschler, 2016).

In flat knitting complete, near-net shaped production of knitted fabrics for technical application, such as sport shoe industry, medical application, or 3D covering and cushioning can be produced (Legner, 2016).

3.6.2 Quality monitoring of knitted fabrics

Quality is interpreted as "uniformity" or homogeneity of the knitted fabric. "Texture imperfections are nontextured patches that locally break the homogeneity of a texture pattern" (Chetverikov, 1988). A defined deviation in quality is called defect. The defective point differs optically by its appearance or material-specific with a changed machine density from the rest of the product. The fabric's quality is evaluated by measuring four main parameters: weights per area, mesh density, fabric feel including softness or stiffness, and elasticity (Au, 2011). A fast, efficient as well as precise quality monitoring of yarn-based textiles is important for industry. In 2010, the waste caused by dead times in production of warp knitted structures has led to a loss of 160,000 €/year/enterprise. Because of the high requirements of the knits' quality, one defect within 4000 m^2 produced fabric is allowed (Strauf Amabile, 2010). By the increase in production speed, e.g., regarding large circular weft-knitting machines, the probability of defect production is increased too, which stays in conflict with the quality requirements (Au, 2011).

Optical quality monitoring can be executed off- or online. The inspection of samples after being produced, often in a downstream process, is called off-line quality monitoring. An inspection during fabric production is called online quality monitoring (Malek et al., 2013). As the online monitoring lead to a higher rate of defect detection than by the trained personnel, the online monitoring system is preferred. The rate of defect detection of trained personnel is around 70%, whereas the off-line monitoring systems can deliver rates up to 100% (Niebel et al., 2013; Jeffrey Kuo et al., 2015). The Institut für Textiltechnik of the RWTH Aachen University, Aachen, in cooperation with the Institute for Imaging and Computer Vision of the RWTH Aachen University, Aachen, Germany, have shown that even horizontal stripe defects in weft-knitted fabrics due to uneven yarn tension settings can be detected off-line with a rate of 97% (Kopaczka et al., 2016).

The online defect detection regarding knitted structures is more challenging than monitoring woven fabrics because of the loop geometry and the overall mesh structure. Therefore fewer amounts of investigations have been executed in the field of online monitoring of weft-knitted structures. There are no sources talking about online defect measuring in knitting. Defect measuring is different to defect detection. Defect detection shows significant differences in the structures' surface or appearance. Defect measuring displays yarn dimensions (length, diameter, etc.)

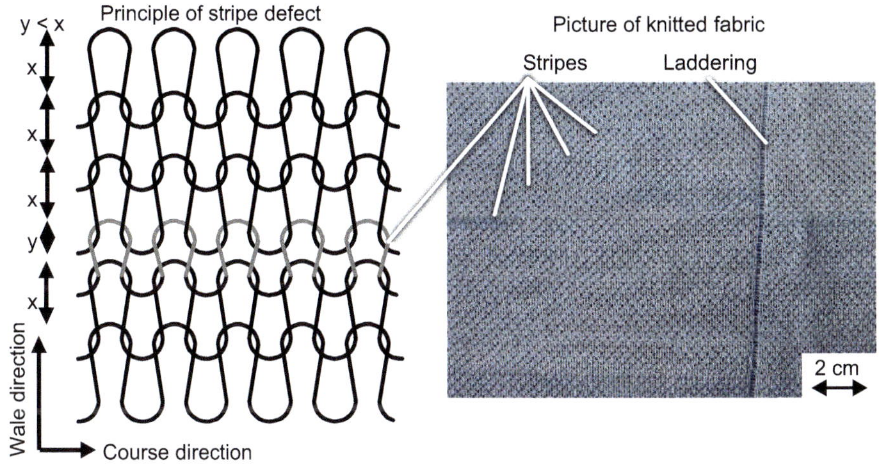

Figure 3.11 Principle and picture of stripe defects (Institut für Textiltechnik der RWTH Aachen University).

within the fabric (Niebel et al., 2013). There are several sources referring to online defect detection during the production of knitted fabrics (Abou-Taleb and Sallam, 2008; KirciTorun and Kircitorun, 2011; Li et al., 2013; Li and Zhang, 2016; Sun and Long, 2011).

The majority of defects in knitting can be divided into two categories: (1) bands and streaks, (2) stitch defects (Au, 2011).

1. Bands and streaks:
   The first category encourages the so-called **barré effect** a continuous horizontal visual barred or striped pattern (Fig. 3.11). These defects are mainly caused by physical, optical or dye differences in the yarns, errors introduced while yarn spinning, or inefficient fabric formation. The last reason can be caused by a wrong setting of sinker height of at least one feeding system, incorrect yarn tension, or variation in fabric take-off tension (Au, 2011).

   The **skew distortion** is the angularly display from the ideal perpendicular angle of wales and courses of the knitted fabric. This defect is mainly caused by errors in yarn twist parameters and yarn twist direction. Moreover, bowing, stop marks, and needle lines belong to the first category of knitting defects (Au, 2011).
2. Stitch defects:
   Dropped stitches, also called **laddering effect**, are caused when loops are casted off during production. This error occurs when the yarn is fed wrongly to the needle because of too loose yarn tension or even yarn break. The process of formation of the laddering effect is described in Fig. 3.12. Cloth press-off and holes also stitch defects and are caused by yarn breaks or weak places in the yarn (Spleiss, 2005; Au, 2011). Also knitting undesired pattern elements such as tuck or miss elements are classified as stitch defects (Au, 2011).

   Defects of weft-knitted fabrics are currently detected by the knitting expert during the knitting process on a random basis. Here, roll material from, e.g., large circular knitting machines, is inspected in an additional and downstream process, therefore, there is no possibility of error prevention but rather error detection and waste disposal.

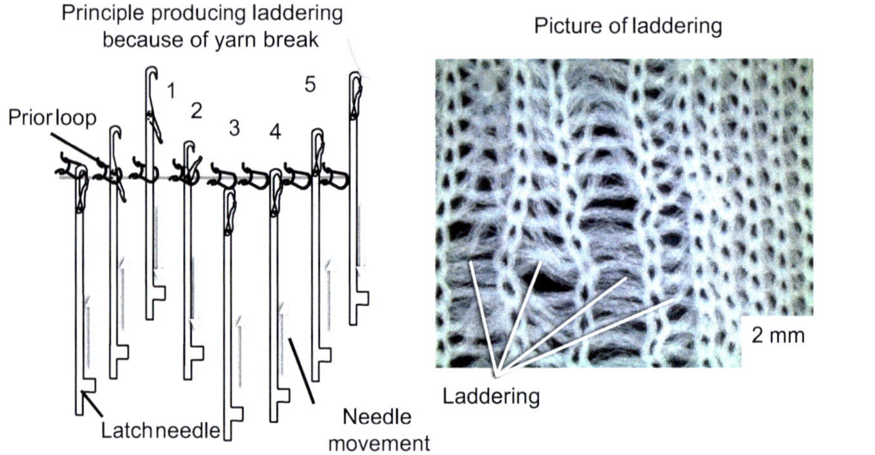

1 – No new yarn fed in
2 – Latch needle closes without yarn fed in
3 – Prior loop is casted off
4 – Latch does not open without yarn encased
5 – No feeding in of new yarn possible–latch remain closed

Figure 3.12 Principle and picture of laddering creation.
Institut für Textiltechnik der RWTH Aachen University.

Defects are caused either by inhomogeneous yarn material mixture, fineness, and tension or by false setting of knitting process parameters such as speed [RPM], sinker height [mm], yarn tension [cN], take-off tension [N], or the width of the spreader [mm] (Saggiomo et al., 2016).

Optical properties of knitted fabrics, such as surface evenness or surface quality, are important quality parameters of knitted fabrics. "The human eye is very sensitive and can detect deviations in patterns as fine as 20 µm" (Schneider, 2015). Regarding technical applications, such as flame-retarding underwear, clothing for fencer skiing underwear, or very fine knitwear (sheer panty hose), a little selective fabric mistake can lead to disqualification of the fabric's mechanical and optical requirements. In particular, errors with a spatial extension in production direction, such as color mistakes, pattern failures, or laddering, are critical because a huge number of wasted running meter result without being detected.

Moreover, repetitive errors, such as stripe defects because of uneven yarn tension of the yarn feeders, are hard to detect—even by trained operators because of the repetitive surface change is hard to differentiate from a (desired) pattern. Therefore, these tasks are more and more substituted by automated vision detection systems by counting or measuring the fabrics characteristics. Despite the practical advantages, there are just less applications in weft-knitting technology so far.

Investigations for online quality monitoring regarding large circular knitting machines exist on research level. In industry these settings are not used, yet. Regarding circular weft-knitting machines two main settings exist: the camera system can be either installed inside or outside the cylindrical knitted tube (Fig. 3.13). Therefore, ring light is being used. Outside the tube, the camera can be installed by the use of a stand to prevent perspective distortion by machine vibrations during production.

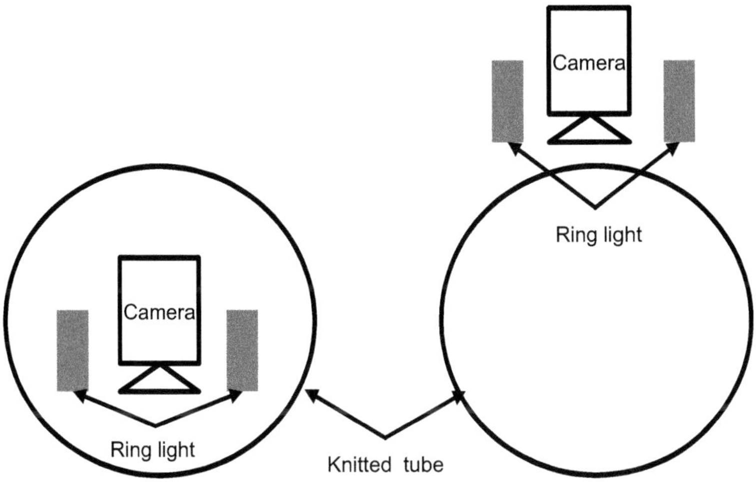

Figure 3.13 Top view of two principle arrangements of online quality monitoring of circular knitted fabrics.

3.7 Conclusion

The spinning and weaving technologies already offer a high degree of mature automation technologies. The next step to improve such existing technologies is to connect the processes within the production and along the textile process chain. To guarantee an overall product quality it is necessary to access all necessary information of the production. Many product quality issues are related to the stability of the whole production process. The knitting process can be divided into the highly productive circular knitting process or the slower but more flexible flat knitting process. The aim is to combine both benefits in one machine to fulfill the increasing demand for mass-customized products in the future. Therefore, the circular knitting process was improved by individual design capabilities. Such a new system will enable unknown flexibility for the circular knitting process with its known productivity advantages. Such a design flexibility requires streamlined interfaces to connect the process to the product design by the use of direct data transfer. In addition, with such a high degree of automation the automated detection of fabric failures is necessary. Therefore, image-based quality monitoring systems are developed and efficiently integrated into the process.

The weaving process already offers a high degree of automation and process stability. By the integration of self-optimizing technologies the process becomes a smart production system. Based on the product requirements the weaving machine automatically adjusts the production parameters according to the measured product parameters. In general the weaving process can run without the attendance of an operator. The financial benefit of low labor costs because of a high amount of machines, which can be operated by one person, has to be compensated by quality monitoring systems to ensure the overall fabric quality. Therefore, imaging systems have been developed to inspect the fabric for typical defects. An advanced and more efficient version of the quality monitoring system is the on-loom imaging system, to detect all relevant fabric defects online at the weaving machine. The advantage is minimization of waste and the possibility to transfer amount and location of defects to subsequent process steps.

Increasing use of automation technologies within these processes of yarn spinning, knitting, and weaving production will also include an implementation of digitalization technologies. The necessary infrastructure on the shop floor needs to be provided and fulfill standard requirements to allow a lead data management. The aggregation of production parameters of yarn production, knitting or weaving, and the analysis of data pattern and structures will increase the understanding of production-related quality through the whole production systems of garment manufacturing. Automation is a key technology and in general necessary to enable added value for the garment manufacturer and also for the customers.

References

Abou-Taleb, H.A., Sallam, A.T.M., 2008. On-line fabric defect detection and full control in a circular knitting machine. AUTEX Research Journal, 8.AS08, 21–29.

Amann & Söhne: Amann product range, 2016. “Mercifil” Yarn, p. 11.

Au, K.F., 2011. Hrsg: Advances in Knitting Technology. Woodhead, Cambridge, UK, Philadelphia, Penn.

Basu, A., September 2009. Yarn structure – properties relationship. Indian Journal of Fibre and Textile Research 34, 287–294.

Bera, S., Mukherjee, I., 2010. Performance analysis of Nelder-Mead and a hybrid simulated annealing for multiple response quality characteristic optimization. In: Proceedings of the International Multi Conference, vol. 3, pp. 1728–1732.

Blobel, V., Lohrmann, E., 1998. Statistical and Numerical Methods for Data Analysis: Statistische und numerische Methoden der Datenanalyse, vol. 1. Vieweg+Teubner Wiesbaden.

Brecher, C., 2011. Integrative Production Technology for High-Wage Countries. Springer Berlin/Heidelberg, pp. 747–1057.

Chen, M., 1996. Computer-based Optimization of the Weaving Process According to Warp Thread Stress and Warp Processing Behaviour (Computergestützte Optimierung des Webprozesses bezüglich Kettfadenbeanspruchung und Kettlaufverhalten). (Dissertation). Universität Stuttgart, Stuttgart.

Chetverikov, D., 1988. Detecting defects in texture. In: 9th International Conference on Pattern Recognition, pp. 61–63.

Derringer, G., Suich, R., 1980. Simultaneous optimization of several response variables. Journal of Quality Technology 12 (4), 214–219.

Gausemeier, J., Rammig, F.J., Schäfer, W., 2009. Design Methodology for Intelligent Technical Systems: Develop Intelligent Technical Systems of the Future. Springer Science and Business Media.

German Industrial Standard DIN 60000.

Gherzi, G., Technische Textilien als Treiber der europäischen Zukunft.

Gloy, Y.-S., Sandjaja, F., Gries, T., May 2015. Model-based self-optimization of the weaving process. Journal of Manufacturing Science and Technology (CIRP) 9, 88–96.

Gries, T., Klopp, K., 2007. Füge- und Oberflächentechnologien für Textilien. Springer Verlag Berlin Heidelberg.

Gries, T., Veit, D., Wulfhorst, B., 2015. Textile Technology. Carl Hanser Munich.

Haller, R., 1928. Chemische Technologie der Baumwolle/Mechanische Hilfsmittel zur Veredlung der Baumwolltextilien, pp. 525–526.

http://www.bmbf.de/pubRD/Umsetzungsempfehlungen_Industrie4_0.pdf.

https://www.lakshmimach.com/.

http://www.muratec.net/tm/index.html.

http://www.rieter.com/en/machines-systems/products/.

https://www.saviotechnologies.com/savio/en/Products/Pages/default.aspx.

http://schlafhorst.saurer.com/de/produkte/.

http://schlafhorst.saurer.com/de/zinser-351/.

Jeffrey Kuo, C.-F., Shih, C.-Y., Huang, C.-C., Wen, Y.-M., 2015. Image inspection of knitted fabric defects using wavelet packets. Textile Research Journal. http://dx.doi.org/10.1177/0040517514553872. Available from: http://trj.sagepub.com/content/early/2015/ 07/07/0040517514553872.abstract.

Kiessling, A., Matthes, M., 1993. Textil-Fachwörterbuch, p. 11.

KirciTorun, T., Kircitorun, T., 2011. Online fault detection system for circular knitting machines: Online-Fehlererkennungssystem für Rundstrickmaschinen, 21. Tekstil ve konfeksiyon dergisi, pp. 164–170.

Kopaczka, M., Ham, H., Kolk, R., Simonis, K., Merhof, D., 2016. Automated enhancement and detection of subtle stripe defects in circularly knitted fabric. ETFA. http://dx.doi.org/10.1109/ETFA.2016.7733678.

Lawrence, C.A., 2003. Fundamentals of Spun Yarn Technology. CRC Press LLC, 2000 N.W. Corporate Blvd, Boca Raton, Florida.

Legner, M., May 10–11, 2016. Endkonturnahe, funktionale Textilprodukte, von der Flachstrickmaschine. H.Stoll AG & Co. KG, Aachen-Dresden-Denkendorf – Deutsches Fachkolloquium Textil. Festhalle Denkendorf.

Li, Y., Zhang, C., 2016. Automated Vision System for Fabric Defect Inspection Using Gabor Filters and PCNN. SpringerPlus, p. 5. S. 1.LZ16.

Li, Y., Ai, J., Sun, C., 2013. Online fabric defect inspection using smart visual sensors. Sensors (Switzerland) 13, 4659–4673.

Malek, A.S., et al., 2013. Optimization of automated online fabric inspection by fast Fourier transform (FFT) and cross-correlation. Textile Research Journal 83, 256–268.

Mutschler, T., 10–11 May, 2016. Change the Game. Terrot GmbH, Aachen-Dresden-Denkendorf – Deutsches Fachkolloquium Textil. Festhalle Denkendorf.

Nawaz, N., Nayak, R., 2015. Seamless garments. In: Nayak, R., Padhye, R. (Eds.). Garment Manufacturing Technology, Elsevier.

Nayak, R., Padhye, R., 2015a. Introduction: the apparel industry. In: Nayak, R., Padhye, R. (Eds.), Garment Manufacturing Technology. Elsevier.

Nayak, R., Padhye, R., 2015b. Garment Manufacturing Technology. Elsevier.

Nelder, J.A., Mead, R., 1965. A simplex method for function minimization. The Computer Journal 7 (4), 308–313.

Niebel, V., Hehl, A., Schulte Südhoff, E., Holtermann, T., Neumann, F., Gries, T., Koßmann, U., Schmitt, R., Schneider, D., Ohm, J.-R., 2013. Integratives Messsystem zur Fehlererkennung während der textilen Flächenherstellung, final report AiF 355 ZN am Institut für Textiltechnik der. RWTH Aachen University, Aachen.

Osthus, T., 1996. Process Optimization and Changeover Time Reduction for Weaving through Automatical Adjustment of Backrest and Warp Stop Motion (Prozessoptimierung und Rüstzeitverkürzung in der Weberei durch automatische Einstellung von Streichbaum und Kettwächterkorb). (Dissertation). Rheinisch Westfälische Technische Hochschule Aachen, Aachen.

Ray, S.C., 2012. Fundamentals and Advances in Knitting Technology. Woodhead. New Delhi.

Saggiomo, M., Gloy, Y.-S., Gries, T., 2016. Reduction of the weaving process set-up time through multi-objective self-optimization. Journal of Textile Science and Engineering 6. http://dx.doi.org/10.4172/2165-8064.1000255. H. 3.

Schneider, D., Gloy, Y.-S., Merhof, D., 2015. Vision-based on-loom measurement of yarn densities in woven fabrics. In: Instrumentation and Measurement, IEEE Transactions on Instrumentation and Measurement, vol. 64, pp. 1063–1074. http://dx.doi.org/10.1109/TIM.2014.2363580. 4.

Schneider, D., 2015. On Loom Fabric Defect Detection. State of the Art and beyond. Techn. Hochsch., Diss., Aachen.

Schrank, V., Hehl, A., Weber, K.-P., 2015. Processes and machines for knitwear production. In: Gries, T., Wulfhorst, B., Veit, D. (Eds.), Textile Technology – An Introduction. Carl Hanser Verlag, Munich.

Simonis, K., Gloy, Y.-S., Gries, T., 2016a. Industrie 4.0 – automation in weft knitting technology. In: IOP Conference Series: Materials Science and Engineering 141. (Ed.), Future Generation/48th International IFKT Congress in Mönchengladbach. International Federation of Knitting Technologists, Germany, June 8–11, 2016. http://dx.doi.org/10.1088/1757-899X/141/1/012014.

Simonis, K., Gloy, Y.-S., Gries, T., January 2016b. Selbstlernende Systeme für die Strickerei von Morgen. Melliand-Newsletter.

Spencer, D., 2001. Knitting Technology – A Comprehensive Handbook and Practical Guide. Woodhead Publishing Limited.

Spleiss, C.K.G., 2005. Fehlerdetektion und -diagnose auf RundstrickmaschinenZürich. ETH.

www.stoll.de; www.santoni.com/circular-electronic-knitting-machines.asp; http://www.merz-maschinenfabrik.de/; www.terrot-g.

www.stoll.de.

Strauf Amabile, M., 2010. Optisches Prüfsystem zur Untersuchung von Fehlern in Wirkwaren. Techn. Hochsch., Diss., Aachen. 2009. Shaker, Aachen.

Sun, Y., Long, H.-R., 2011. Adaptive detection of weft-knitted fabric defects based on machine vision system. Journal of the Textile Institute 102, 823–836.

Tilebein, M., 10–11.05.2016. Textil 4.0: Perspektiven und Gestaltungsansätze, Aachen-Dresden-Denkendorf – deutsches fachkolloquium Textil. Festhalle Denkendorf.

Tuschak, M.A., 10–11.05.2016. Die Spinnstricktechnologie, eine neue Perspektive beim Rundstricken, Aachen-Dresden-Denkendorf – Deutsches Fachkolloquium Textil. Festhalle Denkendorf.

Weber, K.P., Weber, M., 2004. Wirkerei und Strickerei – technologische und bindungstechnische Grundlagen, deutscher Fachverlag.

Automation in fabric inspection

4

Ashvani Goyal
The Technological Institute of Textile & Sciences, Bhiwani, India

4.1 Introduction

In the modern textile industry, visual surface inspection of fabrics plays an important role in quality control of textile and fashion products because defects on the fabric can have a great impact on costs and grading of final products. Fabric defects are responsible for nearly 85% of the second quality items found in the garment industry, which represents a loss in revenue for manufacturers because the second quality product will sell for only 45%–65% of the price of first quality fabric. Therefore, fabric inspection is an utmost priority to prevent delivering inferior quality production.

The traditional methods of manual fabric inspection cannot meet the demand and accuracy sought for good quality garment manufacturing. In recent decades, numerous algorithms based on computer vision have been proposed to address the problem of fabric defect detection. The development of fully automated web inspection system requires robust and efficient fabric defect detection algorithms. Various approaches for fabric defect detection have been proposed in past and on the basis of nature of features from the fabric surfaces; the proposed approaches have been characterized into three categories; statistical, spectral, and model-based.

This chapter discusses the advancements in fabric inspection systems to inspect the quality of fabrics. The principles of various approaches such as statistical, spectral, and model-based approaches for the fabric inspection, which are faster and reliable, are discussed. The research studies on fabric inspection done by several researchers are also highlighted. Various commercial fabric inspection systems such as BarcoVision's Cyclops, Elbit Vision System's I-Tex4, Zellweger Uster's Fabriscan, and Shelton web-Spector for online or on-loom fabric fault analysis are also discussed in this chapter.

4.2 Fabric inspection

Until recently, fabric inspection is carried out by textile experts, who conduct a completely subjective evaluation of fabric products. However, an expert can detect no more than 60% of typical defects on a fabric wider than 2 m or when its moving speed is faster than 30 m/min. In addition, because of factors such as tiredness, boredom, and inattentiveness, the staff performance is often unreliable. As the recent production speeds are faster than ever, manufacturers must be able to identify defects, locate their source, and make the necessary corrections in less time so as to reduce the amount of second quality fabric.

Automation in Garment Manufacturing. http://dx.doi.org/10.1016/B978-0-08-101211-6.00004-5

The increased speed and demand for high-quality fabric have placed a greater strain on the inspection departments of fabric manufacturers. Therefore, the best possibility for objective and consistent evaluation is through the application of an online automated inspection system. The wider application of online automated fabric inspection would seem to offer a number of potential advantages including improved safety, reduced labor costs, the elimination of human error and/or subjective judgment, and the creation of timely statistical product data. Moreover, as they are computer based, these systems do not suffer the drawbacks of human visual inspection.

4.3 Conventional fabric inspection techniques

There are various fabric inspection systems, which are based on the dimensions and severity of faults. Some of the systems widely used for fabric inspection are described in the following section.

4.3.1 Dallas system

Dallas system was specifically developed for knitted fabric in 1970, and this system was approved by Dallas Manufacturers Association. In this system, if any defect was found on a finished garment, the garment would then be termed as a "second." For fabrics, this system defines a second as "more than one defect per 10 linear yards, calculated to the nearest 10 yards." For example, one piece 60 yards long would be allowed to have six defects.

4.3.2 Graniteville "78" system

This system was introduced in 1975 for the field of fabric grading. The system divides defects into major and minor types. The major defect is one, which is very obvious and leads the goods to second quality. The minor defect is one, which may or may not have caused garment to second, depending on its location in the end use item (Rana, 2012). Penalty points are assigned as per Table 4.1.

Table 4.1 Fabric defect grading system in Graniteville "78" system

Length of defect	Penalty points allotted
Up to 9 inches	1
9–18 inches	2
18–27 inches	3
27–36 inches	4

4.3.3 4-point system

The 4-point inspection system is the most common and widely used fabric quality inspection system for all types of fabrics in textile and apparel industry. The 4-point system, also called the American Apparel Manufacturers (AAMA) point grading system for determining fabric quality, is endorsed by the AAMA as well as the ASQC (American Society or Quality Control). The 4-point system assigns 1, 2, 3, and 4 penalty points according to the size and significance of the defect (Table 4.2). No more than 4 penalty points can be assigned for any single defect. Defect can be in either length or width direction, the system remains the same and only major defects are considered. No penalty points are assigned to minor defects. In this system, one should inspect at least 10% of the total rolls in the shipment and make sure to select at least one roll of each color way. Fabric defects are assigned points based on Table 4.2.

Total defect points per 100 square yards of fabric are calculated, if fabric rolls contain less than 40 points per 100 square yards are considered "first" quality and as an acceptance criteria. Fabric rolls containing more than 40 points are considered "seconds."

4.3.4 10-point system

The 10-point inspection system is a point per fault system, which gives a measurable guide to quality grading per roll (Islam, 2015). In 1955s "Ten Points" piece goods evaluation was adapted by the Textile Distributors and National Federation of Textiles. The system assigns penalty points to each defect as in Tables 4.3 and 4.4.

Under the 10-point system, a piece is graded a "first" if the total penalty points do not exceed the total yardage of the piece. A piece is graded a "second" if the total penalty points exceed the total yardage of the piece. This system is bit complicated and difficult to use in practice because points per length are different for warp and weft defects.

Table 4.2 **Fabric defect grading system in 4-point system**

Length of defect	Penalty points allotted
Up to 3 inches	1
3–6 inches	2
6–9 inches	3
Over 9 inches	4
Woven fabric hole	
1 inch or less	2
More than 1 inch	4
Knitted fabric	
Any hole	4

Table 4.3 **Fabric warp defect grading system in 10-point system**

Length of warp defect	Penalty points allotted
Up to 1 inch	1
1–5 inches	3
5–10 inches	5
10–36 inches	10

Table 4.4 **Fabric weft defect grading system in 10-point system**

Length of weft defect	Penalty points allotted
Up to 1 inch	1
1–5 inches	3
5 inches to half the width of fabric	5
Full width	10

4.4 Automatic fabric inspection techniques

In general, images of normal woven fabrics are dominated by a texture, which always exhibits a high periodicity among subpatterns because of the characteristics of the weaving process used in fabric formation. If a defect is present, the local regularity of a fabric will be disrupted, resulting in an imperfection or anomaly against the homogeneous texture. With this observation, to inspect defects with computer vision techniques, a defect in a fabric image is usually defined as any abnormality deriving from the homogeneous texture background. How to describe such anomalies becomes a baseline for researchers to design detection algorithms. Generally, the automatic fabric inspection techniques have been extensively reviewed (Kumar, 2008; Mahajan et al., 2009; Ngan et al., 2011; Malek, 2012) and mainly classified in three categories namely, statistical, spectral, and model-based.

4.4.1 Statistical approach

Statistical approach measures the spatial distribution of pixel values with the objective of separating the image of inspected fabric into the regions of distinct statistical behavior. In this approach, the important assumptions are that the statistics of defect-free regions are stationary, and that these regions extend over a significant portion of inspected images. Statistical methods can be classified into first-order (one pixel), second-order (two pixels), and higher-order (three or more pixels) statistics based on a number of pixels defining the local features (Mahajan et al., 2009).

The first-order statistics estimate properties such as the average and variance of individual pixel values, ignoring the spatial interaction between image pixels, second-, and higher-order statistics on the other hand estimate properties of two or more pixel values occurring at specific locations relative to each other. The defect detection methods employing texture features extracted from fractal dimensions (FDs), first-order statistics, cross-correlation, edged detection, morphological operations, cooccurrence matrix (CM), eigen filters, rank-order functions, and many local linear transforms have been categorized into this class.

4.4.1.1 Autocorrelation function

Autocorrelation technique combines all parts of an image and can be used to characterize repetitive structures (Zhang and Bresee, 1995). It measures the correlation between the image itself and the image translated with a displacement vector. As autocorrelation measure regular textures, it exhibits peaks and valleys. Tolba and Abu-Rezeq (1997) applied a self-organizing feature map to detect and classify automatically the textile defects. For fabric defect detection, Wood (1990) utilized a 2D autocorrelation function to describe the translational and rotational symmetry of an image at plain carpet but no explicit result was given. However, autocorrelation function (Zhang and Bresee, 1995) can misinterpret a fine texture and cannot analyze a texture without a reference frame of tonal primitive.

4.4.1.2 Local linear transform

This approach is closely related to filter-bank analysis methods. It gives a statistical justification for the extraction of texture properties by means of convolution operators (masks). These masks may be considered as local detectors of elementary structures such as defects. Texture properties can be extracted by using several bidimensional transform such as discrete cosine transform (DCT), discrete sine transform (DST), discrete Hadamard transform (DHT), Karhunen–Loeve transform (KLT), and eigen filtering (Kumar, 2008; Mahajan et al., 2009; Harwood et al., 1995).

Unser (1986) tested different local linear transforms for texture classification and found KLT as the best algorithm. Neubauer (1992) has detected fabric defect using texture energy features from low mask on 10×10 pixels windows of inspection images. In his approach, three 5×5 pixels Laws' masks corresponding to ripple, edge, and weave features are used to extract histogram features from every window of the image. These features are then used for the classification of the corresponding window into defect-free of defect class, using a three-layer neural network (NN). Also Ade et al. (1984) compared Laws' filters, KLT, DCT, and DHT for textile defect detection. In online fabric inspection, the local transforms such as DCT or DST could be preferable to eigen filters or KLTs because DCT or DST can be directly obtained from the camera hardware using commercially available chips that perform fast and efficient DCT or DST transforms.

4.4.1.3 Fractal dimension

FD or fractal image analysis can be used occasionally to discriminate between texture defective areas (Anagnostopoulos et al., 2001; Behera, 2004; Gedziorowski and Garcia,

1995; Harwood et al., 1995). The differential box-counting method used differences in computing nonoverlapping copies of a set of images, and the method gave satisfactory results in all ranges of FD. Conci and Proenca (1998) implemented fractal image analysis system using a box-counting approach with an overall detection accuracy of 96% for eight types of fabric defects (Kumar, 2008; Mahajan et al., 2009; Behera, 2004; Gedziorowski and Garcia, 1995). Bu et al. (2009) have tested with seven data sets of 14,378 defect-free and 3222 defective samples of plain and twill fabric of size 256×256 pixels. The detection rates ranged from 94.09% to 98.30%. FD evaluation have some drawbacks that this method does not cover all possible ranges for textiles, that is, any value from 2.0 to 3.0, therefore it is not applicable to many types of textiles. Moreover, the method also has a poor efficiency and high false alarms rate (Kumar, 2008; Mahajan et al., 2009; Conci and Proenca, 2000a,b; Anagnostopoulos et al., 2001).

4.4.1.4 Edge detection

Edge detection is a traditional technique for image analysis. Edges can be detected either as microedges using small edge operator masks or as macro edges using large masks (Davis and Mitiche, 1980). The distribution of amount of edges per unit area is an important feature in the textured images. The amount of gray-level transitions in the fabric image can represent line, edges, spots, ripples, and other spatial discontinuities. Thus these features have been largely employed for conformity testing, assembly inspection, and fabric defect detection. Conci and Proenca (2000a,b) have used Sobel edge detection to detect fabric defects and compared the results with those based on thresholding and FD. Lane (1998) has detailed a systematic approach to detect fabric defect. It is mainly suitable for plain weave fabrics imaged at low resolution. But, the difficulty in isolating fabric defects with the noise generated from the fabric structure results in high false alarm rate and therefore makes them less attractive for textile inspection (Kumar, 2008; Mahajan et al., 2009; Conci and Proenca, 1998, 2000a,b; Unser and Ade, 1984).

4.4.1.5 Cross-correlation

Cross-correlation is used to locate features in one image that appear in another one, and the correlation coefficient can generate a correlation map for defect declaration. The cross-correlation function provides a direct and accurate measure of similarity between two images. Any significant variation in the values of resulting measure indicates the presence of a defect (Kumar, 2008; Mahajan et al., 2009). Bodnarova et al. (1998) have used the correlation coefficient from multiple templates to generate a correlation map for defect declaration. The correlation approach in (Bennamoun and Bodnarova, 1998) yields satisfactory results when detecting imperfections in regularly textured backgrounds, but randomly textured backgrounds do not correlate well and demonstrate a limitation of this approach.

4.4.1.6 Morphological operations

Zhang and Bresee (1995) have detailed on morphological operations for detection of fabric defects. The practical utility of this approach is limited as most of the commonly occurring fabric defects will be missing from the binary image generated from the simple thresholding operation. Detecting defects morphologically on spatially filtered images of fabrics produces better results (Mallik-Goswami and Datta, 2000), particularly when the fabric is fine and contains defect of small size. The mathematical morphology helps describing the geometrical and structural properties of an image (Tunak and Linka, 2005). Because the morphological operations are one of the ideal tools for removing noise, the technique can be profitably exploited for noise removal in spatially filtered images of fabrics. Mallik-Goswami and Datta (2000) illuminated the inspected fabric by a collimated laser beam to obtain its diffraction pattern while the technology spatially filtered noisy image is recorded by a charged coupled device (CCD) camera and converted to a binary one. The noise is then removed using suitable morphological operations with a critically selected structuring element. However, the presented experimental results are on obvious defects (Kumar, 2008).

Mathematical morphology (Serra, 1982) extracts useful components in an image for the geometric representation and description of regional shape (e.g., boundaries and skeletons). It performs operations (Gonzalez and Woods, 2002; Weeks, 1996) such as erosion and dilation, for smoothing, sharpening, and noise removal. The most successful method is an optimal morphological filter designed by Mak et al. (2005, 2009), for plain and twill fabric defect detection. The method reached accuracies of 97.4% (Mak et al., 2005) and 94.87% (Mak et al., 2009) (off-line detection) over 78 images of 256×256 pixels in size, of which there were 39 defect-free and 39 defective images from different defects, resolutions, and textural backgrounds. Mak et al. (2009) further tested their approach on a real-time inspection machine for 276 frames of images sized 768×256 pixels (17 defective images and 259 defect-free images) and achieved 96.7% detection accuracy.

4.4.1.7 Cooccurrence-based features

The CM method, known also as the spatial gray-level dependence method, has been widely used in texture analysis, and it is one of the most popular statistical texture analysis tools for fabric defect detection (Xie, 2008; Kumar, 2008). It is known also as the spatial gray-level dependence (Mahajan et al., 2009). It is based in repeated occurrences of different gray-level configurations in a texture. CM, originally proposed by Haralick and Shanmugam (1973), characterizes texture features as second-order statistics by measuring 2D spatial dependence of the gray values in a CM for each fixed distance and/or angular spatial relationship. He derived 14 features from the CM and used them successfully for characterization of textures. However, only two of these features have been used for the defect detection on fabrics. Rosler (1992) has also developed a real fabric defect detection system, using CM features, which can maintain 95% of the defects as small as 1 mm^2 in size.

Balakrishnan et al. (1998) developed a vision system to identify and classify fabric defects (fabric defect identification and classification system) using the CM with a total cost around $ 5300. For twill fabric, Tsai et al. (1995) applied CM to extract six image features as input parameters for a back-propagation (BP) NN algorithm. A detection success rate of 96% was achieved with 25 images (5 defect-free and

20 defective images). Latif-Amet et al. (2000) proposed the subband cooccurrence matrix method. They achieved a detection accuracy of 90.78% on 36 plain fabric images of 256×256 pixels.

The spatial features of the CM are superior to that of AF because the cooccurrence probabilities can extract more information in one spatial distance, which is the measure between two pixel locations. Two main weaknesses of the CM (Zhang and Bresee, 1995) are poor performance in textures constructed by large-sized primitive and intensive computer requirements because of large number of adjacency pixels in calculation. Despite it is very popular and many studies exploited it as highly accurate technique, the CM features suffer from many drawbacks (Xie, 2008; Kumar, 2008; Mahajan et al., 2009; Anagnostopoulos et al., 2001). It is time-consuming while there is no generally accepted solution for optimizing the displacement vector. In addition, the number of gray levels is usually reduced to keep the size of the CM manageable. For a given displacement vector, a large number of features can be computed, which implies dedicated feature selection procedure. Moreover, this technique is computationally expensive for the demands of a real-time defect inspection system. Finally, the portioning of cooccurrence space and the description of multipixel cooccurrence are inefficient, which should be addressed to achieve the best possible performance for online fabric inspection.

4.4.1.8 Artificial neural network

Artificial neural networks (ANN) are among the fastest and most flexible classifiers used for fault detection because of their nonparametric nature and ability to describe complex decision regions composed of a number of similar elementary processing units (neurons) connected together into a network (Xie, 2008; Kumar, 2008; Mahajan et al., 2009; Kuo and Su, 2003; Malamas et al., 2003; Nayak et al., 2016). Huang and Chen (2001) have used BP NN, with fuzzy logic, to achieve the classification of eight different kinds of fabric defects along with defect-free fabric. Elragal (2006) proposes an automated visual inspection system using adaptive neural fuzzy interface system that can detect and classify knitting machine fabric defects.

M. Shi et al. (2009) described an adaptive image-segmentation method based on a simplified pulse-coupled neural network for detecting fabric defects. They introduced a new parameter called the deviation of the contrast to describe the contrast difference in row and column between the analyzed image and a defect—free image of the same fabric. The fast and noise immune probabilistic neural network classifier has been found to be very suitable for defect detection in homogeneous nonpatterned surfaces with acceptable slight variations, such as textile fabrics (Tolba, 2011). A defect detection accuracy of 99.87% has been achieved with 99.29% recall/sensitivity and 99.91% specificity.

Castilho et al. (2007) implemented a real-time fabric defect detection–based intelligent techniques. They used NNs, fuzzy modeling to obtain a clearly classification for defect detection. The experimental results stated that NN has a faster performance. The used algorithms can be easily implemented online and may be adapted to industrial applications without great efforts. They also proposed new methods for determining threshold values for fabric defect detection using feedforward NN. Behera and Mani (2007) used BP-based NN coupled with the DCT technique to characterize and classify woven fabric defects. Banumathi and Nasira (2012) uses BP-based NN algorithm to calculate the weighted factors and generates the desired classification of fabric defects.

Furferi and Governi (2008) described an artificial vision inspection system for real-time detection and classification of raw material defects. This system based on an ANN approach with 90% detection reliability and an adequate computational time. Many other research works (Mitropulos et al., 1999; Behera and Mani, 2007; Furferi and Governi, 2008; Kuo and Lee, 2003; Kuo et al., 2003; Mursalin et al., 2008; Schmalfuss and Schinner, 1999; Shiau et al., 2000) implemented ANN approach to detect automatically the fabric defects. Its limitations (Egmont-Petersen et al., 2002) include its black-box character, difficulty in coping with abundance of features and concomitant variations in scale, position, and orientation.

Stojanovic et al. (2001) suggested a three-layer BP ANN for low-cost fabric defect detection with off-the-shelf components. It achieved a detection accuracy of 86.2%. Similarly, a cost-effective feedforward NN architecture based on principal component analysis was proposed in Kumar (2003). A three-layer BP NN was proposed by Kuo et al. (2003) for plain white fabric defect detection. From four defect classes, 160 defective images (acquired by 1×4096 high-resolution line-scan camera) were tested with a defect recognition accuracy of 91.88%. Its merit was to model a high-dimensional system by nonlinear regression algorithm. For the same kind of fabric, another BP network (Kuo and Lee, 2003) was presented with a preprocessed filtering step. It was tested on 240 defective images (by an area-scan camera) from four classes and offers 94.38% accuracy. A recent BP NN (Yin et al., 2009) has accomplished 91% and 100% detection success rates for hole (16 images) and oil stain (16 images) of twill fabric, respectively. ANN method (Castilho et al., 2007) achieved over 99.9% accuracy in both off-line (two plain weave fabric) and online defect detections.

4.4.1.9 Bilevel thresholding

Studies of fabric defect detection have been based primarily on gray-level statistical approaches (Zhang and Bresee, 1995). These approaches are direct and simple mean to detect high contrast fabric defects. The occurrence of a defect causes the signal level to rise or fall locally; the presence of a peak or trough then indicates a defect. This defect is detected when the signal crosses a decision threshold. Cho et al. (2005) proposed algorithm for finding defect in textile fabrics with fine web surface, which shows 80% recognition rate on warp and pick float. Stojanovic et al. (2001) have developed a fabric inspection system that uses thresholding, noise removal followed by local averaging to identify eight categories of defects with 86.2% accuracy and 4.3% of false alarm. The advantages of such a technique lie in its ease of implementation. Otherwise, it fails to detect the defects, which appear without altering mean gray level in defect-free areas.

4.4.1.10 Histogram analysis

Histogram analysis and the rank function provide exactly the same information. Histogram analysis is done rather than a point-to-point analysis. Since different images usually become more comparable to one another after histogram equalization because their brightness and contrast are more similar, equalization is usually performed. Zhang and Bresee (1995) used histogram equalization that reassigns gray-level values of pixels to achieve a more uniform gray-level distribution in an image. During this process, individual pixels retain their brightness order, but a more flattened histogram is produced so the brightness and contrast of images are altered. The color information in textured images can also be used to extract color histograms and this has been used in (Boukouvalas et al., 1999; Bergsa et al., 2000) to detect defects. Thilepa and Thanikachalam (2010) applied noise filtering, histogram, and thresholding techniques using Matlab to detect fabric defects with 85% overall efficiency. Despite their simplicity, histogram techniques have proved their low cost and high detection accuracy (Xie, 2008; Behera, 2004).

4.4.1.11 Rank-order functions approach

An image rank function is a simple statistical approach for defect detection based on histogram analysis. It is given by the sequence of gray levels in the histogram when this sequence is sorted in the ascending order (Kumar, 2008). There exists 1:1 correspondence between the rank function and the related histogram, which does not exist between histogram and the image. Therefore the histogram and the rank function provide exactly the same information. However, rank functions are used instead of histograms because of the existence of very efficient definition of rank distances, which can be efficiently computed.

The median filter and other rank-order filters (Tunak and Linka, 2005) such as minimum or maximum are the best known examples of order statistics–based filers. These nonlinear filters are especially useful because of their robustness toward the modifications of the image local properties. Harwood et al. (1995) found that, local rank-order correlations of images with Laws' masks could perform better than the basic convolutions, for suitable image and mask sizes. These more robust measures of correlation are less sensitive to local random pattern and gray-scale variabilities, which are everywhere apparent in large textured images. The fabric texture information regarding spatial distribution and orientation, etc., is not uniquely determined from the knowledge of rank-order functions. Because of such drawbacks the approaches based on rank-order functions or classical histogram analysis have failed to generate any further interest for fabric defect detection.

4.4.1.12 Statistical moments approach

In statistical moments approach, large windows are preferred to gather a statistical sample. Mean, standard deviation, skewness, and kurtosis provide statistical information over a region, whereas the values are used for image segmentation. Abouelela et al. (2005) proposed a method of obtaining texture features by computing the moments in local regions directly from the gray-level image. The used algorithm has successfully segmented binary images containing textures with iso-second-order statistics as well as a number of gray-level texture images. Because of the influence of nonuniform illumination conditions on the image, statistical moments reveal the necessity of a preprocessing step to correct the image illumination inhomogeneities. The main advantage of these techniques is their computational simplicity (Anagnostopoulos et al., 2001).

4.4.1.13 Eigen filters or independent component analysis approach

The eigen filter approaches are useful in separating pairwise linear dependencies, rather than higher-order dependencies, between image pixels (Kumar, 2008). As these filters are of particular interest because they adapt automatically to the class of texture to be treated, Unser and Ade (1984) suggested a flexible texture inspection system based on the evaluation of a sequence of local textural features. Their system presented accurate defect detection with an extremely low probability of false alarms.

Monadjemi (2005) introduced the usage of structurally matched eigen filters to overcome the practical drawbacks of traditional approaches, which require an extensive training stage. Sezer et al. (2004) developed a new methodology for defect detection based on the independent component analysis. This method extracts the feature from the nonoverlapping subwindows of texture images and classifies a subwindow as defective or nondefective according to Euclidean distance between the feature obtained from average value of the features of a defect-free sample and the feature obtained from one subwindow of a test image. It gives good detection results with 96%–97% accuracy.

4.4.1.14 Local binary patterns approach

Usually, a simple local contrast measurement is calculated as a complement to the (Local binary pattern (LBP)) value to characterize local spatial relationships. The (LBP) operator is computationally simple, gives good performance in texture classification, and is relatively invariant with respect to changes in illumination and image rotation (Mallik-Goswami and Datta, 2000). Ojala et al. (1996) described the LBPs as a shift invariant complementary measure for local image contrast. It uses the gray level of the center pixel of a sliding window as a threshold for surrounding neighborhood pixels.

4.4.2 Spectral approach

Spectral approaches are robust and efficient computer vision approaches for fabric defect detection. In these approaches texture is characterized by texture primitives or texture elements and the spatial arrangement of these primitives. Thus, the primary objective of these approaches is firstly to extract texture primitives, and secondly to model or generalize the spatial placement rules. Spectral approaches require a high degree of periodicity thus, it is recommended to be applied only for computer vision of uniform textured materials such as fabrics.

For automated defect detection, such approaches are developed to overcome the efficiency drawbacks of many low-level statistical methods. Therefore, these approaches were rendered as a robust solution for online fabric defect detection. Spectral approaches are more advantageous in terms of computational efficiency but less robust in dealing with random texture images, which cannot be described in terms of primitives and displacement rules. In spectral-domain approaches, the texture features are generally derived from the Fourier transform (FT), Gabor transform, and wavelet transform (WT).

4.4.2.1 Fourier transform

Fourier analysis is a global approach that characterizes the textured image in terms of frequency components. The spatial domain is usually noise sensitive and arduous to locate defects, whereas FT utilizes the frequency domain to characterize the defects. Fourier techniques have desirable properties of noise immunity, translation invariance, and the optimal characterization of the periodic features (Kumar, 2008; Mahajan et al., 2009; Tsai and Huang, 2003). They can be used to monitor the spatial frequency spectrum of a fabric and compare the power spectrum of an image containing a defect with that of a defect-free one. When a defect occurs, the fabric regular structure is changed, so that the corresponding intensity at some specific positions of the frequency spectrum will also change, which could signify the presence of a defect.

Many researchers (Chan and Pang, 2000; Sengottuvelan et al., 2008; Tunak and Linka, 2004, 2005, 2006; Escofet et al., 2001; Rallo et al., 2007; Sakaguchi et al., 2000; Tsai and Hu, 1996; Malek et al., 2011) proposed a simulated fabric model to understand the relationship between the fabric structure in the image space and that in the frequency space. To implement Fourier analysis for fabric defect detection, various methods are available; optical Fourier transforms (OFT) obtained in optical domain by using lenses and spatial filters can be used, but most techniques, digitally implemented, are derived from discrete Fourier transforms (DFT) and/or its inverse discrete Fourier transforms (IDFT), which recovers the images in the spatial domain: classic fast Fourier transforms (FFT) or windowed Fourier transforms (WFT) versions, which have the ability to localize and analyze the features in spatial as well as frequency domain.

Chan and Pang (2000) used the Fourier analysis for fabric defect detection. Seven textural features extracted from the vertical and horizontal frequency components in the Fourier spectrum are used to discriminate four defect types including double yarn, missing yarn, webs, and yarn densities. Tsai and Hsieh (1999) detected fabric defects using a combination of DFT and Hough transform. Based on a global image reconstruction scheme using the FT, Tsai and Huang (2003) presented a global approach for the automatic inspection of defects in randomly textured surfaces as sandpaper. In the restored image obtained by IFT, the homogeneous region in the original image has an approximately uniform gray level, and yet the defective region will be distinctly preserved.

As OFT is relatively easy to implement and fast (Kumar, 2008), Castellini et al. (1996) developed a defect detection system using the measurements of the first- and the zero-order intensities. Ciamberlini et al. (1996a,b) have described the design of spatial filters: a fixed filter adaptable for different types of fabric and a universal spatial filter for the detection of defects in textured materials. Campbell and Murtagh (1998) have detailed a WFT-based method to detect defect on denim fabric samples.

Chan and Pang (2000) used DFT and IDFT to extract seven significant characteristic parameters from the central spatial frequency spectrums. These parameters are then applied using FFT to detect fabric defects (Sengottuvelan et al., 2008; Tunak and Linka, 2004). In addition, Cardamone et al. (2002) used FFT to analyze the woven fabric construction. He et al. (2004) used FT to develop an oblique scanning method, which scans the fabric surface on a running air-jet loom to estimate the fabric fluctuation in the cloth fell during weaving. Mallik-Goswami and Datta (1999) used a joint transform correlator technique, which is an extension of FT analysis and is extremely useful for real-time pattern recognition to identify fabric defects.

Based on FT, Perez et al. (2004) presented an automated analysis system for defect detection in the print process of flocked fabrics with repetitive patterns. Ralló et al. (2007) developed and tested a fully automatic system to inspect a variety of fabrics and defects. The method is achieved by applying Fourier analysis to the image of the sample under inspection, without considering any reference image so that, no prior information about the fabric structure or the defect is required. Based on FT, Weng and Perng (2007) detailed a reliable and computationally efficient two-dimensional (2D) convolution mask to detect irregularities and defects in a periodic 2D signal or image.

Hoffer et al. (1996) presented an OFT to detect and identify the defects on an on-loom NN-based inspection machine. The only explicit detection result, given by Chiu et al. (2002), was Fourier-domain maximum likelihood estimator (FDMLE), which was based on a fractional Brownian motion model for detecting fabric surface defects. Four defective images of size 128 × 128 were shown to be successfully detected by FDMLE. The method was invariant to geometric transformation such as rotation, position shift, gray-level shift, and size rescaling of an image.

FT is simple, fast, and simulates the human visual inspection. Although on the other hand, it has a low-computational complexity and is less sensitivity to noise. The FT is insensitive to the minor modifications to the frequency spectrum caused by local fabric defects. To overcome this drawback, researchers used other approaches, such as Gabor filters and WT. The DFT- and OFT-based techniques are suitable for both global and local defects. Furthermore, the DFT-based approaches are not effective in the fabric images in which the frequency components associated with the homogenous and defective regions are highly mixed together in Fourier domain. It is due to the difficulty in manipulating the frequency components associated with homogenous regions without affecting the corresponding components associated with the defective regions.

The relevant limitation to OFT approach is the laser beam diameter employed to generate the image of the moving fabric. It cannot be too large relative to the spacing of weft and warp yarns in the fabric. Consequently, multiple optical systems are required to cover the width of fabric, which is very costly and complex (Kumar, 2008). Moreover, FT is known to be a computationally expensive method. For instance, the time of 2D DFT is proportional to the square of the image size. Therefore, to reduce the computation time, FFT is used.

4.4.2.2 Gabor filters

Gabor filters are considered to be the most successful approach, of the nonfeature extraction detection schemes, for detecting fabric defects, in that it does not need an explicit feature extraction stage but utilizes a set of optimized Gabor filters and segment defects from the filtered images straightforwardly. The Gabor filter bank has been extensively studied in visual inspection. Researchers have suggested that computer vision systems utilize Gabor filters to more closely mimic the texture recognition abilities of human brains (Kumar and Pang, 2000; Tsai and Wu, 2000). Images captured by the retina are decomposed into several filtered images, each containing varying intensities over a narrow band of frequency and orientation. The neurons in the brain are individually tuned to a particular combination of frequency and orientation, which denotes a channel. These channels, therefore, closely resemble Gabor functions.

Kumar and Pang (2000) perform fabric defect detection using only real Gabor functions. Later in (Kumar and Pang, 2002b), they used a class of self-similar Gabor functions to classify fabric defects. They also investigated defect detection using only imaginary Gabor functions as an edge detector. Bodnarova et al. (2002) applied a Fisher cost function to select a subset of Gabor functions based on the mean and standard deviation of the template feature images to perform textile flaw detection. Shu and Tan (2004) detailed a method of detecting the fabric defects automatically based on multichannel and multiscale Gabor filtering. Experiments on various simulated defect fabric images have shown the effectiveness of this method. This method has accurate location and fine detection of fabric defects.

Han and Zhang (2009) proposed a fabric defect detection method based on Gabor filter masks. The performance is evaluated off-line by using a group of fabric sample images containing many kinds of fabric defects. Their experimental result shows accurate defect detection with low false alarms. In Mak and Peng (2008), Gabor filters are designed on the basis of the texture features extracted optimally from a nondefective fabric image by using a Gabor wavelet network (GWN). Their result exhibits accurate defect detection with low false alarms, showing the effectiveness and robustness of the scheme. Gabor WT is applied to detect the defects in fabrics (Arivazhagan et al., 2006). Zhu et al. (2014) proposed an applicable method for defect detection applied to twill, plain, gingham, and striped fabric. The proposed algorithm firstly uses a Gabor filter to reduce the complexity of the fabric image signal and extract the main texture signal; then, a sparse coding training algorithm is used to find the small scale over complete basis set of defect-free fabric patches. At last, according to this compared distance, the patch is classified as defective or nondefective.

Hou and Parker (2005) investigate a method for detecting defects on textured surfaces using a support vector machines classification approach with Gabor wavelet features. Instead of using all the filters in the Gabor wavelets, an adaptive filter selection scheme is applied to reduce the computational cost on feature extraction while keeping a reasonable detection rate. Their experimental result shows this method can successfully detect and segment defects in texture images. Ogata et al. (2005) suggested a new image visualization technique by an interface of plasma display panel to display an electromagnetic wave shield mesh for twill fabric defect detection. It applied 2D DFT to detect the global defects and an optimal Gabor filter (GF) to segment the local defects.

Recently, Zhang et al. (2010) integrated GFs with a Gaussian mixture model for plain fabric defect detection, but the results were not conclusive (only nine detected images shown) with a classification success rate of 87% from 360 defective images of nine classes of defects. Among the GFs methods, Mak and Peng (2008) achieved the best detection result on a fair amount of testing samples. They applied a GWN to extract optimal texture features from a defect-free image and then a well-tuned real-valued GF was employed for detecting defects. The detection success rates were 96.2% (39 defect-free and 32 defective images sized 256×256 from plain, twill and denim weaving fabrics acquired by a flat-bed scanner) and 97.1% (259 defect-free and 17 defective images sized 768×256 from twill fabric captured by a line-scan camera with front and back lighting). The main drawback of this approach comes from the nonorthogonality of Gabor functions, which results in many correlations of features between the scales.

4.4.2.3 Wavelet transform

In the recent past, multiresolution decomposition schemes based on WT have received considerable attention as alternatives for the extraction of textural features. As the basis functions of FT are sinusoids, WTs (Gonzalez and Woods, 2002) are based on small waves of varying frequency and limited duration called wavelets. WT offers localized information (more local support than FT) from horizontal, vertical, and diagonal directions on any input image. For plain and twill fabrics defect detection, WT (Yang et al., 2002; Tsai and Hsiao, 2001; Tsai and Chiang, 2003; Sari-Sarraf and Goddard, 1999; Han and Shi, 2007; Yang et al., 2004; Truchetet and Laligant, 2004) is commonly used for the feature extraction.

An on-loom fabric inspection system by Sari-Sarraf and Goddard (1998, 1999) proposed to use WT and edge fusion as preprocessing tools to attenuate the background texture and accentuate the defects on sheeting, filament-yarn, and spun-yarn fabrics. It reached an 89% detection success rate over 3700 images of fabrics, containing 26 different kinds of defects. Scharcanski (2005) used the discrete WT to classify stochastic textile texture. Rather using fixed scales, Kim et al. (1999) employed a learning process to choose the wavelet scales for maximizing the defect ability of fabric defects. Latif Amet et al. (2000) extracted cooccurrence and Markov random field (MRF)-based features from WT coefficients for fabric defect detection. Gray-level difference-based features from subbands of the WT were also applied in classifying fabric defects.

Kumar and Gupta (2000) have used mean and variance of "Haar" wavelet coefficient for the identification of surface defects. The fabric texture can also be considered as noise and removed using wavelet shrinkage. Recently, Truchetet and Laligant (2008) gave a detailed review on wavelet analysis in industrial application. On the basis of wavelet and singular signal-characteristic analysis, Guan et al. (2008) presented a new defect detection method based on wavelet characteristics. The detail signal feature after wavelet decomposition of fabric image is extracted, and it is compared with the detail signal feature of normal fabric image decomposition to determine fabric defects. Their experimental result shows the defect detection accuracy is over 92.5%.

Ngan et al. (2005) have developed the method of wavelet preprocesses golden image subtraction (WGIS) for defect detection on patterned fabric. They concluded that the WGIS method provides the best detection result. Dorrity et al. (1996) developed a real-time fabric defect and control system based on fuzzy wavelet analysis. Tsai and Hsiao (2001) detailed with some experimental results, an approach based on selective wavelet coefficients, to reconstruct the fabric image. It enhances the defects to be detected by thresholding in another step. Yang et al. (2002) designed a state-of-the-art technique of an adaptive wavelet-based feature extractor with a Euclidean distance-based detector for plain and twill fabrics. It achieved a detection rate of 97.5% with known defects (480 defect-free and 480 defective samples) and dropped to 93.3% with unknown defects (780 defect-free and 180 defective samples).

Yang et al. (2004) later compared a new discriminative feature extraction (DFE) method with five other WT-based classification methods on nine classes of samples (8 defect and 1 defect-free classes) of plain and twill fabrics. The DFE method outperformed the rest; however, fabric defect classification accuracy slightly decreased to 95.8% for a larger database of plain fabric samples (434 defect-free and 466 defect samples) when compared to that of previous work (Yang et al., 2002). But, after surveying of numerous wavelet-based research works, Truchetet and Laligant (2004) concluded that, wavelet cannot solve all the problems and that there are still a lot of limitations inherent to WT.

It suffered from either image components interference or features correlations between the scales (Kumar, 2008). Xia et al. (2016) modified contourlet transform for a more complex situation to obtain a subtler decomposition of the warp-knitted fabric image. The new contourlet transform is named the nonsubsampled wavelet-packet–based contourlet transform (NWPCT), and it consists of wavelet-packet transform and a nonsub sampled directional filter bank. The result demonstrates that NWPCT has excellent properties to segment out the defects (broken wrap, oil, and width barrier) of warp-knitted fabric.

4.4.2.4 Filtering approach

Some fabric defects that produce very subtle intensity transitions may be difficult to detect using abovementioned spectral approaches. A potential solution to detect such defects is to employ optimal finite impulse response (FIR) filters. An FIR filter has generally more free parameters than an infinite impulse response filter or a Gabor filter and thus offers added advantage of computational ease. Therefore, it offers a large feature separation between the defect-free and the defective regions of the filtered image (Kumar, 2003, 2008; Kumar and Pang, 2002a). The biggest advantage of FIR filters is that they can implement any impulse response, provided it is of finite length.

Kumar (2003) emphasized on smaller spatial masks, as compared to those from optimal Gabor filters, and demonstrated fabric defect segmentation with optimal FIR filters as small as 3×3 or 5×5 mask size. Kumar and Pang (2002a) proposed a linear FIR filter with an optimized energy separation. They investigated the approach performance with the size variation of both optimal and smoothing filters. They concluded that the size of optimal filter has appreciable effect on the performance for the defect detection. These filters can be used to supplement the performance of the existing inspection systems that fail to detect a class of specific defects.

Filtering is utilized in many applications (e.g., image enhancement) and performed (Gonzalez and Woods, 2002) between an image neighborhood and a filter mask. Two kinds of filtering methods (Gonzalez and Woods, 2002) are (1) frequency domain filtering based on FT, and (2) spatial filtering based on direct operations on image pixels. Neubauer (1992) recommended a defect segmentation method based on multiple linear filters (including three separable convolution filters as first-order statistics). Depicting only one fabric sample in poor quality, the true positive rate and the true negative rate of detection were 98.3% and 90.6%, respectively.

An eight-parameter 2D lattice filter (Meylani et al., 1996a,b) was utilized to detect defects on raw fabrics. To reduce the computation complexity for detection, a multiscale differentiation filtering (MSDF) method (Zeng and Hirata, 2002) was suggested with the help of B-spline. The defects from small to large size of 12 plain fabric image sized 256×256 pixels (acquired by camera on real-time air-jet looms) were outlined after detection. The MSDF method was successful in suppressing the background texture, and seemed effective to detect different defects, and had a high sensitivity. Yet, it produced distorted output for large-scale defects. No explicit results were presented (Neubauer, 1992; Meylani et al., 1996a,b; Zeng and Hirata, 2002) and their reliability on a large database was not clear.

4.4.2.5 Wigner distributions approach

The Wigner distribution function is Fourier-like but offers better cojoint resolution than Gabor or difference of Gaussians for cojoint spatial and spatial frequency image representation. This algorithm is effective when implemented for online fabric defect detection but its computation time is prohibitive. However, its utility for unsupervised fabric inspection, in simultaneously detecting defects from a large number of classes, is yet to be demonstrated. The major drawback of this technique (Kumar, 2008) is the presence of interference terms between the different components of the image.

4.4.3 Model-based approach

Model-based texture analysis methods try to capture the process that generated the texture. It tries to model the texture by determining the parameters of a predefined model. Model-based approaches are particularly suitable for fabric images with stochastic surface variations (possibly due to fiber heap or noise) or for randomly textured fabrics for which the statistical and spectral approaches have not yet shown their utility. These approaches often require that the image features at different levels of specificity or detail match one of possible many models of different image classes. This task is very difficult and computationally intensive if the models are complex and if a large number of models must be considered (Malamas et al., 2003). A random field (Jain, 1989) of an image is a stochastic modeling (SM) by a simple function of an array of random variables. In general, SM in image processing can be broadly classified into three classes: covariance, 1D, and 2D models. Autoregressive model belongs to the 1D class. The 2D models include casual, semicasual, and noncasual predictions. Markov random field (MRF) is one of the noncasual predictions.

4.4.3.1 Gauss–Markov random field model

As the brightness level at an image point is dependent on the brightness levels of the neighboring points unless the image is simply random noise, MRFs use a precise model of this dependence. They are able to capture the local (spatial) contextual information in an image. The theory provides a convenient and consistent way for modeling context-dependent entities such as pixels, through characterizing mutual influences among such entities using condition MRF distribution (Xie, 2008; Kumar, 2008; Mahajan et al., 2009; Baykut et al., 1998).

MRFs have been popular for modeling images. It can combine both statistical and structural information in pattern recognition. These models assume that the intensity at each pixel in the image depends on the intensities of only the neighboring pixels. Cohen and Fan (1988) used Gaussian Markov Random Fields (GMRF) to model defect-free textile web. The inspection process was treated as a hypothesis testing problem on the statistics derived from the GMRF model. The images of fabric to be inspected are divided into small windows in inspection process; a likelihood ratio test is then used to classify the windows as nondefective or defective. The testing image was partitioned into nonoverlapping subblocks where each window was then classified as defective or nondefective. Őzdemir and Erçil (1996), Baykut et al. (1998, 2000) implemented GMRF-based defect detection system. They showed that the fifth-order GMRF-based defect detection scheme runs at about 10 times faster than that based on KLT.

4.4.3.2 Poisson model

The stochastic models of some randomly industrial textured materials are based on the nature of the manufacturing process (Kumar, 2008). One example of such material is the fibrous, nonwoven material used for air filtration that is manufactured through adhesive technology. Brzakovic et al. (1995, 1997) discuss a theoretical approach based on a Poissonian model for inspection of web materials. The inspection objective is to quantify the randomness and homogeneity across the material. It was shown that the difference between the theoretical estimated model and actual measurements from the defect-free images is within 10%. Thus a statistical hypothesis testing between these two measurements can also be used to detect the fabric defects.

4.4.3.3 Model-based clustering approach

The problem of locating possible clusters in a data set (image) is a recurrent one with a long history. Campbell et al. (1997, 1999) combined image processing techniques with a powerful new statistical technique to inspect denim fabrics. They detect an alignment pattern in preprocessed images via model-based clustering and use an approximate Bayes factor to assess the evidence for the presence of a defect. Each of the aforementioned defect detection methods has limitations. There are still industrial needs to develop efficient and robust algorithms for online fabric inspection.

4.4.4 Combination of computational methods

It is difficult to perform a robust individual approach that detects all fabric defects with high accuracy. Each technique has some advantages and disadvantages. Therefore, many researchers combined two or more different approaches to give better results and to minimize the computational complexity and enhance the detection capability. Sari-Sarraf and Goddard (1999) described an online automated fabric defect detection system with 100% coverage. The relatively low-cost system is synchronized with the loom motion and produces high-quality fabric images with either front or back lighting. The acquired images were then processed by a segmentation algorithm, which combines WT, image fusion, and the correlation dimension. The approach overall detection rate under realistic conditions was found to be 89%, with 0.2 inches minimum defect size and a false alarm rate of 2.5%.

Röesler (1992) used a combination of two statistical approaches; histograms and CMs to develop a real fabric defect detection system. About more than 50 defective samples were recognizable up to 95%. Chen and Libert (1998) developed a real-time automatic visual inspection system for high-speed plane products. The implemented algorithm combines the connected component labeling, the moment calculation, and the pattern recognition. Jianli and Baoqi (2007) combined discrete WT and BP NN to develop feasible approach for the recognition of fabric defects. Latif-Amet et al. (2000) described an effective algorithm that combines concepts from wavelet theory and CMs to detect fabric defect.

Mak et al. (2009) extracted fabric defects using a pretrained GWN. Then texture features are used to facilitate the construction of structuring elements in subsequent morphological processing to remove the fabric background and isolate the defects. Malek et al. (2013) applied FFT and cross-correlation techniques to examine the structure regularity features of the fabric image in the spatial domain. To improve the efficiency of the technique and overcome the problem of detection errors, further thresholding operation is implemented using a level selection filter. The proposed detection technique is able to detect each fabric defect of a 1.5 mm size with a 100% detection rate even with the deferent added noise levels. Stojanovic et al. (1999) implemented simple and fast binary and statistical algorithms in combination with NN to improve fabric inspection process for reduced number of defect classes under real industrial conditions.

4.4.5 Comparative studies of different model

Because of the huge number of fabric defect detection algorithms and techniques, the need of effective methods to compare between these approaches is very important than before. The statistical and structural approaches have been in favor in terms of the amount of research reported. There are several comparative studies in the literature that evaluate texture analysis methods in applications to fabric defect detection. It must be noted that different studies use different data sets and possibly different parameter settings. Also resolution of the acquired images is an important factor in selecting the suitability of an approach for the defect detection. Therefore comments on the suitability of some approaches, recently cited in the literature, based on the image resolution, computational complexity, and performance would be useful. The high-resolution images are highly suitable for detecting defects with very subtle intensity variations, but their use will require a high volume of online computations for unsupervised defect detection. However, supervised defect detection on these high-resolution images is a real possibility and is therefore suggested for its practical usage.

Ozdemir et al. (1998) compared six texture features, consisting of MRF, KLT, 2D lattice filters, Laws' filters, CMs, and an FFT-based method, for detecting textile defects and found that texture modeling using the ninth-order MRF model gave the best results. Bodnarova et al. (2002) have concluded that the optimal Gabor filters (optimized to detect five types of defects) perform better than gray-level CM, correlation, or FFT-based approaches. Shady et al. (2006) study statistical quantifiers and FT to extract image features for six different knitted fabric defects using a defect-free fabric as a control sample. NN were used as a classifier, using the learning vector quantization algorithm in training the networks. The results showed success in classifying most of the defects. The results of using the FT features extraction approach were slightly more successful than the statistical approach in detecting the free defect and classifying most of the other defects.

Lee (2004), compared the performance of three methods, which utilize matched masks, WT, and NN for fabric defect detection. An evaluation of the performance of the methods was conducted on eight classes of fabric defects (Broken End, Dirty Yarn, Mispick, Netting Multiples, Stack End, Thick Bar, and Wrong Draw) and concluded that, the method employing matched masks proved the most effective in detecting fabric defects followed by NN method. The WT method was the least effective because it was only able to detect effectively certain classes of fabric defects.

Zhang and Bresee (1995) studied and compared two software approaches, based on either gray-level statistics or morphological operations, for detecting and classifying knot and slub defects in solid-shade, unpatterned woven fabrics. It was found that, both methods exhibited similar performance. Although because of the gray-level approach was more noise tolerant, fewer defect-free specimens were falsely determined as defective. Bodmarova et al. (2000) developed a comparative study to examine the suitability of four different detecting algorithms namely, gray-level cooccurrence, normalized cross-correlation, texture-blob detection, and spectral approaches. The correlation approach appeared to be the most promising method for a real-time, high accuracy defect detection algorithm.

Conci and Proença (1998, 2000a,b, 2002) compared the Sobel edge detection with those based on thresholding and FD and found that the use of FD method gives the most reliable results because it correctly detects all defect types with only 2% false alarms while it is faster than the other approaches. Ozdemir and Erçil (1996) compared six texture algorithms: MRF, KLT, 2D lattice filters, Laws' filters, CMs, and FFT, for fabric defect detection. They concluded that, the ninth-order MRF model gives the best results. Cuenca and Cámara (2003) developed a new texture descriptor based on semicover concept and a simplified local measure. They evaluated their method by comparing it with CM, histogram, Gabor Filters, wavelets transforms, and FD algorithms. The results showed a similar or superior performance to more complex approaches but with greatly saving computational cost.

Randen and Husoy (1999) and Chen and Chen (1999) indicate that the Gabor features in most of the cases outperform the other methods regarding the complexity and overall error rate, but the Gabor features suffer from a number of difficulties. A major difficulty of this method is how to determine the number of Gabor channels at the same radial frequency and the size of the Gabor filter window in the application. Although a solid conclusion cannot be drawn to determine the best method for defect detection, it is clearly evident that filtering approaches, in particular Gabor filtering has been more popularly applied in these areas. Finally, Vergados et al. (2001) detailed a description of the state-of-the-art techniques for texture segmentation as well as an evaluation of experimental research and results on the basis of selected algorithms suitable for real-time applications. They concluded that the efficiency of the various methods is strongly related to the nature of the inspected image while an algorithm for real-time applications should be specially designed on the basis of fast computational approaches.

4.5 Commercial automated fabric inspection systems

Traditionally fabric inspection has been carried out manually through human visualization performed by inspectors. This process can be tedious and can add to the production cost. Besides many defects can be missed, and the inspection can be inconsistent where by the output is dependent on the training and the performance and skill level of the human inspectors. In recent years automated machines have been developed based on adaptive and NN systems. Automated fabric inspection systems are designed to find and catalog defects in a wide variety of fabrics including greige, apparel, upholstery, furnishing, dyed, finished, denim, and industrial fabrics. Systems such as the BMSVision Cyclops, Zellweger Fabriscan, Elbit Vision I Tex, and Shelton webSpector have proven to produce consistent and reliable results (Dockery, 2001; Guruprasad and Behera, 2009).

4.5.1 BMSVision Cyclops

What makes the BarcoVision Cyclops system different from the Elbit Vision System (EVS) and Zellweger Uster offerings is that it has a traveling scanning head and can be deployed on the weaving machine itself. I-Tex and Fabriscan both inspect fabric in full width either at the batcher for greige fabrics or at the exit end of a finishing machine. Therefore, Cyclops can prevent the production of off-quality fabric by stopping the weaving process if it detects a serious or running defect. Examples of defects that would prompt Cyclops to stop a loom are running warp defects, recurring filling defects, and a high concentration of local defects (Fig. 4.1).

The Cyclops (Fig. 4.2) has a traveling head that can scan at a speed of 18–54 cm/s. The head includes a camera and illumination system and can prevent production defects by stopping the weaving process. The camera is based on complementary

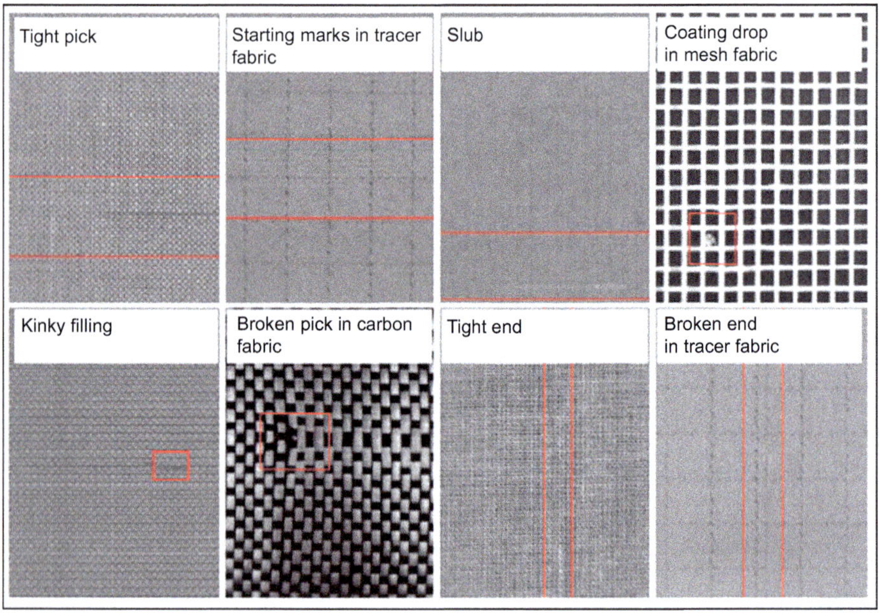

Figure 4.1 Fabric defects detected by Cyclops.
Source: http://www.visionbms.com/sites/default/files/downloads/Cyclops_BRCH_EN_A00511_0.pdf.

Figure 4.2 BMSVision Cyclops.
Source: http://www.visionbms.com/sites/default/files/downloads/Cyclops_BRCH_EN_A00511_0.pdf.

metal–oxide–semiconductor technology. The illumination system has been specifically designed to achieve optimal detection of defects in woven structures. Proprietary algorithms run on a combination of in-house designed processing hardware and an industrial PC to carry out the image processing. The embedded software is the heart of Cyclops and runs on special purpose hardware designed by Barco. The major features of the software include calibration of the camera and illumination, tuning of image processing algorithms for warp/weft density and weave, boundary detection, and JPEG-encoded image storage of fabric defects.

The on-loom fabric inspection system communicates with the microprocessor of the weaving machine. In case a running defect is detected, the on-loom fabric inspection system stops and holds the loom, preventing to continue the production of defective fabric. All detected defects are transmitted to the QualiMaster system. To complete the information, each defect is marked with time, date, and pick stamp. This allows locating the defect in the cloth roll, to generate quality reports by weaver, per shift, per style, and many more.

A typical application of QualiMaster is the "bypass" software. At cloth roll doffing, the system formulates a fabric quality advice. In case, based on the number and concentration of the detected defects, the fabric is judged to be first grade, the system will indicate on the loom's display that the cloth roll can pass the greige cloth inspection and can be sent on directly for further processing. The pieces which have been marked "quality risk" by the bypass software are reinspected and possibly mended on a gray inspection table. QT I/O-unit can interface with the speed control of the inspection table, allowing high-speed forwarding in defect-free zones and slowing down the speed of the table when approaching a mendable defect. This feature significantly increases the efficiency in the mending department.

The latest version of the Cyclops system has now also been optimized for the inspection of carbon, Kevlar, and glass fabrics, which are difficult to inspect because of their specular reflecting nature. A specially designed Cyclops measuring head exploiting the reflection characteristics of these fabrics, together with dedicated software algorithms, guarantee a 100% reliable on-loom inspection system for these high-cost fabrics.

4.5.2 Zellweger Uster Fabriscan

The Zellweger Uster Fabriscan online fabric inspection system like the Cyclops provides automatic quality control directly to the weaving loom. This automatic fabric inspection system from Zellweger Uster uses a NN to learn the characteristics of the flawless fabric (Beri, 2003). The system then detects marks, records, and classifies the faults automatically. The fabric passes over a two-component, illumination module, which allows for an inspection in reflected or transmitted light. The selection of the illumination type depends on the fabric density, the special fault types or the textile process stage at which the inspection is carried out. Depending on the inspection width, there are between 2 and 8 special CCD high-resolution line cameras installed above the light source.

Automatic fabric inspection offers advantages including consistency, speed, and objectivity. The Fabriscan classifies defects in a matrix called Uster Fabriclass, which is similar to the Uster Classimat systems for yarns. This system differentiates between disturbing defects from nondisturbing defects and makes overdetection virtually nonexistent. Fabriscan, can inspect fabric at speeds up to 120 m/min (off-line) and can detect defects down to a resolution of 0.3 mm (Fig. 4.3). The inspection speed of an online system is ~30 m/min. It can handle fabric widths from 110 to 440 cm. The Fabriscan marks the fabric faults by paper tag labels and by ink. In addition to fabric grading, the Fabriscan allows the possibility to determine which faults one wish to

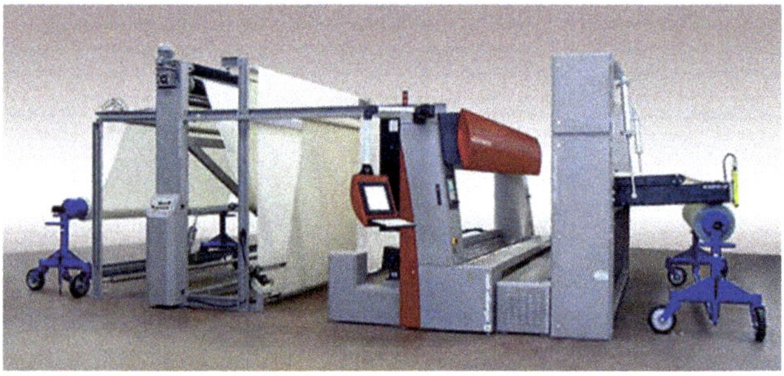

Figure 4.3 Zellweger Uster Fabriscan.
Source: http://www.ttp.com/case-studies/uster.

declare, pass, or repair. Basic weave constructions, cotton, cotton blends, wool and filament yarns, plain greige, denim fabrics, and single piece dyed, unicolored fabrics all can be inspected. However, it cannot measure color shade variations, Dobby and Jacquard designs, pile fabrics (velvets and terry), and knitted goods.

4.5.3 Elbit Vision IQ-Tex system

IQ-TEX4 is an automatic, in-line final inspection solution, and real-time process monitoring system. This latest EVS (IQ-TEX4) innovation utilizes high-resolution color line-scan technology and enhanced Defect Sorting Algorithms, to achieve unparalleled defect detection and interpretation. With the capability to detect defects <0.1 mm in size, at speeds of up to 1000 m/min and fabric widths up to 5 m, the IQ-TEX4 was designed to provide Vision Empowered Monitoring, at each stage of manufacturing. With the capability to capture any visible defect at line speeds, IQ-TEX4 can be used in a variety of positions in the majority of manufacturing equipment. It provides real-time operator alerts for running or repetitive defects as well as creating finished roll inspection maps—driving optimized product yield and high-speed debatcher control (Fig. 4.4). IQ-TEX4 eliminates the most expensive, nonvalue added process in manufacturing by removing the need for slow, costly manual inspection.

IQ-TEX4 combines state-of-the-art, computer-aided, smart vision cameras with patented defect detection algorithms to provide unparalleled inspection in textile fabrics, technical webs, nonwovens, composite materials, and many more. IQ-TEX4 lowers operating costs, optimizes product yield, maximizes raw material utilization, and improves the quality of "first quality" for the majority of running substrates. From process-specific applications to the final inspection of finished products, this solution is tailor-made to meet each customer's individual needs. Its narrow profile and modular frame design allow for seamless integration into virtually any processing machine.

4.5.4 Shelton slow moving vision inspection

The Shelton Vision slow moving web (SMW) vision inspection system meets the need for machine quality monitoring equipment that performs fabric inspection

Figure 4.4 Elbit Vision IQ-TEX4 system.
Source: www.evs.co.il.

to a standard or better than an operator. Because of large number of installations required per factory, SMW vision inspection system has been developed as a cost-effective solution consisting of low-cost high-resolution hardware coupled with sophisticated software and lighting control. For reliable detection of the full range of defects, in particular long running or repeating defects, often more than one light source is required, but to maintain low-unit costs the SMW vision inspection system employs a unique method of multiple light sources operating with a single set of cameras. Moreover, it provides unrivaled inspection performance in terms of defect detection and prevention of value reduction in the fabric forming process because of yarn and fabric manufacturing defects, human error setup, and contamination.

4.6 Conclusion

Fabric is an important component of a garment and fabric quality highly influences the garment quality. In the modern textile and garment industry, visual surface inspection of fabrics plays an important role in quality control because defects on the fabric surface can have a great impact on costs and grading of final products. For a long time the fabric inspection process has been performed with human visual inspection, and thus, the process is insufficient and costly. Therefore, automatic fabric inspection system is developed to offer a number of potential advantages such as improved safety, reduced labor costs, elimination of human error and/or subjective judgment, and the creation of timely statistical product data. In the past two decades, interesting research work relevant to automated fabric inspection has been reported and all researchers interpreted this task as a texture analysis problem.

This chapter has discussed the application of statistical, spectral, and model-based approaches for the fabric inspection, which are faster and reliable. The research work done by several researchers is also discussed in this chapter. Each application has its own advantages and disadvantages, which are also discussed. Only a few existing defect detection methods have real-time implementation. Efficient and effective fabric inspection systems such as BarcoVision's Cyclops, EVS's I-Tex4, Zellweger Uster's Fabriscan, and Shelton webSpector are commercially available for online or on-loom fabric fault analysis. In future, extra effort will be required to develop new computationally efficient methods to apply them to real-time fabric inspection scenarios.

References

Abouelela, A., Abbas, H.M., Eldeeb, H., Wahdan, A.A., Nassar, S.M., 2005. Automated vision system for localizing structural defects in textile fabrics. Pattern Recognition Letters 26 (10), 1435–1443.

Ade, F., Lins, N., Unser, M., July 1984. Comparison of various filter sets for defect detection in textiles. In: International Conference on Pattern Recognition, vol. 1, pp. 428–431.

Anagnostopoulos, C., Vergados, D., Kayafas, E., Loumos, V., Stassinopoulos, G., 2001. A computer vision approach for textile quality control. Computer Animation and Virtual Worlds 12 (1), 31–44.

Arivazhagan, S., Ganesan, L., Bama, S., 2006. Fault segmentation in fabric images using Gabor wavelet transform. Machine Vision and Applications 16 (6), 356–363.

Balakrishnan, H., Venkataraman, S., Jayaraman, S., 1998. FDICS: a vision-based system for the identification and classification of fabric defects. Journal of the Textile Institute 89 (2), 365–380.

Banumathi, P., Nasira, G.M., 2012. Fabric inspection system using artificial neural networks. International Journal of Computer Engineering Science 2, 20–27.

Baykut, A., Ozdemir, S., Meylani, R., Ercil, A., Ertuzun, A., 1998. Comparative evaluation of texture analysis algorithms for defect inspection of textile products. In: Proceedings of the 14th International Conference on Pattern Recognition (ICPR), vol. 2, pp. 1738–1741.

Baykut, A., Atalay, A., Erçil, A., Güler, M., 2000. Real-time defect inspection of textured surfaces. Real-time Imaging 6 (1), 17–27.

Behera, B.K., Mani, M.P., 2007. Characterization and classification of fabric defects using discrete cosine transformation and artificial neural network. Indian Journal of Fibre and Textile Research 32 (4), 421–426.

Behera, B.K., 2004. Image-processing in textiles. Textile Progress 35 (2–4), 127–137.

Bennamoun, M., Bodnarova, A., October 1998. Automatic visual inspection and flaw detection in textile materials: past, present and future. In: IEEE International Conference on Systems, Man, and Cybernetics, 1998, vol. 5. IEEE, pp. 4340–4343.

Bergsa, L., Duffy, N., Laecy, G., Mazo, M., 2000. Industrial inspection using Gaussian functions in a color space. Image and Vision Computing 18, 951–957.

Beri, M.R., 2003. Fabric Inspection That Think for Itself. Retrived from: http://ptj.com.pk/Web%202003/6-2003/uster.htm.

Bodmarova, A., Bennamoun, M., Kubik, K.K., 2000. Suitability analysis of techniques for flaw detection in textiles using texture analysis. Pattern Analysis and Applications 3 (3), 254–266.

Bodnarova, A., Bennamoun, M., Kubik, K.K., October 1998. Defect detection in textile materials based on aspects of the HVS. In: IEEE International Conference on Systems, Man, and Cybernetics, 1998, vol. 5. IEEE, pp. 4423–4428.

Bodnarova, A., Bennamoun, M., Latham, S., 2002. Optimal Gabor filters for textile flaw detection. Pattern Recognition 35 (12), 2973–2991.

Boukouvalas, C., Kittler, J., Marik, R., Petrou, M., 1999. Color grading of randomly textured ceramic tiles using color histograms. IEEE Transactions on Industrial Electronics 46 (1), 219–226.

Brzakovic, D., Vujovic, N., Liakopoulos, A., October 1995. An approach to quality control of texture web materials. In: Proceedings-SPIE the International Society for Optical Engineering. SPIE International Society for Optical, pp. 60–69.

Brzakovic, D.P., Bakic, P.R., Vujovic, N.S., Sari-Sarraf, H., April 1997. A generalized development environment for inspection of web materials. In: Proceedings, 1997 IEEE International Conference on Robotics and Automation, vol. 1. IEEE, pp. 1–8.

Bu, H.G., Wang, J., Huang, X.B., 2009. Fabric defect detection based on multiple fractal features and support vector data description. Engineering Applications of Artificial Intelligence 22 (2), 224–235.

Campbell, J.G., Murtagh, F., 1998. Automatic visual inspection of woven textiles using a two-stage defect detector. Optical Engineering 37 (9), 2536–2542.

Campbell, J.G., Fraley, C., Murtagh, F., Raftery, A.E., 1997. Linear flaw detection in woven textiles using model-based clustering. Pattern Recognition Letters 18 (14), 1539–1548.

Campbell, J.G., Fraley, C., Stanford, D., Murtagh, F., Raftery, A.E., 1999. Model-based methods for textile fault detection. International Journal of Imaging Systems and Technology 10 (4), 339–346.

Cardamone, J.M., Damert, W.C., Phillips, J.G., Marmer, W.N., 2002. Digital image analysis for fabric assessment. Textile Research Journal 72 (10), 906–916.

Castellini, C., Francini, F., Longobardi, G., Tiribilli, B., Sansoni, P., 1996. On-line textile quality control using optical Fourier transforms. Optics and Lasers in Engineering 24 (1), 19–32.

Castilho, H., Gonçalves, P., Pinto, J., Serafim, A., 2007. Intelligent real-time fabric defect detection. Image Analysis and Recognition 1297–1307.

Chan, C.H., Pang, G.K., 2000. Fabric defect detection by Fourier analysis. IEEE Transactions on Industry Applications 36 (5), 1267–1276.

Chen, C.C., Chen, C.C., 1999. Filtering methods for texture discrimination. Pattern Recognition Letters 20 (8), 783–790.

Chen, W.B., Libert, G., 1998. Real-time automatic visual inspection of high-speed plane products by means of parallelism. Real-time Imaging 4 (6), 379–388.

Chiu, S.H., Chou, S., Liaw, J.J., Wen, C.Y., 2002. Textural defect segmentation using a Fourier-domain maximum likelihood estimation method. Textile Research Journal 72 (3), 253–258.

Cho, C.S., Chung, B.M., Park, M.J., 2005. Development of real-time vision-based fabric inspection system. IEEE Transactions on Industrial Electronics 52 (4), 1073–1079.

Ciamberlini, C., Francini, F., Longobardi, G., Poggi, P., Sansoni, P., Tiribilli, B., August 1996a. Weaving defect detection by Fourier imaging. In: Lasers, Optics, and Vision for Productivity in Manufacturing I. International Society for Optics and Photonics, pp. 9–18.

Ciamberlini, C., Francini, F., Longobardi, G., Sansoni, P., Tiribilli, B., 1996b. Defect detection in textured materials by optical filtering with structured detectors and self-adaptable masks. Optical Engineering 35 (3), 838–844.

Cohen, F.S., Fan, Z., April 1988. Rotation and scale invariant texture classification. In: Proceedings, 1988 IEEE International Conference on Robotics and Automation. IEEE, pp. 1394–1399.

Conci, A., Proença, C.B., 1998. A fractal image analysis system for fabric inspection based on a box-counting method. Computer Networks and ISDN Systems 30 (20), 1887–1895.

Conci, A., Proença, C.B., 2000a. A comparison between image-processing approaches to textile inspection. Journal of the Textile Institute 91 (2), 317–323.

Conci, A., Proença, C.B., 2000b. A computer vision approach for textile inspection. Textile Research Journal 70 (4), 347–350.

Conci, A., Proença, C.B., July 2002. A system for real-time fabric inspection and industrial decision. In: Proceedings of the 14th International Conference on Software Engineering and Knowledge Engineering. ACM, pp. 707–714.

Cuenca, S.A., Cámara, A., September 2003. New texture descriptor for high-speed web inspection applications. In: ICIP 2003. Proceedings. 2003 International Conference on Image Processing, vol. 3. IEEE, pp. III-537–540.

Cyclops Automatic On-loom Fabric Inspection. Retrieved from: http://www.visionbms.com/sites/default/files/downloads/Cyclops_BRCH_EN_A00511_0.pdf.

Davis, L.S., Mitiche, A., 1980. Edge detection in textures. Computer Graphics and Image Processing 12 (1), 25–39.

Dockery, A., 2001. Automated Fabric Inspection: Assessing the Current State of the Art. Techexchange.com.

Dorrity, J.L., Vachtsevanos, G., Jasper, W., 1996. Real-time fabric defect detection and control in weaving processes. National Textile Center Annual Report 113–122.

Egmont-Petersen, M., de Ridder, D., Handels, H., 2002. Image processing with neural networks—a review. Pattern Recognition 35 (10), 2279–2301.

Elragal, H.M., March 2006. Neuro-fuzzy fabric defect detection and classification for knitting machine. In: Radio Science Conference. NRSC 2006. Proceedings of the Twenty Third National. IEEE, pp. 1–8.

Escofet, J., Millán, M.S., Ralló, M., 2001. Modeling of woven fabric structures based on Fourier image analysis. Applied Optics 40 (34), 6170–6176.

Furferi, R., Governi, L., 2008. Machine vision tool for real-time detection of defects on textile raw fabrics. Journal of the Textile Institute 99 (1), 57–66.

Gedziorowski, M., Garcia, J., 1995. Programmable opticaldigital processor for rank order and morphological filtering. Optics Communications 119 (1), 207–217.

Gonzalez, R.C., Woods, R.E., 2002. Digital Image Processing, second ed. Prentice Hall.

Guan, S., Shi, X., Cui, H., Song, Y., 2008. Pacific-asia workshop on computational intelligence and industrial application. PACIIA 266–370.

Guruprasad, R., Behera, B.K., June 2009. Automation fabric inspection system. Indian Textile Journal 119 (9). Available from: http://www.indiantextilejournal.com/articles/FAdetails.asp?id=2131.

Han, Y., Shi, P., 2007. An adaptive level-selecting wavelet transform for texture defect detection. Image and Vision Computing 25 (8), 1239–1248.

Han, R., Zhang, L., May 2009. Fabric defect detection method based on Gabor filter mask. In: GCIS'09. WRI Global Congress on Intelligent Systems, vol. 3. IEEE, pp. 184–188.

Haralick, R.M., Shanmugam, K., 1973. Textural features for image classification. IEEE Transactions on Systems, Man, and Cybernetics (6), 610–621.

Harwood, D., Ojala, T., Pietikäinen, M., Kelman, S., Davis, L., 1995. Texture classification by center-symmetric auto-correlation, using Kullback discrimination of distributions. Pattern Recognition Letters 16 (1), 1–10.

He, X., Taguchj, Y., Sakaguchi, A., Matsumoto, Y.I., Toriumi, K., 2004. Measuring cloth fell fluctuation on a weaving machine. Textile Research Journal 74 (7), 576–580.

Hoffer, L.M., Francini, F., Tiribilli, B., Longobardi, G., 1996. Neural networks for the optical recognition of defects in cloth. Optical Engineering 35 (11), 3183–3190.

Hou, Z., Parker, J.M., January 2005. Texture defect detection using support vector machines with adaptive Gabor wavelet features. In: WACV/MOTIONS'05 Volume 1. Seventh IEEE Workshops on Application of Computer Vision, vol. 1. IEEE, pp. 275–280.

Huang, C.C., Chen, I.C., 2001. Neural-fuzzy classification for fabric defects. Textile Research Journal 71 (3), 220–224.

IQ-TEX4 Automatic Web Inspection. Retrieved from: https://www.evs.co.il/UserFiles/File/EVS/EVS%20products%20brochures/EVS_IQ_TEX4_pages.pdf.

Islam, M., 2015. 10 point System for Fabric Inspection in Garment Industry. Available from: http://textilelearner.blogspot.in/2015/07/10-points-system-for-fabric-inspection.html.

Jain, A.K., 1989. Fundamentals of Digital Image Processing. Prentice-Hall, Inc.

Jianli, L., Baoqi, Z., 2007. Identification of fabric defects based on discrete wavelet transform and back-propagation neural network. Journal of the Textile Institute 98 (4), 355–362.

Kim, S., Lee, M.H., Woo, K.B., 1999. Wavelet analysis to fabric defects detection in weaving processes. In: ISIE'99. Proceedings of the IEEE International Symposium on Industrial Electronics, vol. 3. IEEE, pp. 1406–1409.

Kumar, A., Gupta, S., January 2000. Real time DSP based identification of surface defects using content-based imaging technique. In: Proceedings of IEEE International Conference on Industrial Technology, vol. 2. IEEE, pp. 113–118.

Kumar, A., Pang, G., 2000. Fabric defect segmentation using multichannel blob detectors. Optical Engineering 39 (12), 3176–3190.

Kumar, A., Pang, G.K., 2002a. Defect detection in textured materials using optimized filters. IEEE Transactions on Systems, Man, and Cybernetics, Part B (Cybernetics) 32 (5), 553–570.

Kumar, A., Pang, G.K., 2002b. Defect detection in textured materials using Gabor filters. IEEE Transactions on Industry Applications 38 (2), 425–440.

Kumar, A., 2003. Neural network based detection of local textile defects. Pattern Recognition 36 (7), 1645–1659.

Kumar, A., 2008. Computer-vision-based fabric defect detection: a survey. IEEE Transactions on Industrial Electronics 55 (1), 348–363.

Kuo, C.F.J., Lee, C.J., 2003. A back-propagation neural network for recognizing fabric defects. Textile Research Journal 73 (2), 147–151.

Kuo, C.F.J., Su, T.L., 2003. Gray relational analysis for recognizing fabric defects. Textile Research Journal 73 (5), 461–465.

Kuo, C.F.J., Lee, C.J., Tsai, C.C., 2003. Using a neural network to identify fabric defects in dynamic cloth inspection. Textile Research Journal 73 (3), 238–244.

Lane, J.S., June 1998. US Patent No. 5,774,177.

Latif-Amet, A., Ertüzün, A., Erçil, A., 2000. An efficient method for texture defect detection: sub-band domain co-occurrence matrices. Image and Vision Computing 18 (6), 543–553.

Lee, T.C., 2004. Fabric Defect Detection by Wavelet Transform and Neural Network. University of Hong Kong.

Mahajan, P.M., Kolhe, S.R., Patil, P.M., 2009. A review of automatic fabric defect detection techniques. Advances in Computational Research 1 (2), 18–29.

Mak, K.L., Peng, P., 2008. An automated inspection system for textile fabrics based on Gabor filters. Robotics and Computer-Integrated Manufacturing 24 (3), 359–369.

Mak, K.L., Peng, P., Lau, H.Y.K., December 2005. Optimal morphological filter design for fabric defect detection. In: ICIT 2005. IEEE International Conference on Industrial Technology. IEEE, pp. 799–804.

Mak, K.L., Peng, P., Yiu, K.F.C., 2009. Fabric defect detection using morphological filters. Image and Vision Computing 27 (10), 1585–1592.

Malamas, E.N., Petrakis, E.G., Zervakis, M., Petit, L., Legat, J.D., 2003. A survey on industrial vision systems, applications and tools. Image and Vision Computing 21 (2), 171–188.

Malek, A.S., Drean, J.-Y., Bigue, L., Osselin, J.-F., June 8–10, 2011. Online fabric defect detection by fast Fourier transform and cross-correlation. In: 11th World Conference AUTEX, Mulhouse, France, 1056.

Malek, A.S., Drean, J.Y., Bigue, L., Osselin, J.F., 2013. Optimization of automated online fabric inspection by fast Fourier transform (FFT) and cross-correlation. Textile Research Journal 83 (3), 256–268.

Malek, A.S., 2012. Online Fabric Inspection by Image Processing Technology (Doctoral dissertation). Université de Haute Alsace-Mulhouse.

Mallik, B., Datta, A.K., 1999. Defect detection in fabrics with a joint transform correlation technique: theoretical basis and simulation. Textile Research Journal 69 (11), 829–835.

Mallik-Goswami, B., Datta, A.K., 2000. Detecting defects in fabric with laser-based morphological image processing. Textile Research Journal 70 (9), 758–762.

Meylani, R., Ertuzun, A., Ercil, A., October 1996a. A comparative study on the adaptive lattice filter structures in the context of texture defect detection. In: ICECS'96., Proceedings of the Third IEEE International Conference on Electronics, Circuits, and Systems, vol. 2. IEEE, pp. 976–979.

Meylani, R., Ertuzun, A., Erçil, A., September 1996b. Texture defect detection using the adaptive two-dimensional lattice filter. In: Proceedings, International Conference on Image Processing, vol. 3. IEEE, pp. 165–168.

Mitropulos, P., Koulamas, C., Stojanovic, R., Koubias, S., Papadopoulos, G., Karayanis, G., January 1999. Real-time vision system for defect detection and neural classification of web textile fabric. In: Proceedings of SPIE-the International Society for Optical Engineering, vol. 3652, pp. 59–69.

Monadjemi, A., 2005. Towards Efficient Texture Classification and Abnormality Detection (Doctoral dissertation). University of Bristol.

Mursalin, T.E., Eishita, F.Z., Islam, A.R., 2008. Fabric defect inspection system using neural network and microcontroller. Journal of Theoretical and Applied Information Technology 4 (7), 560–570.

Nayak, R., Kanesalingam, S., Wang, L., Padhye, R., 2016. Artificial intelligence: technology and application in apparel manufacturing. In: TBIS-APCC 2016. Binary Information Press, Textile Bioengineering and Informatics Society, pp. 648–655.

Neubauer, C., August 1992. Segmentation of defects in textile fabric. In: Pattern Recognition. Proceedings, 11th IAPR International Conference on Conference A: Computer Vision and Applications, vol. I. IEEE, pp. 688–691.

Ngan, H.Y., Pang, G.K., Yung, S.P., Ng, M.K., 2005. Wavelet based methods on patterned fabric defect detection. Pattern Recognition 38 (4), 559–576.

Ngan, H.Y., Pang, G.K., Yung, N.H., 2011. Automated fabric defect detection—a review. Image and Vision Computing 29 (7), 442–458.

Ogata, N., Fukuma, S., Nishikado, H., Shirosaki, A., Takagi, S., Sakurai, T., December 2005. An accurate inspection of PDP-mesh cloth using Gabor filter. In: ISPACS 2005. Proceedings of 2005 International Symposium on Intelligent Signal Processing and Communication Systems. IEEE, pp. 65–68.

Ojala, T., Pietikäinen, M., Harwood, D., 1996. A comparative study of texture measures with classification based on featured distributions. Pattern Recognition 29 (1), 51–59.

Ozdemir, S., Ercil, A., November 1996. Markov random fields and Karhunen-Loeve transforms for defect inspection of textile products. In: EFTA'96. Proceedings, 1996 IEEE Conference on Emerging Technologies and Factory Automation, vol. 2. IEEE, pp. 697–703.

Ozdemir, S., Baykut, A., Meylani, R., Ercil, A., Ertüzün, A., 1998. International Conference on Pattern Recognition, pp. 1738–1740.

Perez, R., Silvestre, J., Munoz, J., 2004. Defect detection in repetitive fabric patterns. In: Proceeding of Visualization, Imaging and Image Processing, September, 6–8.

Ralló, M., Millán, M.S., Escofet, J., 2007. Referenceless segmentation of flaws in woven fabrics. Applied Optics 46 (27), 6688–6699.

Rana, N., 2012. Fabric inspection systems for apparel industry. Indian Textile Journal 122 (11), 57.

Randen, T., Husoy, J.H., 1999. Filtering for texture classification: a comparative study. IEEE Transactions on Pattern Analysis and Machine Intelligence 21 (4), 291–310.

Rosler, R.N.U., 1992. Mellind Texilberichte 73, 292.

Sakaguchi, A., Kim, H., Matsumoto, Y.I., Toriumi, K., 2000. Woven fabric quality evaluation using image analysis. Textile Research Journal 70 (11), 950–956.

Sari-Sarraf, H., Goddard, J.S., June 1998. Robust defect segmentation in woven fabrics. In: Proceedings. 1998 IEEE Computer Society Conference on Computer Vision and Pattern Recognition. IEEE, pp. 938–944.

Sari-Sarraf, H., Goddard, J.S., 1999. Vision system for on-loom fabric inspection. IEEE Transactions on Industry Applications 35 (6), 1252–1259.

Scharcanski, J., 2005. Stochastic texture analysis for monitoring stochastic processes in industry. Pattern Recognition Letters 26 (11), 1701–1709.

Schmalfuss, H., Schinner, K.L., 1999. Automatic fabric inspection on the loom. Melliand Textilberichte International Textile Reports 80, 499–500.

Sengottuvelan, P., Wahi, A., Shanmugam, A., 2008. Automatic fault analysis of textile fabric using imaging systems. Research Journal of Applied Sciences 3 (1), 26–31.

Serra, J., 1982. Image Analysis and Mathematical Morphology, vol. 1. Academic press.

Sezer, O.G., Ertüzün, A., Erçil, A., 2004. Independent component analysis for texture defect detection. Pattern Recognition and Image Analysis 14 (2), 303–307.

Shady, E., Gowayed, Y., Abouiiana, M., Youssef, S., Pastore, C., 2006. Detection and classification of defects in knitted fabric structures. Textile Research Journal 76 (4), 295–300.

Shi, M., Jiang, S., Wang, H., Xu, B., 2009. A simplified pulse-coupled neural network for adaptive segmentation of fabric defects. Machine Vision and Applications 20 (2), 131–138.

Shiau, Y.R., Tsai, I.S., Lin, C.S., 2000. Classifying web defects with a back-propagation neural network by color image processing. Textile Research Journal 70 (7), 633–640.

Shu, Y., Tan, Z., June 2004. Fabric defects automatic detection using Gabor filters. In: WCICA 2004. Fifth World Congress on Intelligent Control and Automation, vol. 4. IEEE., pp. 3378–3380.

Slow Moving Vision Inspection. Retrieved from: https://www.sheltonvision.co.uk/visual-inspection-systems/slow-moving-web-solutions/.

Stojanovic, R., Mitropoulos, P., Koulamas, C., Koubias, S., Papadopoulos, G., 1999. Vision Inspection of Web Textile Fabric. QCAV, France.

Stojanovic, R., Mitropulos, P., Koulamas, C., Karayiannis, Y., Koubias, S., Papadopoulos, G., 2001. Real-time vision-based system for textile fabric inspection. Real-time Imaging 7 (6), 507–518.

Thilepa, R., Thanikachalam, M., 2010. A paper on automatic fabrics fault processing using image processing technique in MATLAB. Signal and Image Processing: An International Journal (SIPIJ) 1 (2), 88–99.

Tolba, A.S., Abu-Rezeq, A.N., 1997. A self-organizing feature map for automated visual inspection of textile products. Computers in Industry 32 (3), 319–333.

Tolba, A.S., 2011. Neighborhood-preserving cross correlation for automated visual inspection of fine-structured textile fabrics. Textile Research Journal 81 (19), 2033–2042.

Truchetet, F., Laligant, O., October 2004. Wavelets in industrial applications: a review. In: Proc. SPIE, vol. 5607, pp. 1–14.

Truchetet, F., Laligant, O., 2008. Review of industrial applications of wavelet and multiresolution-based signal and image processing. Journal of Electronic Imaging 17 (3), 031102.

Tsai, D.M., Chiang, C.H., 2003. Automatic band selection for wavelet reconstruction in the application of defect detection. Image and Vision Computing 21 (5), 413–431.

Tsai, D.M., Hsiao, B., 2001. Automatic surface inspection using wavelet reconstruction. Pattern Recognition 34 (6), 1285–1305.

Tsai, D.M., Hsieh, C.Y., 1999. Automated surface inspection for directional textures. Image and Vision Computing 18 (1), 49–62.

Tsai, I.S., Hu, M.C., 1996. Automatic inspection of fabric defects using an artificial neural network technique. Textile Research Journal 66 (7), 474–482.

Tsai, D.M., Huang, T.Y., 2003. Automated surface inspection for statistical textures. Image and Vision Computing 21 (4), 307–323.

Tsai, D.M., Wu, S.K., 2000. Automated surface inspection using Gabor filters. The International Journal of Advanced Manufacturing Technology 16 (7), 474–482.

Tsai, I.S., Lin, C.H., Lin, J.J., 1995. Applying an artificial neural network to pattern recognition in fabric defects. Textile Research Journal 65 (3), 123–130.

Tunák, M., Linka, A., 2004. Applying spectral analysis to automatic inspection of weaving density. In: 11th International Conference on Structure and Structural Mechanics of Textiles (STRUTEX 2004). Liberec, vol. 6, pp. 133–140 No. 7.12.

Tunák, M., Linka, A., November 2005. Planar anisotropy of fibre systems by using 2D Fourier transform. In: Proceedings of the 12th International Conference (STRUTEX), Liberec, Czech Republic.

Tunák, M., Linka, A., November 2006. Simulation and recognition of common fabric defects. In: Proceedings of 13th International Conference on Structure and Structural Mechanics of Textiles, Liberec, Czech Republic, pp. 363–370.

Unser, M., Ade, F., 1984. Feature extraction and decision procedure for automated inspection of textured materials. Pattern Recognition Letters 2 (3), 185–191.

Unser, M., 1986. Local linear transforms for texture measurements. Signal Processing 11 (1), 61–79.

Uster fabriscan. Retrieved from: http://www.ttp.com/case-studies/uster.

Vergados, D., Anagnostopoulos, C., Anagnostopoulos, I., Kayafas, E., Loumos, V., Stassinopoulos, G., 2001. An Evaluation of Texture Segmentation Techniques for Real-Time Computer Vision Applications, Advances in Automation, Multimedia and Video Systems and Modern Computer Science. WSES Press, pp. 332–335.

Weeks, A.R., 1996. Fundamentals of Electronic Image Processing. SPIE Optical Engineering Press, Bellingham, pp. 316–414.

Weng, Y.S., Perng, M.H., May 2007. Periodic pattern inspection using convolution masks. In: MVA, pp. 544–547.

Wood, E.J., 1990. Applying Fourier and associated transforms to pattern characterization in textiles. Textile Research Journal 60 (4), 212–220.

Xia, D., Jiang, G., Ma, P., 2016. Warp-knitted fabric defect segmentation based on the non-subsampled wavelet-packet-based Contourlet transform. Textile Research Journal 86 (19), 2043–2055.

Xie, X., 2008. A review of recent advances in surface defect detection using texture analysis techniques. ELCVIA Electronic Letters on Computer Vision and Image Analysis 7 (3), 1–22.

Yang, X.Z., Pang, G.K., Yung, N.H., 2002. Discriminative fabric defect detection using adaptive wavelets. Optical Engineering 41 (12), 3116–3126.

Yang, X., Pang, G., Yung, N., 2004. Discriminative training approaches to fabric defect classification based on wavelet transform. Pattern Recognition 37 (5), 889–899.

Yin, Y., Zhang, K., Lu, W., 2009. Textile flaw classification by wavelet reconstruction and BP neural network. Advances in Neural Networks–ISNN 2009, 694–701.

Zeng, P., Hirata, T., October 2002. On-loom fabric inspection using multi-scale differentiation filtering. In: 37th IAS Annual Meeting. Conference Record of the Industry Applications Conference, vol. 1. IEEE, pp. 320–326.

Zhang, Y.F., Bresee, R.R., 1995. Fabric defect detection and classification using image analysis. Textile Research Journal 65 (1), 1–9.

Zhang, Y., Lu, Z., Li, J., 2010. Fabric defect detection and classification using Gabor filters and Gaussian mixture model. Computer Vision–ACCV 2009, 635–644.

Zhu, Q., Wu, M., Li, J., Deng, D., 2014. Fabric defect detection via small scale over-complete basis set. Textile Research Journal 84 (15), 1634–1649.

Artificial intelligence and its application in the apparel industry

5

Rajkishore Nayak[1], *Rajiv Padhye*[2]
[1]RMIT University Vietnam, Vietnam; [2]RMIT University, Melbourne, VIC, Australia

5.1 Introduction

Wearing clothing by human being started since the time immemorial and is a feature of all human societies now. The type of clothing in a specific society mainly depends on social, cultural, and geographic factors. As the world population is ever growing, the demand for clothing, fashion accessories, and other textiles is also growing. The growing global demand is fulfilled by the textile and clothing manufacturers mainly located in developing countries such as Bangladesh, Cambodia, India, and Vietnam (Nayak and Padhye, 2014). The demand for cheap clothing and rising labor cost in developed countries in addition to globalization has resulted in the shifting of manufacturing of these products to developing counties (Nayak and Padhye, 2015).

Garment manufacturing from the day of it's invent is a labor-intensive process, which has not done much progress toward automation (Nayak and Padhye, 2015). The manufacturing of textile, clothing, and fashion accessories in developing countries is still done by manual practices due to the high cost of installing automated tools and equipment. However, due to high demand on garment quality and increased consumer awareness is leading to the use of automated tools and equipment in recent years during garment manufacturing. Every year, increased numbers of garment manufacturing industries are adopting the automation process to facilitate speed of production and improved quality of products. The automation in garment production is becoming a reality due to the technical developments and the use of modeling and simulation (Nayak and Padhye, 2011).

In this era of information technology, artificial intelligence (AI) has revolutionized the field of engineering, physics, medicine, and management (Vas, 1999). Traditional mathematical models are used to solve problems or in the decision making process, which is the key principle of AI (Nayak et al., 2016). AI can provide superior solutions to various problems due to its heuristic and intelligent characteristics. Significant results such as improving quality, increasing productivity, and lowering production cost can be achieved with the help of AI (Shamey and Hussain, 2003; Wong et al., 2004).

5.1.1 History of artificial intelligence

Around three decades ago, the main emphasis on apparel production was solely on the volume of production, which has now changed a lot. The recent focus is on achieving

Automation in Garment Manufacturing. http://dx.doi.org/10.1016/B978-0-08-101211-6.00005-7

good quality, reduced cost, appropriate information management, statistical process control, just-in-time manufacturing, and computer-integrated manufacturing in addition to the volume of production (Nayak and Padhye, 2014; Guo et al., 2011). AI is increasingly used in apparel design, pattern construction, production planning, marker planning, sewing floor, sales forecasting, and supply chain. A number of AI techniques such as expert system, neural network (NN), fuzzy logic (FL), genetic algorithm (GA), evolution strategy (ES), artificial immune system (AIS), and multiagent system (MAS) are used in clothing manufacturing.

AI is a field of computer science that can simulate characteristics of human intelligence and human sensory capabilities (Moon and Ngai, 2008). AI systems can provide superior solutions over classical systems due to their heuristic and intelligent nature. For example, it is too difficult to use classical systems to get global optima for the assembly line balancing problem, which can be easily achieved by the use of the GA. Conventional computers lack the ability to learn, and this restricts them to operate only under the conditions for which they are programmed. In essence the limitations of computers can be summarized as (Nayak et al., 2016):

- They are not intelligent,
- They have no self-awareness, and
- They have no understanding of the environment.

In contrast although humans are slow and unreliable compared to computers, they possess the intelligence and understanding that allow them to solve infinitely varied problems. AI is the creation of machines that can simulate human intelligence and work accordingly. The heuristic and intelligent nature of AI assists in providing superior solutions over the classical techniques.

Apparel manufacturing is labor-intensive, which is characterized by low-fixed capital investment; a wide range of product designs and, hence, input materials; variable production volumes; high competitiveness; and often high demand on product quality (Nayak and Padhye, 2014). Although the manufacturing process is associated mainly with apparels and household linens, it is also used in a variety of industries and crafts such as upholstery, shoemaking, sailmaking, bookbinding, and the production of varieties of sporting goods. Sewing is the fundamental process with ramifications into a variety of textile arts and crafts, including tapestry, quilting, embroidery, appliqué, and patchwork (Nayak and Padhye, 2015).

The apparel manufacturing process evolved as an art and underwent several technical changes (Nayak and Padhye, 2014). The technological advancements in the apparel industry include the use of computerized equipment (especially in design, patternmaking and cutting), three-dimensional (3D) scanning technology, integration of wearable technology, advanced material transport systems, automation, and use of robotics or AI. The recent important development involves increasing use of robotics or AI to transport components and materials within the plant or facilitate the production, which helps in improving the production efficiency. However, the apparel industry especially the sewing technology has remained significantly less automated compared to many other manufacturing industries.

5.1.2 Current status of artificial intelligence

The clothing production process has been described in Fig. 5.1. It starts from the conceptualization phase, passes through design development, manufacturing, supply chain, and retailing till it reaches to the consumers (Nayak et al., 2015a). In the conceptualization stage a designer conceptualizes a theme based on the forecasting of trends in the color, fabrics, silhouette, and trims. The concept is then translated into mood boards as abstract images. Inspiration boards are also prepared from the images and trims to represent the final theme.

The abstract images are subsequently transformed into real fashion illustrations by using computer-aided design (CAD) software, which is the first stage in the design development process (Kennedy, 2015). In this stage the color, silhouette, fabrics, and trims are finalized. The selection of the color palette is done from the mood board, considering the future fashion trends based on forecasting. The selected colors are matched with the pantone shades for a closest match. The right type of fabric is selected to produce the silhouette. Trims and embellishments are selected to complement the design and which are cost competitive. Subsequently, the initial garment is created in white muslin and draped on mannequin to visualize the appearance of the garment (Nayak and Padhye, 2017).

Once the design development has been finalized, the garment production process starts. The garment production process involves fabric spreading, cutting, bundling, sewing, pressing, inspection, and packing (Fig. 5.2) (Nayak and Padhye, 2015). Fabric is the major component of a garment and it is the input material for

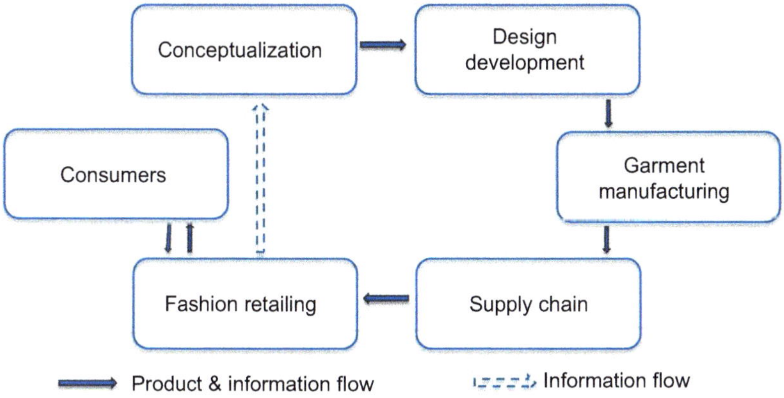

Figure 5.1 Production processes involved in the clothing manufacturing.

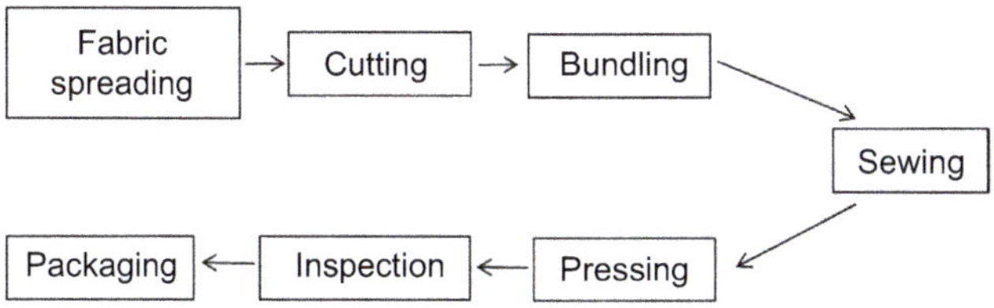

Figure 5.2 Process sequence of garment manufacturing.

many garment manufacturing industries. Once the fabric is received, the quality is inspected, stored for some time, and then fabric is spread for cutting. Depending on the garment design, several components are cut using various cutting equipment. The cut components are bundled and fixed with a bundle ticket, which is then passed to the sewing floor.

The garment components are sewn by skilled operators, pressed and the finished garments are inspected for quality (Nayak and Padhye, 2015). Then the garments are packed and sent to the retailers by own or third party logistics providers. The retail point is the place where the consumers buy their product. Today's consumers are much conscious on garment style, fit, quality, and price. Majority of these parameters are governed during the production steps as described in Fig. 5.2. Hence, during these processes, various tasks are needed to be controlled by the managing staffs, which may be difficult in several instances. The use of AI can help to control these problems effectively for decision making, cut order planning, marker making, production planning, supply chain management (SCM), and retailing (Nayak et al., 2016).

Apparel manufacturers have to produce a diverse product mix as consumers are difficult to understand and predict (Pfahl and Moxham, 2014). Their choice is unstable and unpredictable, and there is a wide variation in their demographics and physiographics. The product quality depends on several factors related to yarn manufacturing, fabric preparation (weaving and knitting), fabric chemical processing, and garment manufacturing. Hence, all these factors can be better controlled by the application of AI in the whole process of apparel manufacturing. Although there is some automation, the apparel industries are still far behind the other sectors and rely on manual intervention. AI is gaining impetus over the last two decades, in the apparel industry in different areas. The automation of various instruments by the application of AI in spreading, cutting, sewing, and material handling can reduce the production cost and minimize faults.

The production of textiles and clothing involves a large number of variables relating to the material and process (Nayak and Padhye, 2015). As there is high variability in raw materials in addition to the multiple stages of operation, it is hard to precisely control the process parameters to achieve a desired output. Until now, establishing a proper relationship between these variables and the properties of a fabric depends on human expertise. In many instances there are chances of error involved with human working as it is a difficult task to always remember such a large number of variables and apply the knowledge for accurate property prediction. This is possible by the application of AI as the developments in computation and simulation have created various systems to deal with multiple variables. The application of AI can now deal with a large range of datasets during training to establish an effective relationship between the variables and the product properties (Nayak et al., 2016). Therefore, over the last decade the use of AI is rapidly growing in textile and clothing manufacturing industries for various applications.

This chapter focuses on the types of AI techniques that are employed in the garment manufacturing process. Various applications of AI in the textile industry such as fiber, yarn, and fabric production are described. It also covers the applications of AI in apparel industry such as pattern design, production planning and control (PPC), marker making, sewing automation, sales forecasting, fashion recommendation, SCM, and retail. The challenges faced by the AI and the future directions are also discussed.

5.2 Types of artificial intelligence

AI can be broken into several different disciplines. Each of them is unique but often they intermix to accomplish a specific task. Till date there are several types of artificial neural networks (ANNs) that can be employed to solve a specific problem. The selection of a right type of ANN depends on the characteristic features of the problem and the number of variables influencing the parameter that need to be investigated (Nayak et al., 2016). The different disciplines of AI that are mainly used in garment manufacturing process are as follows:

- Expert system,
- NN,
- FL,
- ES,
- AIS,
- Generalized regression NN,
- GA.

5.2.1 Expert system

Expert system is a knowledge-intensive computer program, which solves a problem that normally requires human expertise (Nayak et al., 2016). It operates in the same manner as a human expert. Expert systems are made of two parts: a knowledge base and an inference engine. The program in the expert system incorporates the knowledge of the expert consisting of facts and heuristics by storing it in the form of rules. The inference engine is the software used to search and interpret the rules in the knowledge base by logical deduction. Facts are the information that is widely available and generally agreed on by the experts. Heuristics are the rule of thumb derived from personal experience and judgment of individual experts. Production rules of IF/THEN form are the easiest way of coding the knowledge. "IF" portion contains the conditions and "THEN" portion contains the actions. By prompting the right questions and considering the user's reply, an expert system decides which element of knowledge to use as the basis for further questioning until the intended goal is reached. This goal is the information, which the expert is expected to provide.

Expert systems can be built in two ways. One way is to develop the computer code by using programming languages such as PROLOG and LISP or by using any of the high-level programming languages such as BASIC and FORTRAN. The other way is to use expert system building tools called expert system shells. These are specially designed programs by specialists in computer science. MYCIN is the first expert system developed in the world. This expert system has been used by physicians to diagnose certain forms of bacterial blood infections and to determine suitable treatments for various diseases.

5.2.2 Neural network

Mankind is distinguished from other animals by their ability to think and their intelligence. Animals react adaptively to the changing environment by their nervous system

Figure 5.3 Image of a simple neuron used in artificial neural network .

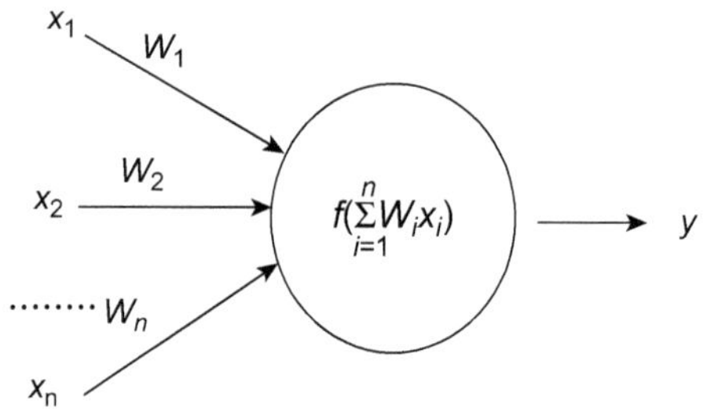

(Nayak et al., 2016). Similar responses are produced by the appropriate simulation of the nervous system. The nervous system is built up by simple units: The Neurons. Human brain consists of 1011–1014 neurons. These neurons receive any process and transmit electrochemical signals via synaptic connections. ANN behaves similar to a human brain in a digital computer. ANN may be applied where traditional techniques have failed to give satisfactory results, or where small improvement in modeling performance can make a significant difference in operational efficiency.

ANN is the representation of biological NN composed of many artificial neurons, which are linked together by special network architecture and have the ability to learn from examples. Though ANN was established before the advent of computers, it has evolved into a powerful tool for solving complex problems only recently. In AI, the input signal is converted to output signal by the neurons obeying some transfer function. This output becomes an input to the subsequent neuron. A simple neuron is shown in Fig. 5.3.

In the above figure $x_1, x_2, \ldots, x_n$ are inputs; $W_1, W_2, \ldots, W_n$ are the weights, and y is the Output. The single neuron converts the input units to output units according to the weights and the transfer function. The single neuron acts in a linear behavior. This is also known as single layer perception. All most all the NNs are based on multilayer perception (MLP) consisting of an agglomerate of neurons interconnected in some specific architectural way. Fig. 5.4(a) shows the typical NN with one hidden layer and Fig. 5.4(b) with multiple hidden layers between the input and the output. The selection of the number of hidden layers is important. AI may follow a linear function or a nonlinear (sigmoid) function.

In MLP the units are arranged in distinct layers and each unit receives weight input from each unit in previous layer. Initially the output is calculated with initial set of weights. Then weights are modified according to the expected output by reducing the difference between the actual and expected output values. Each input and output pairs would be presented to the network in turn and small changes made to weights every time. The presentation of all the training pairs to the network is known as an epoch. Training may require many hundreds of thousands of epochs and could take many hours even in a fast computer.

When the training is complete the values of the weights are fixed. During the recall phase the weights are not modified but the network can be used to predict the output values when set of input values are given. If the network has been trained correctly then it will be able to do this accurately for all the training data and also for other input values that were not included in the training set. This important property is called generalization. The training of the NNs is a complex activity and it can be done by back

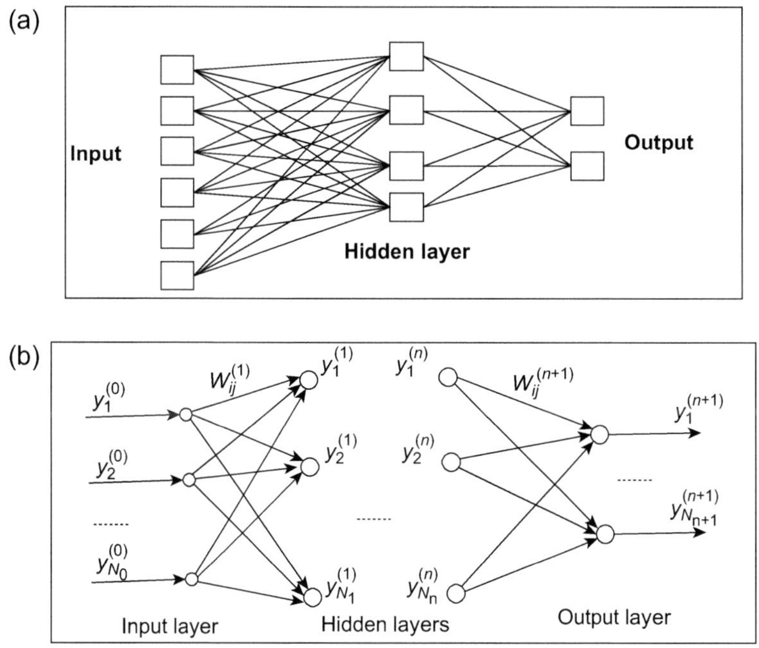

Figure 5.4 (a) Neural network (NN) with one hidden layer and (b) NN with multiple hidden layers.

propagation (BP), counter propagation, standard conjugate gradient, dynamic learning vector quantization, etc. Among these the BP is widely used.

5.2.3 Fuzzy logic

FL is used where the behavior is not linear (e.g., the trend in today's dyeing industry is to use highly concentrated dyeing liquor that do not behave in a linear fashion due to interaction between the dye molecules). In FL, IF/THEN reasoning is used when applying a fuzzy inference system. Adaptive–neuro-fuzzy inference systems (ANFIS) are used where the BP method is implemented like work for modeling or fitting of spectroscopic data. The number of membership functions (MFs) has a clear influence on the ANFIS' training results. The root-mean-square error is determined as the difference when the number of MFs increases.

ANNs based on FL have shown a high success rate in AI applications in many fields including the garment manufacturing. FL-based ANN systems have been applied in various fields of textile clothing research such as prediction of the thermal environment, thermophysiological comfort (Hamdi et al., 1999; Wong et al., 2003; Hui et al., 2004), recipe prediction, color solutions, supply chain, and retail management.

5.2.4 Evolution strategy

ES is an optimization technique in computer science, which is based on the theory of adaptation and evolution. ES is one of the artificial evolution methodologies, which acts as search operator by using natural problem-dependent representations. Similar to the evolutionary algorithms (EA), the operations in ES are applied in a loop. A generation is known by iteration of the loop and during the operations, the sequence of generations is performed till it meets with a termination criterion.

5.2.5 Artificial immune system

AISs are computational algorithms that work on the principles of the vertebrate immune system. AIS is inspired by theoretical immunology and observed immune functions, which is an adaptive system based on principles and models. AIS can be used to solve problems in mathematics, engineering, and information technology by abstracting the structure and function of the immune system to computational systems. AISs are based on biologically inspired natural computing and widely used in the area of AI. AISs are applied for complex problem solving in computing, which are inspired by theoretical immunology and observed immune functions.

5.2.6 Generalized regression neural network

A generalized regression neural network (GRNN), mainly used in function approximation, is a one-pass learning algorithm consisting of highly parallel structures. Based on the nonparametric regression, GRNNs represent an improved technique in the AI field. The algorithms used in GRNN provide a smooth transition from one observed value to another in spite of small number of data. GRNNs are selected for AI applications due to the following reasons:

- GRNNs take very less time to design and training despite they need higher number of nodes in their architecture, and
- GRNNs can closely approximate any desired function with finite discontinuities.

5.2.7 Genetic algorithm

Both the constrained and unconstrained optimization problems can be solved by GA, which is a method based on a natural selection process that mimics biological evolution (Nayak et al., 2016). In GA, a population of individual solutions is repeatedly modified by the algorithm. GA is a part of the larger class of EA, which generates solutions for optimization problems using techniques inspired by natural evolution, such as inheritance, mutation, selection, and crossover. GA is an iterative procedure that maintains a population of chromosomes representing different possible solutions to an optimization problem. Each individual iteration is called a generation and in each generation, the fitness of each chromosome is evaluated by a suitable fitness function. By this approach, the optimal solution can be obtained in various fields of apparel manufacturing such as apparel design, marker planning, PPC, production scheduling, fabric and other materials location, and replenishment decision making in apparel supply chain.

5.2.8 Other artificial intelligence approaches

Various other approaches such as genetic programming (GP), MAS, and simulated annealing (SA) have also been used in various sections of apparel manufacturing. GP is an EA, which is inspired by biological evolution and based on computer programs that perform a user-defined task. MAS is composed of multiple interactive intelligent agents, those possessing the characteristics of interaction and cooperation and can help the agents execute different tasks with autonomous functions on receipt, transfer and delivery of information simultaneously. SA approach is based on a generic probabilistic metaalgorithm, which is inspired by the annealing process used in metallurgy.

5.3 Applications of artificial intelligence in apparel industry

A detailed outline of apparel manufacturing process starting from fiber till the consumers has been given in Chapter 1, and the complete process of garment manufacturing is described in Fig. 5.2. AI can be used in various processes of textile production such as fiber grading, prediction of yarn properties, fabric fault analysis, and dye recipe prediction. Similarly, AI can be applied in all the stages (preproduction, production, and postproduction) of garment manufacturing. Garment manufacturing involves processes such as conceptualization, design development, PPC, spreading, cutting, bundling, sewing, pressing, and packaging. Some of the major applications of AI in textile and garment manufacturing are discussed in this section. Out of several types of AI as discussed above, ANN is widely used in garment manufacturing mainly in following fields:

1. Prediction of mechanical properties,
2. Classification and grading,
3. Identification and analysis of faults,
4. Process control and online monitoring, and
5. SCM and retailing.

The following section describes the application of AI in various production processes involved with garment manufacturing.

5.3.1 Application of artificial intelligence in fiber and yarn production

Textile fibers are the basic raw material for the production of clothing and other textiles. As there are many different types of textile fibers, it is often difficult to identify an unknown fiber by visual inspection. The traditional practices of fiber identification are based on destructive tests using flame or chemicals. The recent advancements include the use of optical microscope, Fourier transform infrared and Raman spectroscopy (Nayak et al., 2012). AI can also be used to identify and grade textile fibers according to their color and other properties such as fineness, length, uniformity ratio, tenacity, and effect of spinning performance on yarn properties (Jasper and Kovacs, 1994; Chattopadhyay and Guha, 2004).

Among the several methods of yarn manufacturing, ring spinning is the robust and versatile for a wide range of yarn production. Operational parameters such as ring size, traveler type, spinning speed influence the spinning performance and yarn properties (linear density, amount and direction of twist, yarn strength, and uniformity). As several independent variables are involved during yarn manufacturing, the application of AI is rather difficult to cover the entire range of parameters. The assessment of interactive contribution of each independent variable and interpolating and extrapolating the experimental observations or mill measurements is also difficult. There have been several applications of AI in yarn manufacturing that includes virtual modeling of yarn from fiber properties (Yin and Yu, 2007), prediction of yarn tensile properties (Majumdar and Majumdar, 2004; Üreyen and Gürkan, 2008), prediction of yarn unevenness (Demiryürek and Koç, 2009), and yarn engineering (Majumdar et al., 2006).

Two different spinning industries can use the same fiber to produce a specific yarn count as per the established norms. However, the spinning performance of the yarns usually varies from one to the other industry. Therefore, empirical models need to be developed to address all the known processing variables existing in different spinning industries. Furthermore, it is also needed to generalize this information for an individual mill so that the yarn quality is predicted accurately. Conventional techniques involve several limitations due to strict assumptions of normality, linearity, and variable independence. On the other hand, ANNs are based on universal approximations, they can learn directly from the data being modeled.

5.3.2 Application of artificial intelligence in fabric production

The major raw material for a clothing industry is fabric. The quality of the fabric influences the quality of the garment, productivity, and the ease with which garments can be manufactured. The fabrics are selected based on the type of the garment and their end use applications (Kumar et al., 2016). The fabric specifications for making any garment can be classified as primary and secondary. The physical dimensions are considered to be the primary, whereas the fabric reaction to external forces is considered to be the secondary. From consumer perspective, garment appearance, comfort, and durability are the important parameters. AI can be applied to control these parameters and achieve the fabric specifications as discussed below.

5.3.2.1 Predicting fabric properties

In today's competitive market, clothing manufacturers are trying to reduce the lead time, which influences the product development process. The fabric manufacturers have to work hard to meet the quality requirements during production and deliver the fabric on time. Hence, the use of modern computational and simulation tools can help to reduce the time to predict the fabric behavior. For example, AI can be used to predict the fabric properties before manufacturing with the help of neuro-fuzzy or other approaches by using the fiber, yarn, and fabric constructional data. While applying AI, it is essential to establish a proper linear and nonlinear relationship between the input fiber and yarn parameters and the property of the fabric need to be predicted. However, the application of AI can be very expensive for the fabric manufacturers, which can increase the cost of production.

Several researches have been done to predict fabric performance properties, comfort, and durability by the application of AI. For example, Behera and Mishra (2007) used the ANN to predict the aesthetic and functional properties of worsted suiting fabrics. The ANN was able to predict the fabric properties accurately from fiber properties and fabric constructional features. There was a good correlation between the predicted results and experimental data. Beltran et al. (2005) investigated fabric pilling propensity by using ANN and establishing the relationships among fiber, yarn, and fabric properties. This model was used for pure wool knitted fabrics to predict the pilling performance. Similarly, Unal et al. (2012) used regression analysis and ANN models to predict the properties of single jersey fabric. Murrells et al. (2009) investigated the spirality of fully relaxed single jersey fabrics by using ANN.

The other applications include the application of NN and image processing technology to classify woven fabric patterns (Jeon et al., 2003); application of Taguchi methodologies to predict the strength of woven fabrics (Zeydan, 2008); application of BP algorithm-based (Behera and Karthikeyan, 2006) ANN for the prediction of initial load-extension behavior of plain woven fabrics and their derivatives (Hadizadeh et al., 2009); prediction of drape profile of cotton woven fabrics using multiple regression-based ANN (Pattanayak et al., 2011); engineering the drapability of textile fabrics (Stylios and Powell, 2003); and development of ANN-embedded expert system for the design of canopy fabrics (Behera and Karthikeyan, 2006).

There have been growing interest of consumers on the clothing sensorial and thermophysiological comfort properties. Several research organizations around the world are involved with the research and development of clothing comfort properties, relating to the fabric feel (sensorial comfort) and transportation of heat and sweat from body, while wearing clothing (thermophysiological comfort). Both the comfort aspects (sensorial and thermophysiological) can be investigated in laboratory by the use of appropriate equipment such as Kawabata evaluation system (KES) to predict fabric sensorial properties and a set of other equipment to predict fabric air permeability, thermal resistance, water vapor resistance, and moisture management properties.

In addition to the above listed equipment, AI can also be applied to investigate the comfort properties. While sensorial comfort is considered, fabric can be classified according to their hand value by the application of AI (Wong et al., 2003; Hui et al., 2004; Park et al., 2000). Generally, the prediction of hand value is subjective based on the experts' opinion after touching the fabric. However, there have been several researches for predicting fabric hand by using AI. The results can be used for selecting right type of fabrics according to the season or type of clothing. A fuzzy neural network (FNN) can predict the total hand value of outer fabrics (Halttunen et al., 1999). While developing a NN for hand prediction from fabric properties, it is essential to establish a relationship between the fabric properties and hand attributes (Hui et al., 2004).

There have been wide area of application of AI in predicting fabric thermophysiological comfort properties (Luo et al., 2007; Wang et al., 2005). For example, Wang et al. (2005) developed a FL-based system to predict fabric thermal comfort properties. The subjective perceptions of thermal comfort were based on simulated results of thermal and moisture sensations, while wearing the cloth. A modeling system consisting of a series of published mathematical models was used to simulate heat and moisture exchanges between human skin and clothing. In this study the thermoregulatory system of human body and the neuropsychological responses were taken into consideration.

Luo et al. (2007) proposed a FNN to predict the clothing thermal comfort, while designing functional textile systems. This model was based on the human body's core and skin temperatures. The feed-forward BP network was used for developing the FNN. The use of FNN helped to reduce the amount of training data. Furthermore, for acquiring, representing, and using the knowledge of the domain expert, FNN helped to achieve a human–machine friendly knowledge representation scheme. A good correlation was established between the experimental results and the results predicted by the FNN. NN systems are also used to predict the thermal resistance of fabrics (Bhattacharjee and Kothari, 2007; Alibi et al., 2012).

AI can be used to predict fabric end use applications based on fabric properties. For example, Chen et al. (2001) used a NN to predict the end uses of fabrics from its properties at low stress such as extension, shear, bending, compression, surface friction, and surface roughness. These parameters were measured by using the KES instruments. The fabric end use application was determined by the NN with an estimated error rate of 0.07 during the prediction.

5.3.2.2 Color solutions

Color is the first element of design to which the consumers respond, hence it is one of the important features of textiles. Consumers select or reject the clothing or other fashion accessories on the basis of the color appeal. Hence, for getting the right color, precise quality control during dyeing and printing is essential, which can affect the volume of sales. Both the dyeing and printing process should ensure that the required color fastness, depth of shade, color matching, and surface characteristics have been achieved. These parameters are influenced by the dye and fabric combinations and the chemical rules governing them. Deviation of these parameters from the allowable limit may lead to reprocessing or rejection of the whole fabric batch.

The use of AI can resolve these issues, which can be used for recipe prediction; process control during dyeing and printing; color matching; and evaluation of the final dyed or printed fabric. This process involves the training of the NN with a set of input (color values) and output data (concentrations of each dye). Once trained, the NN can be used to calculate the concentrations for any set of unknown colored samples.

One of the applications of AI for color solution is during the fiber blending stage, while the roving is converted to yarn. The roving is converted into yarn, and subsequently the yarn is converted into fabric by weaving or knitting. As there is no chemical compound used in this stage, the color effect is not influenced by any chemical rules. However, the intimacy of mixing of different types of fibers rather physically governs the color. Hence, the color production depends on the nature of the fibers and the compatibility of the blend. The use of AI can assist in predicting the color produced when fibers of different colors are mixed together. In the case of a homogenous blend, the prediction of color can be performed more accurately by using theoretical and empirical models.

AI can be used for color matching of fabrics and shade sorting. It can be used for the true color production by predicting the concentration of dyes from their spectrophotometric absorbance. For example, Senthilkumar (2007) designed a NN for predicting the CIE L*a*b* values for cotton fabric dyed with vinyl sulphone reactive dye (Fig. 5.5). A multilayer feed-forward network was used to predict the required shade of color. The CIE L*a*b* values of the fabric were predicted by this NN, and the results were accurate with about 2.0% of the average error. For other dyes and for the input and output parameters beyond the training range, the NN also showed similar error.

It is very important to achieve the required depth of shade of the fabric after dyeing is finished. Correct recipe and accurate timing during dyeing are essential to achieve a good dyed fabric. When the difference in shade between the dyed good and control exceeds the tolerance limit, then the whole lot is either reworked or rejected. Statistical tools such as multiple regression analysis can be used to predict the duration of dyeing. However, the application of ANN has been found to be more accurate than the as multiple regression analysis (Senthilkumar and Selvakumar, 2006; Jasper et al., 1993).

Other application of AI in color solution involves: application of ANN and linear regression (LR) approaches to predict the color of laser-treated denim fabrics

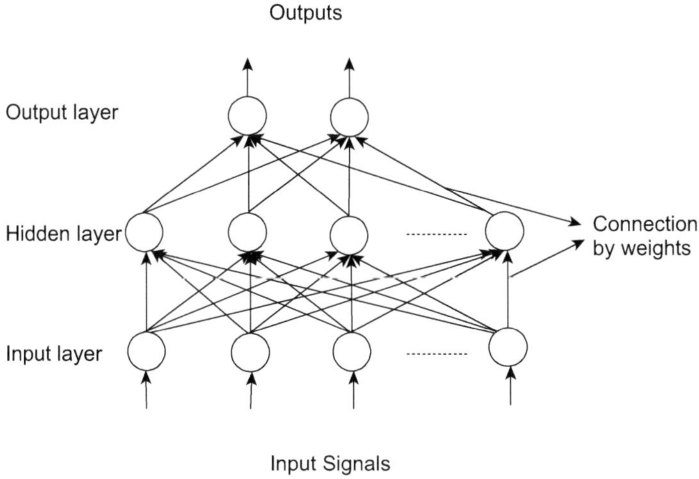

Figure 5.5 Artificial neural network for color prediction (Senthilkumar, 2007).

(Hung et al., 2014); application of FNN in classifying dyeing defects in fabrics (Huang and Yu, 2001); application of ANN to predict the color properties of 100% cotton fabric treated with laser (Hung et al., 2011); application of ANN to predict the efficiency of dye decolorization with UV/$K_2S_2O_8$ (ultraviolet/potassium peroxydisulfate) (Soleymani et al., 2011); and the use of BP NN for computerized color separation system for printed fabrics (Kuo et al., 2007).

5.3.2.3 Fabric fault detection

A poor quality fabric can result in substandard garments as well as reduces the productivity during garment manufacturing. Any defect in the fabric is passed into the final garment, which can result in the rejection of the garment. Hence, it is essential to check the quality of the fabric before manufacturing the garment (Vijayan and Jadhav, 2015). Generally, fabric inspection is performed by skilled workers using lighted tables or equipment. This process is rather slow and many times can allow faults to pass to the garment. Furthermore, the efficiency of the fabric inspectors will be reduced quickly with fatigue. However, the use of AI can perform this task at a faster rate, with much higher accuracy and without fatigue.

AI can be applied for fault inspection and grading of the fabrics that are received in a garment industry (Tsai and Hu, 1996). The image of the faults are stored in the system and compared with the captured image (Fig. 5.6) (Shady et al., 2006). If a difference is observed, the defect is identified and the inspector calculates the fault points.

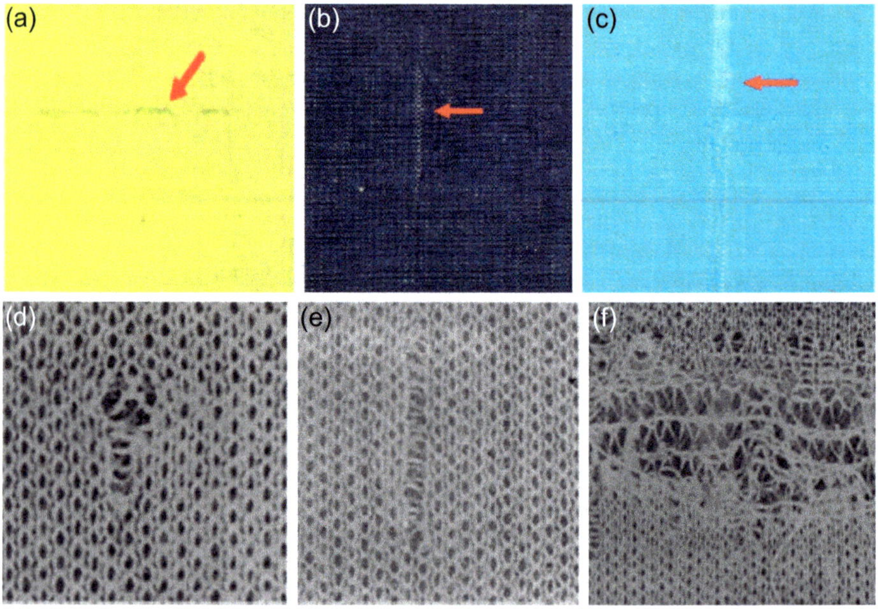

Figure 5.6 Different fabric defects inspected (*arrow* indicates defects) by artificial intelligence: (a) gout, (b) warp float, (c) draw back, (d) hole, (e) dropped stitches, and (f) press-off (Ngan et al., 2011).

Furthermore, during the fabric manufacturing a pretrained NN can be used to detect and mark the fabric faults in the loom or knitting machine by charge coupled device (CCD) cameras as shown in Fig. 5.7. This can be used as a process control tool and save the time and cost of fabric checking at latter stages. The grading of fabrics can be done on the basis of frequency and severity of the faults by the application of AI. Fabric quality is dependent of the fiber type, yarn properties, and fabric structure. For different types of clothing, the fabric quality requirements are different. AI can be used for appropriate fabric selection depending on the end use application of the clothing.

Various applications of AI in fabric fault inspection includes application of BP algorithm-based ANN to recognize fabric defects (Tsai and Hu, 1996); detection and classification of knitted fabric faults based on image analysis and ANN (Shady et al., 2006); detection and classification of defects with ANN using two kinds of optical patterns (Tilocca et al., 2002); detection of fabric faults using a feed-forward-based NN (Kumar, 2003); classification of woven and knitted fabric based on FL and BP learning algorithm-based ANN (Huang and Chen, 2001); develop an automated visual inspection method using the wavelet transform to detect defects on patterned fabric (Ngan et al., 2005); and detection of faults in woven fabrics by ANN-trained structuring elements (Chandra et al., 2010).

5.3.3 Application of artificial intelligence in garment manufacturing

The garment manufacturing process is becoming more automated to cater the increasing demand of consumers, reduce the number of faults, and keep the production cost low. AI is increasingly used to predict the performance of a sewn seam, designing of clothing, in PPC, in various sewing operations, and in quality control. AI can be applied for intelligent manufacturing of clothing to predict the clothing properties after a particular process. So it can be used for suitable garment designing by fabric engineering and monitoring the garment manufacturing processes.

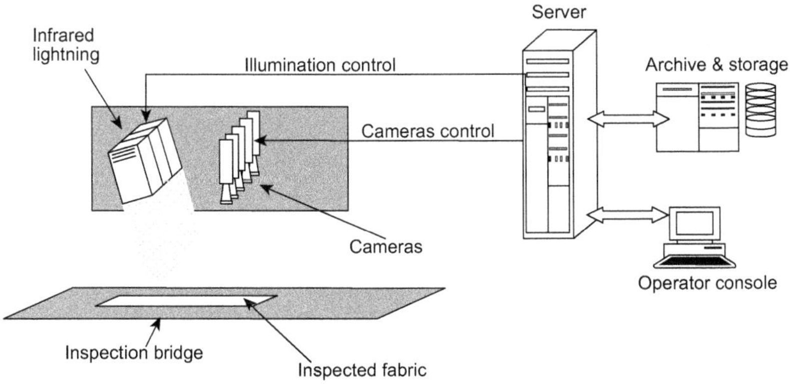

Figure 5.7 Configuration of a typical automatic inspection system (Abouelela et al., 2005).

5.3.3.1 Performance of sewn seam

In sewn garments seams and stitches are used to join two or more pieces of fabric together. The ease of seam formation and the performance of the seam are the important parameters, which are judged by the term known as "sewability." Fabric low-stress mechanical properties such as tensile, shear, bending, and surface can affect the sewability. AI system can be used to find the sewability of different fabrics during garment production (Nayak et al., 2010, 2013). Fabric mechanical properties affect the performance during spreading, cutting, and sewing. For example, Chen et al. (2001) used fabric mechanical properties as input parameter to evaluate their performance during garment manufacturing.

A good quality seam is essential for a good quality garment. Variations in seam quality can be caused by improper setting of sewing machine or by human error. The sewing machine parameters need to be optimized depending on the fabric properties and type of seam before starting the sewing operation. Fabrics parameters such as stretch, distortion, needle cutting index, and other mechanical properties govern the efficiency of sewing, which need to be detected before optimization. Similarly, a skilled operator is essential to produce a good quality garment free from sewing faults.

The performance of a sewn seam depends on the type of fabrics and sewing thread combination; seam and stitch type; and sewing conditions, which includes needle size, stitch density, and the sewing machine condition. The performance properties of the seam are evaluated by seam puckering, seam slippage, and yarn severance, which can be predicted by AI. Seam puckering is detected using BP algorithm by taking fabric areal density, thickness, and bending stiffness in warp and weft direction as input. In addition, two networks can be used for the prediction of puckering. The process involves pattern reorganization by a laser beam followed by training of the network. The seam performance properties can be accurately predicted from the fabric manufacturing parameters by training the network with sufficient number of replicated experimental data.

Hui et al. (2007) and Hui and Ng (2009) applied ANN to investigate the performance of fabrics during in garment manufacturing based on seam puckering, seam flotation, and seam efficiency. The ANN model was very accurate in detecting fabric faults although a small amount of training data set was used. Several other researches on the prediction of seam performance include: application of ANN for automated quality control of textile seams (Bahlmann et al., 1999); application of ANN to predict the performance of fabrics during sewing (Hui and Ng, 2005); and the application of ANN for modeling the seam strength and elongation at break of fabrics (Yildiz et al., 2013).

5.3.3.2 Computer-aided design systems

One of the important steps in garment manufacturing is patternmaking, where paper patterns are made by the designers and subsequently digitized to a computer (Kennedy, 2015). Several two dimensional (2D) patterns are prepared for a garment, which are the basic blocks of a 3D garment. Various CAD software are used in the garment industry for patternmaking, digitizing, grading, and marker planning. The CAD software helps in achieving high productivity and improved quality. The designers involved in clothing designing create numerous designs by using the CAD software. However, the CAD software cannot be used to automatically generate clothing patterns or designs for a specific garment style. In addition, in many garment industries the traditional method of garment pattern generation is still done by experienced designers and does not include the use of CAD, although there is the scope of using AI in pattern generation.

Several researches have been done to implement the AI that can help to develop basic clothing patterns automatically. For example, Inui (1996) had developed an AI integrated CAD system (combination apparel CAD and GA) that can be used to search apparel designs that the system users prefer. The search process involves the man–machine interaction cycles, where the user assesses the examples produced by the systems. An IGA (interactive genetic algorithm)-based fashion design support system was developed by Kim and Cho (2000) for women's dress design. This system can predict the optimal fashion design by using human preferences and emotions. Virtual Reality Modeling Language (3D Studio MAX) and OpenGL were used to enhance the system interface.

Fang and Ding (2008) developed an expert knowledge-based system to automatically prepare the basic patterns. Similarly, an interactive CAD system was designed by Hu et al. to create various patterns for a garment. This CAD system allows the collaboration of different designers and experts to best utilize the expert system to create a specific design. The limitations that existed in iterative GA were investigated by Gong et al. (2007), who proposed a novel iterative GA with multipopulation adaptive hierarchy for clothing design solutions. The novel iterative GA was successfully applied in fashion design, which suggested that the IGAs are feasible and efficient.

During patternmaking, the factor "ease allowance" is considered to provide comfort and facilitate body movement to the wearer. The ease allowance depends on the type of fabric and wearer's body shape and body dimensions. Appropriate value of ease allowance is essential for successful pattern designing with right fit and comfort to the wearer. This area has drawn the attention of large number of researchers. For example, Chen et al. (2006) developed and investigated mathematical models to establish a good relationship between ease allowance for jean type trousers and the related elements (such as garment constructional parameters, wearer's requirements on comfort, and wearer's body sizes). An intelligent model was developed to deal with the clothing pattern creation in terms of sensory evaluation of wearers. This model helped in producing new values of ease allowance depending on the wearer's body shape and size to facilitate the wearer comfort.

This work on the ease allowance was further extended in another study (Chen et al., 2009). A relationship was established between the wearer's key body positions and their movements. The aggregate ease allowance was estimated by utilizing an ordered weighted averaging operator. Jean type trousers were used to investigate the effectiveness of the research and for validating the results. Although it was successful, additional research work are needed to validate other garment types for the effectiveness in ease allowance.

To meet customers' demand, while selecting apparels for different occasions, Lin (2007) developed an intelligent design system that can help designers to design an innovative apparel styling for customers. GA-based system was used for the searching operation to search for different types of clothing combinations. ANN was used to construct the classifier through learning and knowledge acquisition to evaluate the mechanism of classifying the combinations of evolved clothing wearing styles. This system can help a customer with no prior knowledge in fashion design to select the design schema, which is the best solution for the demand. The limitation of this research was that it considered only several simple color types.

Nowadays the consumers want to have proper fitting of the garments they purchase to their body. However, the mass produced garments may lack in fitting in some body parts (Nayak et al., 2015b). The proper fitting of a garment is controlled during the patternmaking process by designing appropriate pattern sizes and shapes. Hu et al. (2009) used the data on customer's body size to develop an intelligent system for designing optimal garment component sizes to predict the fit of the garments. Precise values of body measurements were used to generate the exact body dimensions for proper fitting of the garments.

CAD systems are used in garment manufacturing for creating designs, patternmaking, and grading operations. Several attempts have been made by researchers to integrate AI with CAD systems to generate designs automatically. For example, an intelligent system was developed by Zeng et al. (2008) to assist the designers in decision making process. The objectives were to predict the fashion messages conveyed by the design elements in the garment and to extract design elements for real garment design.

Experienced designers are needed for appropriate pattern design of different clothing styles. However, the AI system can be used to provide the expert knowledge of experienced designers (Fang and Ding, 2008). Hu et al. (2008) designed an expert system that combines the knowledge of experts and designers. It provided valuable suggestion on specific design by interactive design functionalities. To set up a huge anthropometrical database for designing the garments that fit the wide ranges of variation in body type is time-consuming and uneconomical. To avoid that AI can be remarkably used for prediction of shirt pattern from 3D body measurements by multiple LR.

During the garment manufacturing process the 2D fabric is converted in to a 3D garment. The nature of the fabric influences the drapability of the 3D garment, which is rather a difficult task. The application of AI can help to predict the drapability and model to create 3D virtual garment designs. An intelligent system was developed by Fan et al. (2001) to predict the drape images for different clothing styles for women using different fabrics. The intelligent system searches from the database consisting of typical drape images of various garment styles and finds the most identical drape image with the specific fabric properties and feature dimensions. The intelligent system worked satisfactory in predicting the drapability of fabric for women's dress.

5.3.3.3 Production planning and control

PPC coordinates between various departments of production so that delivery dates are met and customers'orders are delivered on time. Various research activities focused on the problems related to PPC and avoid the bottlenecking (Wong et al., 2005; Guo et al., 2008). Majority of the studies were based on the problems in PPC relating to sewing floor such as fixing the machine layout, line balancing in sewing, and managing operators in the sewing floor. AI can be used to solve or optimize the problem of machine layout, operation assignment, and sewing line balancing. This can help in achieving the objectives of PPC.

AI based decision support system was used in decision making to determine the most appropriate manufacturing plant for a particular customer order (Chen et al., 2005). A GA-based real-time segmentation for rescheduling was developed by Wong et al. (2005) to deal with the PPC-related problems in sewing floor during: (1) marker making, (2) fabric spreading, (3) cutting, and (4) bundling.

In another study a GA-based system was designed by Guo et al. (2008) for production scheduling for each production order to appropriate assembly lines. This system was designed by considering various factors for production delay, production uncertainties (such as processing time, orders, and lead arrival times), and other bottlenecking.

5.3.3.4 Final garment inspection

The inspection of finished and semifinished garments during their production is essential to get fewer rejections. The final quality of a finished garment depends on the sewing quality and other faults present in it. The final quality inspection of finished garments is mainly done by experienced people, which is very time-consuming and often subjective in nature. The results of the inspection are influenced by the physical and mental condition of the inspector. Therefore, automated inspection devices are essential to achieve increased efficiency and accurate results. Although limited studies have been done, automated inspection can be performed by the use of AI and image processing for inspection of the quality of finished garments (Yuen et al., 2009a).

During the garment production, each process (cutting, sewing, and pressing) plays a vital role influencing the quality of the finished garment. The quality of the semifinished products should be inspected at each of these processes before the final inspection. The finished garments are inspected as per their specifications, overall appearance, faults, and sizing and fit. In detail the finished garments are inspected for stitching quality, mismatched plaids or stripes along the seam, puckered seam or extra material caught in seams, uneven seam along hems, and many other faults that can arise during the garment production.

The application of AI in final garment inspection includes: automatic classification of general faults in shirt collars (for monocolored materials) using machine vision (Norton-Wayne, 1990); application of AATCC (American Association of Textile Chemists and Colorists) wrinkle rating for evaluation of wrinkle by using a laser sensor (Kim, 1999; Mori and Komiyama, 2002); detection and classification of stitching defects using wavelet transform and BP NN (Wong et al., 2009; Yuen et al., 2009b); seam pucker evaluation by using self-organizing mapping (Mak and Li, 2008); and designing of a smart hanger for garment inspection (Yuen et al., 2008). In manufacturing the seamless garments (Nawaz and Nayak, 2015), AI can be used to detect faults online. The image of the final garment can be captured and compared with the standard and any variation from the standard is reported as a fault that can be mended at that time or a marking is done where the fault occurs.

5.3.3.5 Application in supply chain

SCM in fashion includes the flow of fibers, yarns, fabrics, garments, trims, and accessories in between different production points or to retail. It also involves the storage and control of all the above listed materials including the flow of information. SCM integrates various business processes, activities, information, and resources for creating value for the customers. Appropriate SCM can manage the cost and business competitiveness. Although there have been a wide application of AI in supply chain activities of other goods, there are several areas in fashion supply chain where there is limited application.

AI-based models can be used for integrating and sharing information at any point in SCM (Lo et al., 2008; Au et al., 2006). Various AI approaches can be used for supply chain planning (Chiu and Lin, 2004); supply chain demand forecasting (Carbonneau et al., 2008); optimization of supply chain network (Yu et al., 2011); managing the logistics in textile supply chain (Chandra and Kumar, 2000); sample management (Choy et al., 2009); effective inventory management (Paul and Azeem, 2011); and inventory replenishment (Dong and Leung, 2009; Pan et al., 2009).

5.3.3.6 Application in retailing

Fashion retailing establishes the link between the manufacturers of fashion goods with the consumers. Over the last two decades or so, fashion retailing has become one of the most competitive retail sectors due to technological advancements and behavioral changes of consumers toward fast fashion. There are several areas in retailing such as sales forecasting (Yu et al., 2011; Sun et al., 2008; Wong and Guo, 2010; Choi et al., 2014); fashion retail forecasting (Au et al., 2008; Xia et al., 2012); style suggestion to consumers (Hsu et al., 2009); customer relationship management (Ngai et al., 2009; Groves and Valsamakis, 1998); demand forecasting (Fumi et al., 2013); determining customer satisfaction (Goode et al., 2005); and fashion coordination; where AI application is ever increasing. Sales forecasting in fashion has become more challenging now due to volatility of demand as it depend on several factors. Historical data on sales in combination with the style, color, and garment size can be used for sales forecasting (Sztandera et al., 2004; Thomassey et al., 2005; Thomassey and Fiordaliso, 2006).

AI suggestion systems can be used for selection of appropriate style and design combination for consumers (Ding and Xu, 2008; Hsu et al., 2009). In several instances it is very hard to identify the subtle differences between two different styles. AI can be used to identify the differences and similarities between two or more different styles. For example, Koehl et al. (2008) used FL-based AI to characterize the dissimilarity and inclusion between fashion themes of fabric products. The dissimilarity between fashion themes identified by AI can be used to analyze the relevancy of the input space, which can establish relationship between consumer's fashion choices and the technical parameters of fabric products.

In addition it can be also used to select the right clothing combination depending on the occasion. In fashion coordination, AI can be used to provide improved service to customers and increased sales. Professional advice can be given to consumers on style selection based on price. Today's consumers are more aware on the comfort features than before. AI can also be used for selection of right type of garment for providing necessary comfort including the appearance, which can be used by the customers (Wong et al., 2004; Luo et al., 2007; Nayak et al., 2009).

While making the purchase decision of a garment, the consumers evaluate the fabric performance by using their sensory perceptions, which is more convenient and direct. A fabric advisory system was developed by Lau et al. (2006) to assist fashion designers in selecting appropriate fabric in fashion product development by using 14 fabric hand descriptions as inputs.

In addition to the fabric hand, consumers also try to inspect the texture of the fabric. Fabric texture is the quality of a textile surface. Texture can be rough or smooth and it greatly influences the design process of a garment. Several researches have been done to investigate the texture of a fabric. For example, to produce optimal results in generating fabric texture, Muni et al. (2006) developed an interactive texture generation process using GP. The GP-based texture generation process was really helpful in generating textile texture. However, the GP-based approach was not suitable for generating color texture of textiles.

Jung et al. (2003) proposed the Fashion Design Recommender Agent System based on customer's representative sensibility and preference by applying two-way combined filtering technologies. A database was established by extracting the representative sensibility adjective from user's sensibility and preference about textiles to create textile designs. Although the system helped in recommending textile designs to a customer with similar choice on textiles, the accuracy of prediction by the agent system was not high, which needs further improvement for reliability and practical applications.

5.4 Challenges and future directions of artificial intelligence

5.4.1 Challenges faced by artificial intelligence

Although the use of AI is increasing rapidly in apparel production starting from raw material selection to the final product, there are several challenges faced by the technology. In fiber selection and classification, the accuracy of classification needs to be improved by adopting more powerful learning strategies. For the accurate predictions of number of end down in yarn manufacturing and the occurrence of neps, more work is needed for the simulation of AI. Further developments are needed to accurately map the yarn tenacity at higher spinning speeds. Similarly, modeling and simulation tools need to be developed to accurately detect the yarn hairiness.

There are limitations in the range of constructional parameters that can be used for training the AI. Hence, to improve the correlation of the actual values and the results predicted by AI, the constructional parameters need improvement. The accuracy of predicting fabric properties depends on the quantity and quality of data used in the training of AI. For the prediction of fabric properties accurately, different ANN configurations can be tried in addition to adequate qualitative training data. The coefficient of variation (CV) of the training data also plays a big role in providing accurate results. Higher is the CV of the input training data, the higher is the chance of getting errors in the output results.

It is always essential to have good yarn quality for producing fabrics with lower amount of defects. The application of ANN plays a crucial role in the quality control of yarns and hence the fabric quality. Hence, there should be wider application of ANN during various operations of yarn manufacturing such as roving formation and yarn spinning. Furthermore, monitoring the fabric quality online by the use of CCD during the production process can help to improve the fabric quality. Although many faults can be easily identified by the use of ANN, there are several faults such as missing picks and missing ends, reed mark, finger marks to name a few, which cannot be detected accurately. Hence, further research is needed to investigate these faults by the use of AI tools such as neuro-fuzzy ES, GRNN for accuracy of the predicted results.

The application of AI is lower in knitting compared to weaving. This can be attributed to the higher speed and design of the circular knitting machine. Research and development can help to achieve online monitoring of fabric faults at higher speeds. In this context, the application of AI need to be extended to the 3D space from its current application in 2D space and the results of the AI prediction in higher speeds need to be validated. The application of AI in 3D pattern prediction is not standardized yet. The research and development is not enough to study all garment sizes and the virtual fit pattern.

Limited amount of research has been done to address the clothing comfort issues due to complexities involved in training the NNs by using the fabric properties to predict clothing comfort. One of the major challenges of applying AI in predicting clothing comfort is the high amount of fabric parameters that influence clothing comfort properties. Other challenges involve long computational time needed to handle a large amount of data, determination of adequate model inputs, choice of suitable network architecture, and validation of the predicted results.

5.4.2 Future directions of artificial intelligence

Application of AI in predicting the properties of textile and clothing is rapidly increasing. Several research groups around the world are focusing to apply various AI tools and optimize the prediction of fabric and clothing properties. The future research and development should focus on various challenges faced by AI to establish intelligent systems to predict the performance properties of textile and clothing. The future research on the application of AI in textiles and clothing industries will focus on:

1. Improving the method of data collection such as capturing online data during the production process for the training of ANN,
2. Enhancing the feature-extraction processes before the data can be fed to an ANN,

3. Improving the prediction capability of ANN by extrapolation of the system, and
4. Improving the man–machine interaction during ANN applications.

There have been the highest number of investigations on the application of AI in woven textiles than other structures. The number of research is increasing in knitted fabric. However, limited research has been done on the application of AI in nonwoven, seamless garments, and composites. The application of computational tools is limited to feed-forward and FL networks. Hence, the applicability of computational tools such as pulse propagation network, tree network, and support vector machines need to be investigated in future research.

The application of AI has not received wider acceptance in the production floor. Although the automation is increasing, the AI application is limited. This might be due to the fact that there are limited textile-specific software, high cost, and availability of cheap labor. This may become a reality in future and increased research and development will help to achieve the desired results. In future the application of AI may be involved in the decision making process.

Although SCM is a wide area in garment manufacturing, limited research has been done to address the issues and risks in SCM with the application of AI. The application of AI is increasing but at a much slower pace compared to other industries such as automotive. Some of the leading brands are applying AI in their supply chain and logistic network. Research and development can help to increase the application by providing more solutions and reducing the cost. The future research should focus on the application of AI to optimize SCM network for solving decision making and inventory management problems.

5.5 Conclusion

In this modern era, AI is being used in many areas to solve various problems with intelligence similar to human being. The application of AI was not widely accepted in the labor-intensive clothing production. However, the global competitive environment and a target to achieve low cost of production are the main reasons for the AI's wider applications in apparel industry starting from material selection and sourcing, through manufacturing till retailing. AI can be used in various processes of textile production such as fiber grading, prediction of yarn properties, detection of fabric faults, and dye recipe prediction. Similarly, AI can be applied in all the stages of garment production such as preproduction, production, and postproduction operations. Developed countries have already started using AI to improve quality of garment, enhanced customer service, and hence increased sales. Much progress is undergoing in AI rapidly and in near future it will become an important tool for the garment manufacturers for enhancing quality, increasing production, lowering operating costs, and exercising in house control over production, leading to quick response and just-in-time concept. The application of AI in garment manufacturing has a bright future similar to other areas of application.

Sources of further information

There are several research organizations and academic institutions that are involved in the research and development of AI in different application fields. In addition, several journals also publish the recent research work in the field of AI. The list of research organizations and journals relating to the development in AI is given in the following section.

Research organizations

- School of Computer Science & Engineering, University of Washington, AC101 Paul G. Allen Center for Computer Science & Engineering, Box 352350, 185 Stevens Way, Seattle WA 98195-2350, Phone: (206) 543-1695, Fax: (206) 543-2969
- MIT's Computer Science and Artificial Intelligence Laboratory (CSAIL), The Stata Center, Building 32 32 Vassar Street Cambridge, MA 02139 USA, CSAIL HQ Reception Tel: 617-253-5851, CSAIL HQ Fax: 617-258-8682
- Machine Intelligence Research Institute, 2030 Addison St Fl 7, Berkeley CA 94704
- Stanford Artificial Intelligence Laboratory, 353 Serra Mall, Stanford, CA 94305, Tel: 650.725.3860, Fax: 650.725.3358
- Computer Science Department, Carnegie Mellon University, Gates Center 6105, Pittsburgh PA 15213-3891, Phone: 412-268-2565
- Artificial Intelligence Research Center, AIST Tokyo Waterfront, 2-3-26, Aomi, Koto-ku, Tokyo 135-0064, Japan
- Istituto Dalle Molle di studi sull'intelligenza artificiale, Galleria 2, Via Cantonale 2c, CH-6928 Manno, Tel: +41 (0)58 666 66 66, Fax: 41 (0)58 666 66 61
- German Research Center for Artificial Intelligence, DFKI GmbH, Trippstadter Strasse 122, D-67663 Kaiserslautern, Tel: +49 (0)631/205 75-0, Fax: +49 (0)631/205 75-5030
- School of Computer Science, University of Waterloo, 200 University Avenue West, Waterloo, Ontario, Canada N2L 3G1, Tel: 519-888-4567 x33293, Fax: 519-885-1208
- Centre for and Robotics, Defence Research & Development Organization, DRDO Complex, C.V. Raman Nagar, Bangalore-560 093
- The Austrian Research Institute for Artificial Intelligence, Freyung 6/6, A-1010 Vienna, Austria, Tel: (+43-1) 5336112-0, Fax: (+43-1) 5336112-77
- Artificial Intelligence Center, 333 Ravenswood Avenue, Menlo Park, CA 94025-3493, United States, Tel: +1 (650) 859-2000
- Computer Science Department, University of Texas at Austin, 2317 Speedway, Stop D9500, Austin, TX 78712, Phone: 512.471.7316
- Department of Computer Science, University of Rochester, 734 Computer Studies Building, P.O. Box 270226, Rochester, NY 14627
- Department of Computing Science, University of Alberta, 2-32 Athabasca Hall, Edmonton, Alberta, Canada T6G 2E8, Tel: 1 (780) 492-2285, Fax: 1 (780) 492-6393

Journals to read

- Journal of Artificial Intelligence Research (www.jair.org)
- Artificial Intelligence Research (http://www.sciedupress.com/journal/index.php/air)
- Artificial Intelligence An International Journal (https://www.journals.elsevier.com/artificial-intelligence/)

- International Journal of Robotics Research (http://journals.sagepub.com/home/ijr)
- Journal of Memory and Language (https://www.journals.elsevier.com/journal-of-memory-and-language)
- Autonomous Robots (http://www.springer.com/engineering/control/journal/10514)
- Fuzzy Optimization and Decision Making (https://link.springer.com/journal/10700)
- Pattern Recognition (https://www.journals.elsevier.com/pattern-recognition)
- Cognitive Science (http://onlinelibrary.wiley.com/journal/10.1111/(ISSN)1551-6709)
- Neural Networks (https://www.journals.elsevier.com/neural-networks/)
- Journal of Intelligent Manufacturing (http://www.springer.com/business+%26+management/operations+research/journal/10845)
- Engineering Applications of Artificial Intelligence (https://www.journals.elsevier.com/engineering-applications-of-artificial-intelligence/)
- International Journal of Intelligent Systems (http://onlinelibrary.wiley.com/journal/10.1002/(ISSN)1098-111X)
- IEEE Transactions on Neural Networks and Learning Systems (http://ieeexplore.ieee.org/xpl/RecentIssue.jsp?punumber=5962385)
- Intelligent Service Robotics (http://www.springer.com/engineering/control/journal/11370)

References

Abouelela, A., et al., 2005. Automated vision system for localizing structural defects in textile fabrics. Pattern Recognition Letters 26 (10), 1435–1443.

Alibi, H., et al., 2012. A neural network system for prediction of thermal resistance of knit fabrics. Special Topics and Reviews in Porous Media: An International Journal 3 (1).

Au, K., Wong, W., Zeng, X., 2006. Decision model for country site selection of overseas clothing plants. The International Journal of Advanced Manufacturing Technology 29 (3–4), 408–417.

Au, K.-F., Choi, T.-M., Yu, Y., 2008. Fashion retail forecasting by evolutionary neural networks. International Journal of Production Economics 114 (2), 615–630.

Bahlmann, C., Heidemann, G., Ritter, H., 1999. Artificial neural networks for automated quality control of textile seams. Pattern Recognition 32 (6), 1049–1060.

Behera, B., Karthikeyan, B., 2006. Artificial neural network-embedded expert system for the design of canopy fabrics. Journal of Industrial Textiles 36 (2), 111–123.

Behera, B., Mishra, R., 2007. Artificial neural network-based prediction of aesthetic and functional properties of worsted suiting fabrics. International Journal of Clothing Science and Technology 19 (5), 259–276.

Beltran, R., Wang, L., Wang, X., 2005. Predicting the pilling propensity of fabrics through artificial neural network modeling. Textile Research Journal 75 (7), 557–561.

Bhattacharjee, D., Kothari, V.K., 2007. A neural network system for prediction of thermal resistance of textile fabrics. Textile Research Journal 77 (1), 4–12.

Carbonneau, R., Laframboise, K., Vahidov, R., 2008. Application of machine learning techniques for supply chain demand forecasting. European Journal of Operational Research 184 (3), 1140–1154.

Chandra, C., Kumar, S., 2000. An application of a system analysis methodology to manage logistics in a textile supply chain. Supply Chain Management: An International Journal 5 (5), 234–245.

Chandra, J.K., Banerjee, P.K., Datta, A.K., 2010. Neural network trained morphological processing for the detection of defects in woven fabric. The Journal of the Textile Institute 101 (8), 699–706.

Chattopadhyay, R., Guha, A., 2004. Artificial neural networks: applications to textiles. Textile Progress 35 (1), 1–46.

Chen, R., et al., 2005. A decision support system based on genetic algorithm for garment production. In: Proceedings of the Fourth International Conference on Information and Management Sciences.

Chen, Y., et al., 2006. Estimation of ease allowance of a garment using fuzzy logic. In: Fuzzy Applications in Industrial Engineering. Springer, pp. 367–379.

Chen, Y., et al., 2009. Optimisation of garment design using fuzzy logic and sensory evaluation techniques. Engineering Applications of Artificial Intelligence 22 (2), 272–282.

Chen, Y., Zhao, T., Collier, B., 2001. Prediction of fabric end-use using a neural network technique. Journal of the Textile Institute 92 (2), 157–163.

Chiu, M., Lin, G., 2004. Collaborative supply chain planning using the artificial neural network approach. Journal of Manufacturing Technology Management 15 (8), 787–796.

Choi, T.-M., et al., 2014. Fast fashion sales forecasting with limited data and time. Decision Support Systems 59, 84–92.

Choy, K.L., et al., 2009. A RFID-case-based sample management system for fashion product development. Engineering Applications of Artificial Intelligence 22 (6), 882–896.

Demiryürek, O., Koç, E., 2009. Predicting the unevenness of polyester/viscose blended open-end rotor spun yarns using artificial neural network and statistical models. Fibers and Polymers 10 (2), 237–245.

Ding, Y., Xu, Y., 2008. Intelligent optimal selection of garment sizes by using immune algorithm and AHP method. Journal of the Textile Institute 99 (3), 281–286.

Dong, A., Leung, S., 2009. A simulation-based replenishment model for the textile industry. Textile Research Journal 79 (13), 1188–1201.

Fan, J., et al., 2001. Predicting garment drape with a fuzzy-neural network. Textile Research Journal 71 (7), 605–608.

Fang, J.-J., Ding, Y., 2008. Expert-based customized pattern-making automation: part I. Basic patterns. International Journal of Clothing Science and Technology 20 (1), 26–40.

Fumi, A., et al., 2013. Fourier analysis for demand forecasting in a fashion company. International Journal of Engineering Business Management 5, 30.

Gong, D.-W., et al., 2007. Interactive genetic algorithms with multi-population adaptive hierarchy and their application in fashion design. Applied Mathematics and Computation 185 (2), 1098–1108.

Goode, M.M., et al., 2005. Determining customer satisfaction from mobile phones: a neural network approach. Journal of Marketing Management 21 (7–8), 755–778.

Groves, G., Valsamakis, V., 1998. Supplier-customer relationships and company performance. The International Journal of Logistics Management 9 (2), 51–64.

Guo, Z., et al., 2008. Genetic optimization of order scheduling with multiple uncertainties. Expert Systems with Applications 35 (4), 1788–1801.

Guo, Z., et al., 2011. Applications of artificial intelligence in the apparel industry: a review. Textile Research Journal. p. 0040517511411968.

Hadizadeh, M., Jeddi, A.A., Tehran, M.A., 2009. The prediction of initial load-extension behavior of woven fabrics using artificial neural network. Textile Research Journal 79 (17), 1599–1609.

Halttunen, M., et al., 1999. Applicability of FTIR/PAS depth profiling for the study of coated papers. Vibrational Spectroscopy 19 (2), 261–269.

Hamdi, M., Lachiver, G., Michaud, F., 1999. A new predictive thermal sensation index of human response. Energy and Buildings 29 (2), 167–178.

Hsu, C., Lee, T., Kuo, H., 2009. Applying fuzzy theory based data mining to establish the female sizing systems for garment production. In: Proceedings of the 8th WSEAS International Conference on Applied Computing and Applied Computational Science, Hangzhou, China, World Scientific and Engineering Academy and Society.
Hu, Z.-H., et al., 2008. An interactive co-evolutionary CAD system for garment pattern design. Computer-Aided Design 40 (12), 1094–1104.
Hu, Z.-H., et al., 2009. A hybrid neural network and immune algorithm approach for fit garment design. Textile Research Journal 79 (14), 1319–1330.
Huang, C.-C., Chen, I.-C., 2001. Neural-fuzzy classification for fabric defects. Textile Research Journal 71 (3), 220–224.
Huang, C.-C., Yu, W.-H., 2001. Fuzzy neural network approach to classifying dyeing defects. Textile Research Journal 71 (2), 100–104.
Hui, C., et al., 2004. Neural network prediction of human psychological perceptions of fabric hand. Textile Research Journal 74 (5), 375–383.
Hui, P.C., et al., 2007. Application of artificial neural networks to the prediction of sewing performance of fabrics. International Journal of Clothing Science and Technology 19 (5), 291–318.
Hui, C., Ng, S., 2005. A new approach for prediction of sewing performance of fabrics in apparel manufacturing using artificial neural networks. Journal of the Textile Institute 96 (6), 401–405.
Hui, C.L., Ng, S.F., 2009. Predicting seam performance of commercial woven fabrics using multiple logarithm regression and artificial neural networks. Textile Research Journal 79 (18), 1649–1657.
Hung, O., et al., 2011. Using artificial neural network to predict colour properties of laser-treated 100% cotton fabric. Fibers and Polymers 12 (8), 1069–1076.
Hung, O., et al., 2014. Artificial neural network approach for predicting colour properties of laser-treated denim fabrics. Fibers and Polymers 15 (6), 1330–1336.
Inui, S., 1996. A combined system of computer aided design and genetic algorithm for apparel designing. Sen'i Gakkaishi 52 (11), 605–611.
Jasper, W.J., Kovacs, E.T., 1994. Using neural networks and NIR spectrophotometry to identify fibers. Textile Research Journal 64 (8), 444–448.
Jasper, W.J., Kovacs, E.T., Berkstresser IV, G.A., 1993. Using neural networks to predict dye concentrations in multiple-dye mixtures. Textile Research Journal 63 (9), 545–551.
Jeon, B.S., Bae, J.H., Suh, M.W., 2003. Automatic recognition of woven fabric patterns by an artificial neural network. Textile Research Journal 73 (7), 645–650.
Jung, K.-Y., Na, Y.-J., Lee, J.-H., 2003. FDRAS: fashion design recommender agent system using the extraction of representative sensibility and the two-way combined filtering on textile. In: Database and Expert Systems Applications. Springer.
Kennedy, K., 2015. Pattern construction. In: Nayak, R., Padhye, R. (Eds.), Garment Manufacturing Technology. Elsevier, UK.
Kim, H.-S., Cho, S.-B., 2000. Application of interactive genetic algorithm to fashion design. Engineering Applications of Artificial Intelligence 13 (6), 635–644.
Kim, E.H., 1999. Objective evaluation of wrinkle recovery. Textile Research Journal 69 (11), 860–865.
Koehl, L., et al., 2008. Analysis and identification of fashion oriented industrial products using fuzzy logic techniques. In: 3rd International Conference on Intelligent System and Knowledge Engineering, 2008. ISKE 2008. IEEE.
Kumar, S., et al., 2016. Designing and development of denim fabrics: part 1-study the effect of fabric parameters on the fabric characteristics for women's wear. Journal of Textile Science and Engineering 6 (265), 1–5.

Kumar, A., 2003. Neural network based detection of local textile defects. Pattern Recognition 36 (7), 1645–1659.

Kuo, C.-F.J., Su, T.-L., Huang, Y.-J., 2007. Computerized color separation system for printed fabrics by using backward-propagation neural network. Fibers and Polymers 8 (5), 529–536.

Lin, J.-J., 2007. Intelligent decision making based on GA for creative apparel styling. Journal of Information Science and Engineering 23 (6).

Lau, T.W., et al., 2006. A new fuzzy approach to improve fashion product development. Computers in Industry 57 (1), 82–92.

Lo, W.-S., Hong, T.-P., Jeng, R., 2008. A framework of E-SCM multi-agent systems in the fashion industry. International Journal of Production Economics 114 (2), 594–614.

Luo, X., et al., 2007. A fuzzy neural network model for predicting clothing thermal comfort. Computers and Mathematics with Applications 53 (12), 1840–1846.

Majumdar, P.K., Majumdar, A., 2004. Predicting the breaking elongation of ring spun cotton yarns using mathematical, statistical, and artificial neural network models. Textile Research Journal 74 (7), 652–655.

Majumdar, A., Majumdar, P., Sarkar, B., 2006. An investigation on yarn engineering using artificial neural networks. Journal of the Textile Institute 97 (5), 429–434.

Mak, K., Li, W., 2008. Objective evaluation of seam pucker on textiles by using self-organizing map. IAENG International Journal of Computer Science 35 (1), 47–54.

Moon, K., Ngai, E., 2008. The adoption of RFID in fashion retailing: a business value-added framework. Industrial Management and Data Systems 108 (5), 596–612.

Mori, T., Komiyama, J., 2002. Evaluating wrinkled fabrics with image analysis and neural networks. Textile Research Journal 72 (5), 417–422.

Muni, D.P., Pal, N.R., Das, J., 2006. Texture generation for fashion design using genetic programming. In: Control, Automation, Robotics and Vision, 2006. ICARCV'06. 9th International Conference on. 2006. . IEEE.

Murrells, C.M., et al., 2009. An artificial neural network model for the prediction of spirality of fully relaxed single jersey fabrics. Textile Research Journal 79 (3), 227–234.

Nawaz, N., Nayak, R., 2015. Seamless garments. In: Nayak, R., Padhye, R. (Eds.), Garment Manufacturing Technology. Cambridge, UK, pp. 373–383.

Nayak, R., et al., 2009. Comfort properties of suiting fabrics. Indian Journal of Fibre and Textile Research 34, 122–128.

Nayak, R., et al., 2013. Sewability of air-jet textured sewing threads in denim. Journal of Textile and Apparel, Technology and Management 8 (1), 1–11.

Nayak, R., et al., 2015a. RFID in textile and clothing manufacturing: technology and challenges. Fashion and Textiles 2 (1), 1–16.

Nayak, R., et al., 2015b. The role of mass customisation in the apparel industry. International Journal of Fashion Design, Technology and Education 1–11 (ahead-of-print).

Nayak, R., et al., 2016. Artificial intelligence: technology and application in apparel manufacturing. In: TBIS-APCC 2016. Binary Information Press, Textile Bioengineering and Informatics Society.

Nayak, R., Padhye, R., 2017. Manikins for Textile Evaluation. Elsevier.

Nayak, R., Padhye, R., 2011. Application of modelling and simulation in smart and technical textiles. In: Patanaik, A. (Ed.), Modeling and Simulation in Fibrous Materials: Techniques and Applications. Nova Science, pp. 259–286.

Nayak, R., Padhye, R., 2014. Introduction: the apparel industry. In: Nayak, R., Padhye, R. (Eds.), Garment Manufacturing Technology. Elsevier, pp. 1–18.

Nayak, R., Padhye, R., 2015. Garment Manufacturing Technology. Elsevier.

Nayak, R., Padhye, R., Gon, D.P., 2010. Sewing performance of stretch denim. Journal of Textile and Apparel, Technology and Management 6 (3), 1–9.

Nayak, R., Padhye, R., Fergusson, S., 2012. Identification of Natural Textile Fibres. Handbook of Natural Fibres, p. 1.

Ngai, E.W., Xiu, L., Chau, D.C., 2009. Application of data mining techniques in customer relationship management: a literature review and classification. Expert Systems with Applications 36 (2), 2592–2602.

Ngan, H.Y., et al., 2005. Wavelet based methods on patterned fabric defect detection. Pattern Recognition 38 (4), 559–576.

Ngan, H.Y., Pang, G.K., Yung, N.H., 2011. Automated fabric defect detection—a review. Image and Vision Computing 29 (7), 442–458.

Norton-Wayne, L., 1990. Automated garment inspection using machine vision. In: IEEE International Conference on Systems Engineering, 1990. IEEE.

Pan, A., et al., 2009. Optimal reorder decision-making in the agent-based apparel supply chain. Expert Systems with Applications 36 (4), 8571–8581.

Park, S.-W., et al., 2000. Applying fuzzy logic and neural networks to total hand evaluation of knitted fabrics. Textile Research Journal 70 (8), 675–681.

Pattanayak, A.K., Luximon, A., Khandual, A., 2011. Prediction of drape profile of cotton woven fabrics using artificial neural network and multiple regression method. Textile Research Journal 81 (6), 559–566.

Paul, S., Azeem, A., 2011. An artificial neural network model for optimization of finished goods inventory. International Journal of Industrial Engineering Computations 2 (2), 431–438.

Pfahl, L., Moxham, C., 2014. Achieving sustained competitive advantage by integrating ECR, RFID and visibility in retail supply chains: a conceptual framework. Production Planning and Control 25 (7), 548–571.

Senthilkumar, M., Selvakumar, N., 2006. Achieving expected depth of shade in reactive dye application using artificial neural network technique. Dyes and Pigments 68 (2), 89–94.

Senthilkumar, M., 2007. Modelling of CIELAB values in vinyl sulphone dye application using feed-forward neural networks. Dyes and Pigments 75 (2), 356–361.

Shady, E., et al., 2006. Detection and classification of defects in knitted fabric structures. Textile Research Journal 76 (4), 295–300.

Shamey, R., Hussain, T., 2003. Artificial intelligence in the colour and textile industry. Review of Progress in Coloration and Related Topics 33 (1), 33–45.

Soleymani, A.R., Saien, J., Bayat, H., 2011. Artificial neural networks developed for prediction of dye decolorization efficiency with $UV/K_2S_2O_8$ process. Chemical Engineering Journal 170 (1), 29–35.

Stylios, G.K., Powell, N.J., 2003. Engineering the drapability of textile fabrics. International Journal of Clothing Science and Technology 15 (3/4), 211–217.

Sun, Z.-L., et al., 2008. Sales forecasting using extreme learning machine with applications in fashion retailing. Decision Support Systems 46 (1), 411–419.

Sztandera, L.M., Frank, C., Vemulapali, B., 2004. Prediction of women's apparel sales using soft computing methods. In: Knowledge-Based Intelligent Information and Engineering Systems. Springer.

Thomassey, S., Fiordaliso, A., 2006. A hybrid sales forecasting system based on clustering and decision trees. Decision Support Systems 42 (1), 408–421.

Thomassey, S., Happiette, M., Castelain, J.M., 2005. A short and mean-term automatic forecasting system—application to textile logistics. European Journal of Operational Research 161 (1), 275–284.

Tilocca, A., et al., 2002. Detecting fabric defects with a neural network using two kinds of optical patterns. Textile Research Journal 72 (6), 545–550.

Tsai, I.-S., Hu, M.-C., 1996. Automatic inspection of fabric defects using an artificial neural network technique. Textile Research Journal 66 (7), 474–482.

Unal, P.G., Üreyen, M.E., Mecit, D., 2012. Predicting properties of single jersey fabrics using regression and artificial neural network models. Fibers and Polymers 13 (1), 87–95.

Üreyen, M.E., Gürkan, P., 2008. Comparison of artificial neural network and linear regression models for prediction of ring spun yarn properties. I. Prediction of yarn tensile properties. Fibers and Polymers 9 (1), 87–91.

Vas, P., 1999. Artificial-Intelligence-Based Electrical Machines and Drives: Application of Fuzzy, Neural, Fuzzy-Neural, and Genetic-Algorithm-Based Techniques, vol. 45. Oxford University Press.

Vijayan, A., Jadhav, A., 2015. Fabric Sourcing and Selection. Garment Manufacturing Technology, p. 109.

Wang, Z., Li, Y., Wong, A.S.W., 2005. Simulation of clothing thermal comfort with fuzzy logic. Elsevier Ergonomics Book Series 3, 467–471.

Wong, A., et al., 2003. Neural network predictions of human psychological perceptions of clothing sensory comfort. Textile Research Journal 73 (1), 31–37.

Wong, W.K., et al., 2009. Stitching defect detection and classification using wavelet transform and BP neural network. Expert Systems with Applications 36 (2), 3845–3856.

Wong, W., Guo, Z., 2010. A hybrid intelligent model for medium-term sales forecasting in fashion retail supply chains using extreme learning machine and harmony search algorithm. International Journal of Production Economics 128 (2), 614–624.

Wong, A., Li, Y., Yeung, P., 2004. Predicting clothing sensory comfort with artificial intelligence hybrid models. Textile Research Journal 74 (1), 13–19.

Wong, W., Leung, S., Au, K., 2005. Real-time GA-based rescheduling approach for the pre-sewing stage of an apparel manufacturing process. The International Journal of Advanced Manufacturing Technology 25 (1–2), 180–188.

Xia, M., et al., 2012. Fashion retailing forecasting based on extreme learning machine with adaptive metrics of inputs. Knowledge-Based Systems 36, 253–259.

Yildiz, Z., et al., 2013. Use of artificial neural networks for modelling of seam strength and elongation at break. Fibres and Textiles in Eastern Europe 101 (5), 117–123.

Yin, X., Yu, W., 2007. The virtual manufacturing model of the worsted yarn based on artificial neural networks and grey theory. Applied Mathematics and Computation 185 (1), 322–332.

Yu, Y., Choi, T.-M., Hui, C.-L., 2011. An intelligent fast sales forecasting model for fashion products. Expert Systems with Applications 38 (6), 7373–7379.

Yuen, C., et al., 2008. Application of smart system to textile industry: preliminary design of a smart hanger for garment inspection. Journal of the Textile Institute 99 (6), 569–580.

Yuen, C., et al., 2009a. A hybrid model using genetic algorithm and neural network for classifying garment defects. Expert Systems with Applications 36 (2), 2037–2047.

Yuen, C., et al., 2009b. Fabric stitching inspection using segmented window technique and BP neural network. Textile Research Journal 79 (1), 24–35.

Zeng, X., Ge, X., Bruniaux, P., 2008. Developing an intelligent system for supporting fashion design decisions. In: Proceedings of the 8th International FLINS Conference. World Scientific.

Zeydan, M., 2008. Modelling the woven fabric strength using artificial neural network and Taguchi methodologies. International Journal of Clothing Science and Technology 20 (2), 104–118.

Automation in spreading and cutting

6

Ineta Vilumsone-Nemes
University of Novi Sad, Novi Sad, Serbia

6.1 Introduction

In very competitive global market, fashion companies are constantly forced to offer greater product selection, deliver goods faster, and reduce their costs. For several decades the cost savings were reached by reducing operating costs—using lower cost labor and advanced manufacturing tools to improve production efficiency. However, over the past years input costs of textile and garment manufacturers such as labor, energy, and raw materials, are constantly increasing. In the main manufacturing countries, such as China, Vietnam, Indonesia, Bangladesh, Burma, and Cambodia, the minimum wage, inflation, utility costs, currency exchange are rising fast, often faster than work efficiency gains.

Besides the growing labor problem, textile manufacturers face constant increase in raw material costs. The rise of raw cotton and wool prices has increased the demand for alternate man-made fibers. The increased demand for man-made fibers, and rising oil prices have boosted the prices of synthetic fibers such as polyester and nylon. Although the material prices are rising and they (fibers) account for more than half (i.e., about 50%–70%) of the total product costs, there is always a growing pressure to keep the cost of garments low.

The other challenging facts are the rising labor and transportation costs. Hence, it has become a real challenge to keep the cost of garment low. Now, more than ever, the time has come to make good and well-grounded changes in work organizations and material savings directly in the garment manufacturing process, to maintain the trend of low cost. In this context, a cutting room has become the most vital place in an apparel enterprise, as exactly its management and production processes ensure smooth flow or work to all further garment manufacturing processes and determine fabric consumption efficiency.

This chapter highlights the role of automation in spreading and cutting processes, which plays vital role in garment quality. This also focuses on the fault registration by the use of automated machines. Various automatic cutting machines such as laser, knife, and water-jet are discussed in this chapter. Furthermore, the automatic pattern matching process is illustrated by matching of images. The advancements of fusing technologies are also covered in this chapter. The future directions section includes the further scope in the automation in spreading and cutting including the online assistance during the production processes.

Automation in Garment Manufacturing. http://dx.doi.org/10.1016/B978-0-08-101211-6.00006-9

6.2 The role of automation in textile material spreading and cutting

A cutting room is a separate area in a production enterprise where garment components are cut from fabrics. Traditionally, the most part of the activities in the cutting room are performed manually. However, the manual spreading and cutting equipment and work methods used cannot ensure high productivity and work quality. During the working in spreading and cutting sections of a garment industry, operators have both heavy physical load, and mental load, which can negatively influence work efficiency and quality. Unlike other processes in clothing manufacturing, garment components' cutting is very much dependent on the skills, experience, and decision making of the cutting room workers. The immense variety of textile materials, their quality variability and the constant need to minimize the usage of materials, force the cutting room specialists to work creatively in every new situation.

The first changes in the cutting room came with the development of mass production, when a simple spreading machine was invented to carry a roll of fabric over the table. From that time, the spreading and cutting equipment have changed drastically reducing the human resource importance in the cutting room more and more. As it is well know that in many other industries, for several decades, die cutting machines were used to increase cutting efficiency and quality. However, they had limited flexibility and could not be used to process frequently changing styles. Die cutting presses were replaced with a new type of numerically controlled machines that performed a continuous cut by means of a specialized cutting device, which moved around the profile of the object. In the late 1960s, H. Joseph Gerber invented the first fully automated multi-ply textile cutting system (the GERBERcutter S-70).

Nowadays, due to powerful software, high-tech equipment and advanced services, the cutting room has become the most advanced department in an apparel manufacturing enterprise. The automated management and manufacturing processes can successfully react on the latest tendencies of the global economy and changes in the apparel market.

6.3 Automation in cutting room work process organization

The efficient and well-planned operations of a cutting room ensure a smooth flow of work to all further manufacturing processes and controls the utilization of fabric. In traditional management system, material requirement planning, material inventory, and lay planning processes have been separate work steps fulfilled in different departments. These traditional work methods have several serious disadvantages. Often specialists doing the job are weakly informed and connected with the manufacturing process. They have insufficient knowledge about production; the departments are located far from real production sites, even abroad. A lot of paperwork, manual calculations and unavailability of easily accessible database, complicates the work process and reduces the product quality. As a result, companies are forced to purchase certain amount of extra raw material and accept the changes in the delivery dates of finished goods.

By the help of automation, the cutting room work process are fully changed. Specialized management software schedules jobs for production, select optimal raw material and utilize remnants, generate cut plans and nests, track production activities, and provide management reports for analysis and future planning. All these described processes can be done simultaneously, in high speed, providing an optimal solution in every work step.

6.3.1 Automated cutting room management system

A centralized database is the main part of the computerized management system. It ensures possibility to view and edit style, inventory, labor, costing, an order, cut planning, and manufacturing information. The data are transparent and easily available for all members of authorized staff. There is no need to enter identical information more than one time. Once a change is made it is instantly available for all the database users. During work process all data are accumulated and a company gradually gains knowledge and valuable experience about its material consumption, costs, and their optimization options.

Often the management system also includes *material requirement planning software*. Optimal material purchase is calculated using a set of special techniques and data from the database—previous bills of material and production orders, as well as, current inventory data.

The management systems include *two inventory systems*: finished goods inventory and raw material inventory. The finished goods inventory system provides the ability to enter incoming stock from manufacturing orders or a raw material purchase orders. The raw materials inventory system tracks everything not covered by the finished goods inventory. The system is able to transfer and maintain inventory levels between different warehouses and contractors. At any point in time the company knows the current status of its inventory. The system ensures information about raw materials on hand, ordered raw materials, raw materials in work process, available inventory amounts, as well as, raw material utilization, and raw material requirements.

Based on inventory data, cutting orders or finished goods requirements, *material purchasing software* can automatically generate a purchase order for needed items and post them directly to suppliers. Thus materials are delivered just before their processing. There is no need to do long term purchases of regularly used materials, such as, interlinings, elastic bands, packing material, and labels.

Management systems for apparel industry are developed by companies such as: AIMS Technology, AMS, Apparel Business Systems, Apparel Data Solutions, ASAP Apparel Software, Assyst, Jomar Softcorp International, Jonar Systems, Lectra, Olotech, Optitex, Plataine, Polygon Software, Reach Technologies, Texbase, TradeStone Software, and others.

6.3.2 Automated lay planning

The part of the management system coordinating work process in a cutting room is a cut planning software. It links together ERP (enterprise resource planning), fabric management system, computer-aided design (CAD) and computer-aided manufacturing (CAM), exchanges information in between these systems, and creates the best solutions to cut material for manufacturing orders.

Cut planning software is developed by companies such as: Lectra (Optiplan), Polygon Software (Cut Planning), AMS (Cut Plan), Option Systems (Cutting Room Planning), Optitex (Cut Plan), Assyst (Lago), FK Group (Future Cut planner), Plataine (Cut Order Planning), Reach Technologies (Reach Cut Planner), and others.

6.3.2.1 Methodology

Cut planning software imports customer orders from internal or any external system. It runs different cutting plan scenarios (markers and their combinations) to see their impact to the fabric usage, cutting time, productivity, and choose the best of them. The system is designed to reuse already existing suitable nests from maker library and only after that sends requests for new markers directly to CAD system. Defining which markers are necessary for the order, it creates the optimal cutting plan for one or multiple factories considering specific product and manufacturing characteristics (fabric qualities, technological limits of the spreading and cutting processes). To perform the created cutting plan, the system firstly selects (using material inventory data) fabric rolls that can be 100% consumed then it takes those pieces that result in the least end and width loss. If reusable fabric remnants appear, they can be used, giving preference to the smallest pieces. The system can determine where to lay and how to cut selected fabric to achieve most effective fabric savings.

Then the system generates optimized cutting schedule in sync with sewing plan, print manufacturing reports, and send the orders to the cutting room. Spreading and cutting operations are monitored at each stage of the cutting process (using barcode scanners). Reports are provided to monitor production throughput and efficiencies across multiple plants. Fabric utilization reports can be used for efficient material requirement planning of further manufacturing processes.

Most important steps of automated cut planning process are as follows: running of different planning scenarios, establishing the marker processing time, performing marker calculations, spreading planning, and processing of manufacturing reports.

6.3.2.2 Running of different planning scenarios

After the user fills up all necessary order data (order quantity for each fabric type, fabric, initial marker, and spreading settings), the program tries all possible size combinations in markers. Based on a model information and marker library (data from previously used the same or similar styles) the program estimates length and efficiency of every yet uncreated marker. Finally the program selects and displays the best marker combinations for certain order.

6.3.2.3 Establishing the marker processing time

The yet uncreated markers are classified by their importance depending on the number of sizes in a marker, lays in a spread and garment pieces produced from the marker.

More time to find the best fabric consumption is given to progress more important markers. Marker progressing time can be determined:

- Automatically—the program distributes the time for each marker considering the marker size;
- Semiautomatically—the user can indicate how much time he/she wants the program to spend for each marker.

6.3.2.4 Performing marker calculations

Getting markers ready back from CAD, the program obtains the exact length and the efficiency of every performed marker. The fabric amount needed to produce the order is calculated now using data of markers length. The available info is also used to calculate statistics regarding average fabric use per product, per fabric type or total, fabric input, total average efficiency, etc.

Working with Lectra cut planning software (Optiplan) and also with their spreading and cutting equipment, the program can calculate: garment costs, fabric costs, manufacturing time (spreading time, cutting time, bundling time), and total costs. This information can be obtained for every markers to choose the best solution.

6.3.2.5 Spreading planning

Trying to respect the maximum number of fabric plies in the lay, program generates all spreads. Fabrics with similar properties are grouped together for one spread to reduce spreading time. If disproportioned number of layers (very small number) appears for separate spreads, the program can make automatic balancing.

6.3.2.6 Processing of manufacturing reports

During and at the end of planning process several reports could be generated: marker making, spreading, and cutting instructions as well as fabric use reports. To avoid mistakes, barcode scanning can be used to deliver data for automated spreading. The barcode can be used also to load a marker for a cutting process. Fabric report is send to warehouse to know how much fabric has to be taken to spreading. It can also go down to a roll level. Then planning will be done for every roll separately, finding its best usage.

6.4 Automated spreading methods and machines

Manual spreading is time and labor consuming. With the development of mass production, it could no longer provide the necessary productivity, therefore, the need arose for specialized machines that were capable of carrying out spreading at a much higher speed. The first stage spreading machines only carried a roll of fabric over the table. Nowadays, with the help of computer technologies, spreading process has become fully automated. Automated spreading systems have significantly increased the productivity of the spreading process and reduced the work load of the spreading operator, but have not altered its main work principles.

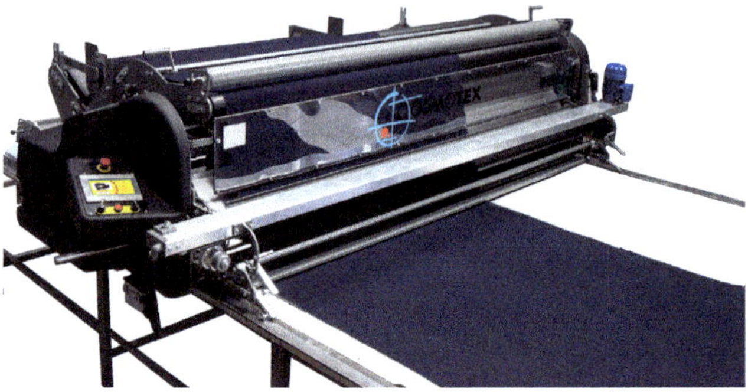

Figure 6.1 The automatic spreading machine, Apolo Smart 300 by Cosmotex.

Several companies produce automated spreading machines for a variety of textiles and uses such as: Lectra (France), Gerber (United States), Kuris (Germany), Bullmer (Germany), Phillips (UK), Eastman (United States), Unicraft Corporation (United States), Cosmotex (Spain), Oteman (Spain), Morgan Tecnica (Italy), FK group (Italy), B.K.R. Italia (Italy), Caron Technology (Italy), Ozbilim (Turkey), Tukatech (India), Shimaseiki (Japan), and Oshima (Taiwan).

6.4.1 Automatic spreading machine and its main parts

Multi-ply spreading is performed by an automatic spreading machine, which provides fabric feeding and transportation over the spreading table and ensures tension free spreading. The main parts of the spreading machine are as follows: a fabric spreader truck, a fabric feed system, an automatic cutting device, an end catcher, an operator stand panel, an encoder system, and a control panel (see Fig. 6.1).

A spreader truck ensures the transportation of a fabric roll above a spreading table in the lengthways and transverse directions. It consists of two main parts—a body and a turret.

- *The body of a spreader truck* ensures lengthways transportation of a fabric roll above a spreading table. The body of the truck also carries several special spreading machine devices: a cutter, a zigzag spreading device, a tubular fabric spreading device, and a fabric tearing device.
- The *turret* ensures transverse transportation of the roll during the spreading process to adjust the fabric ply to achieve a perfect alignment of one fabric edge on the table.

A fabric feed system rolls material from a fabric setting bar or a special cradle (see Fig. 6.2) and moves it to the spreading table. The feed system is synchronized to the speed of the spreader and ensures fabric tension control to provide tension free spreading. During the spreading process, the fabric feed system automatically adjusts the feeding speed of the material and measures its length.

During the spreading process, *the cutting device* moves together with the spreader truck. When the lay of fabric is fully spread, a round knife automatically moves across the table and cuts it off. The cutting device is sharpened automatically. It has an adjustable cutting speed. An automatic height detection sensor ensures the minimum

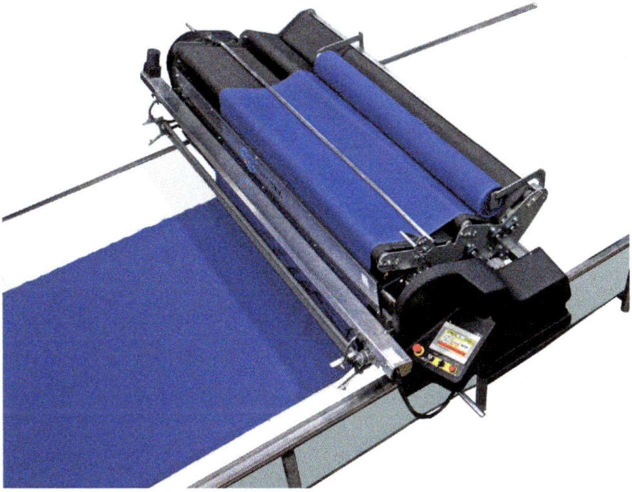

Figure 6.2 A fabric feeding system of an automatic spreading machine.

distance between the cutting device and the top of the spread. The spreading machine can be also equipped with a material tearing device.

A special platform allows the operator to ride alongside the table during the spreading process. It can be equipped with an adjustable seat. However, the panels where the operator stands during the work process are more comfortable and, therefore, more widely used.

The spreading process is driven and controlled by *an encoder*. Using a special belt with metal denticles fixed to one side of the spreading table, the encoder system counts the number of denticles and recalculates them in the distance (meters, inches) from the start point. This defines the placement coordinates of the spreader truck on the spreading table at a given moment. A spread can be started and finished at any point on the spreading table. Semiautomated machines have a special mechanical stopping device, which stops the machine by friction at the end of a ply.

An interactive control panel is used to set up parameters and to program the spreading process. A spreading machine may perform different spreading programs: basic spreading, block spreading, or step spreading. A spreading machine can ensure automatic counting of plies during the spreading process. Usually the machine will show the number of plies that remain to be spread.

Spreading machines are designed to work with fabrics of different weight and quality and can be categorized according to spread fabric properties—diameter, weight, width of the fabric roll, and fabric manufacturing way—opened or tubular (Fig. 6.3). However, there is the last tendency used by brands to categorize spreading machines in accordance with the type of the fabric to be spread (woven, knitted, technical) or the sector of application (apparel textiles, automotive, furniture, others).

6.4.2 Automated spreading process

The spreading process may be performed in semiautomated and fully automated modes.

Figure 6.3 The automatic spreading machine Apolo Smart 300 by Cosmotex equipped with several special devices to lay tubular fabrics.

6.4.2.1 Semiautomated spreading process

In the semiautomated spreading process, the operator moves along the spreading table (walking or riding on a stand panel) and follows the spreading process. The operator smoothes the surface of the lay, identifies faults in the spread fabric, and decides whether to leave faults in the spread or to cut them out. The operator uses a manually operated speed control handle to change the spreading speed and to reduce the speed in problematic areas, or even to stop the spreading process if it is necessary to define the location of a fault and to cut it out.

6.4.2.2 Fully automated spreading process

The fully automated spreading process is used for high-quality materials, which are easily spread. An operator sets the necessary parameters (the length of the lay, the spreading speed, the fabric tension, etc.). The spreading machine automatically performs the following operations: lays the fabric in the required length of the spread, cuts the material at the end of every ply, counts the number of plies, and stops after laying the required number of plies.

6.4.3 Automated fabric fault registration

Automated spreading systems can also determine fabric faults during the spreading process in a similar manner to that of manual spreading processes. Fault registration system is used to carry out this process. During the spreading process, faults are identified visually by a spreading operator. Using a joystick and laser beam, the operator marks the fault and determines its position on a ply and also on the marker on the screen. The operator can assess whether the fault affects any cut component and makes the decision whether or not to leave the fault in the spread. The spreading machine can automatically cut off the unusable part of the fabric and record it. Then the spreading machine moves to the place of the corresponding splice mark on the cutting table to continue the spreading process.

6.5 Automated fabric pattern matching

Spreading intricate pattern fabrics, the coordination of patterns for all the cut components is an important task. Three reasons: pattern matching necessity, variability in fabric structures, and incompleteness of the work methods used, make the spreading and cutting processes very complicated. Manual processing of intricate pattern fabrics is still widely used in the industry because of its low expenses, and in long period, well-developed and proved work principles and methods are established. Usually sectioned markers are performed directly on a fabric, using real size pattern pieces printed on paper. Spreading and cutting are performed in two work steps. During the first step a lay of plies matching the fabric pattern is created. Rough cutting is done leaving certain fabric allowances around cut components. During the second work step the roughly cut components are respread to ensure more precise pattern coordination and the fine cutting is performed. Finally, the fabric allowances that had ensured shifting of roughly cut components to increase pattern matching accuracy are cut off and wasted. Precise pattern matching is obtained, which is time and labor consuming work process. Besides, the operator performing the process requires expertise in cutting intricate pattern fabrics and, because of it, work productivity and pattern matching accuracy is dependent on the skills and work experience of the operator.

To improve processing of the styles from intricate pattern fabrics, several companies have automated it developing specialized pattern matching hardware and software. Using the new work methods, semiautomated and fully automated pattern matching, can be achieved in several different ways.

6.5.1 *Semiautomated work methods*

In semiautomated way, styles from intricate pattern fabrics can be processed by:

- projecting images of the pattern pieces directly on the fabric and performing pattern matching and marker making on the fabric ply; and
- generating an on-screen fabric design image and performing pattern matching and marker making on the image of the fabric pattern.

Projecting images of the pattern pieces on the fabric: Using an overhead projector, a full size pattern piece is projected onto the fabric as it is spread onto a cutting surface. The pattern piece can be moved or rotated to match with the fabric design. It is possible also to distort slightly the shape of the pattern piece and to cut the component in accordance with irregularities in the fabric design. The cutting process is performed by a single-ply cutter.

In this way, cutting of intricate pattern fabrics can be done by using "CutWorks Matching" software by Gerber, "Visual Nest" by Morgan Tecnica, "VisionCUT" by

Kuris and overhead projectors and nesting software by FK Group, Aeronaut, Shima Seiki, Elitron, and others.

Advantages of the method: (1) high pattern matching accuracy can be achieved for every fabric ply; (2) although the marker is made of a fabric ply, possibility to use projected images of pattern pieces significantly ease and speed up the work process; (3) possibility to distort slightly pattern pieces allows to conform them to the design of every certain fabric ply; and (4) as a marker is performed for every fabric ply, material faults can be easily avoided and there is no need to add material allowance around the pattern piece thus the highest material utilization is ensured comparing with other methods.

Disadvantages of the method: (1) styles with simple shapes and few components are not typical for apparel manufacturing; (2) placing (projecting) of pattern pieces on material (nesting) has to be done for every fabric ply; (3) the method does not ensure pattern matching and cutting of multi-ply spreads; and (4) single-ply cutters are not typically used in apparel manufacturing.

Generating an on-screen image of the fabric pattern: Fabric design image is scanned, imported, or generated as a grid representing fabric pattern and a marker is performed on a PC screen, placing pattern pieces on the fabric image. Certain movement allowances are added around the pattern pieces in the marker to perform final corrections of the marker directly before the material cutting. When a fabric ply is laid on a cutting surface, by use of special camera fixed in the cutting head and pattern matching software, the corrections in the previously created marker are realized by shifting and slightly distorting the pattern pieces to compensate for existing pattern irregularities in each piece of fabric. Cutting is performed by a single-ply cutter.

In this way, pattern matching and cutting is done by using "SiliconEye Machine vision" and "Cyclops Machine Vision" by Aeronaut, "Visual Nest" by Morgan Tecnica, "Matching B&S" and "Match It" by Bullmer, a overhead camera and "PathCut" by Pathfinder, old versions of "In Vision" by Gerber and "Mosaic" by Lectra.

Advantages of the method: (1) performance of markers on PC screen speed up the work process; (2) high pattern matching accuracy can be achieved for every fabric ply; and (3) possibility to shift and slightly distort pattern pieces in the marker allows to conform it to the design of every certain fabric ply.

Disadvantages of the method: (1) the shifting and slight distortions of pattern pieces have to be performed for every fabric ply; (2) the cutting head is used for both, to scan fabric ply pattern and to cut material, therefore, while geometry of the pattern pieces in the marker is corrected, cutting can not be performed; and (3) the operator has to perform the corrections in the initial marker.

6.5.2 Fully automated pattern matching

The most advanced, latest generation pattern matching systems "AutoMatch" by Gerber, "Mosaic" by Lectra, "Multi-array scanner system" by Bullmer, "VisionCut" by Gemini, and "Match++" by Optitex ensure highest level of automation in the work process. A high-resolution digital camera creates an image of the fabric pattern, while the fabric from the roll is advanced on the cutting table. In real time the specialized software analyzes the fluid image of fabric pattern—its exact position and distortions—and, in accordance with it, recalculates geometry of pattern pieces, modifies markers, and automatically launches cutting process. The modified markers are developed gradually. The new part of the nest is created on the bases of the newly scanned part of fabric, while the previous one is being cut. Comparing with the previous generation scanning and matching systems, which use a camera fixed in a cutting head, the cutting speed is increased dramatically.

Advantages of the method: fabric ply image is created in real time gradually; shifting, rotation, and distortions of original geometry of pattern pieces in the marker are performed automatically in real time, while the fabric ply is advanced on the cutting surface; minimal training and skills are necessary for an operator as most part of actions are performed automatically; and the continuous work process without any interruptions is ensured feeding fabric from a roll.

Disadvantages of the method: only single-ply cutting can be performed.

6.5.3 Multi-ply pattern matching and cutting

This section describes about the pattern matching and cutting processes of multi-ply patterns. Similar way pattern matching and nesting is done using a pattern matching and cutting system "Match-It" developed by Topcut-Bullmer. In this case one more work surface—a matching segment is involved in a work process. The matching segment with the movable camera is used to obtain image of the real fabric ply.

To ease pattern matching process, the system uses initial markers. They are available in a database and used for several fabrics of similar typical design (striped, checked) and with pattern repeats of similar size. Performing the work process, an appropriate initial marker is superimposed onto the fabric ply image on a PC screen and the pattern pieces are shifted (within movable allowances) to match the patterns to the actual design. Only the necessary corrections in the initial marker are made to get the final marker.

Multi-ply spreads on the matching segment or on a spreading table are performed manually, laying every fabric ply onto pins to match the designs with the patterns of the first basic fabric ply. The ready fabric spread is transported to the cutter by a conveyor. Cutting can be performed by a single-ply or multi-ply cutting system. While the cutting process is performed, necessary corrections in the initial marker can be performed on the matching segment to get the final marker for the next fabric spread.

Advantages of the method: (1) the use of the initial markers save time and speed up pattern matching and nesting; (2) if the system involves a separate matching segment, it provides a continuous work process thereby ensuring higher productivity; and (3) both single- and multi-ply pattern matching and cutting can be performed.

Disadvantages of the method: (1) the use of the initial markers restricts free placement of pattern pieces on a material and thus reduces marker efficiency; (2) to use a separate matching segment, an additional area in a cutting room is necessary; (3) performance of pattern matching in a low-ply and high-ply spreads using pins is performed manually; and (4) significantly complicating work process and reducing its productivity.

6.6 Automated cutting methods and cutting systems

Traditional multi-ply cutting by manual cutting equipment is still often used at small enterprises for its low production costs. The manual method can be used to cut normal fabrics in multi-ply, as well as, to cut intricate fabrics, when automated cutting cannot be used. However, the work process is time and labor consuming and cannot ensure high cutting accuracy and productivity. The cutting is performed in several sequential steps such as: dividing a spread into sections, the rough cutting of components, fine cutting of components, placing notches, and drill marks in cut components, using different cutting equipment (straight and round knife machines, band knife machines, a drill machine). The cutting quality is influenced by: the displacement of fabric plies in a spread, the shape of a cutting device, the movement of the cutting blade, the ease of movement of a cutting machine, or parts of a spread during the cutting process.

From 1970s, apparel industry has been using automated cutting methods, which brought about significant changes in garment manufacturing processes. Automated multi-ply knife cutting systems are highly efficient while working with large diversity of styles and large orders, and cutting components with complex shapes. Using fully different work methods and cutting tools, they reduce or even eliminate most serious disadvantages of manual cutting process. The displacement of fabric plies in the spread, while cutting is avoided by the help of a vacuum system, which compresses the fabrics plies during the cutting. The high precision of the knife movement helps to increase the cutting quality significantly. The productivity of automated cutting process is higher than that of the automated material spreading and sewing processes. Many important changes in the management of manufacturing processes have arisen from this as several spreaders and sewing lines have to operate with one cutter.

During the last decade new changes have happened in the apparel industry. Because of market demand to raise diversity of styles, reducing their quantities, small orders have become the reality of the industry. Their cutting by a powerful multi-ply cutter or manually is very much inefficient. Therefore, the single-ply knife cutters, which can process single fabric ply and low spreads slowly increase their importance in apparel manufacturing.

There are many companies which produce automated knife cutting systems for textiles, such as: Lectra (France), Gerber (United States), Eastman (United States), Bullmer (Germany), Kuris (Germany), Novocut (Germany), Oteman (Spain), Morgan Tecnica (Italy), Elitron (Italy), FK group (Italy), Tecno-Systems (Italy), B.K.R. (Italy), S.M.R.E. Engineering (Italy), Atom (Italy), Zund (Switzerland), Blackman & White (UK), Autometrix (United States), Gemini (Romania), Kimla (Poland), Shimaseiki (Japan), Takatori Corporation (Japan), Tukatech (India), Iecho (China), Oshima (Taiwan), Aeronaut (Australia), and Pathfinder (Australia).

6.6.1 Automated cutting systems

Automated cutting systems are used to process a wide variety of materials. Despite differences in the cutting tools and materials to be cut, the work principles and main parts involved in automated cutting process are similar. These are: a cutting device and a carriage in which the cutting device is fixed, a crossbar (gantry, beam, cutting bridge) that carries the carriage across the cutting surface, a working surface, a control panel to control the cutting process (see Fig. 6.4), and a nesting and cutter control software.

Two synchronized servo motors move the beam (crossbar) along the length of the cutting table (the X axis), while the third servo motor moves the carriage on the beam across the width of the cutting table (the Y axis) and a fourth rotates (the C axis) the cutting tools during the cutting process. If the cutting tool also moves vertically (the Z axis), an additional motor is used to complete the cutting process. To increase productivity, the automated cutting system can be equipped with two cutting heads (fixed on two crossbars) to perform synchronized and simultaneous cutting process.

The cutting process is carried out on static or conveyorized cutting tables. Using the static table the cutting process is completed in one step on the fixed surface. The conveyorized table has a movable surface that ensures a continuous cutting process and increased productivity. The most part of cutting tables are equipped with a special vacuum system to hold the material plies fixed during the entire cutting process.

The cutting process is controlled by the use of a control panel. Before starting the process, the operator sets the main cutting parameters. Using advanced cutters, the operator have to set up only the properties of the material, while the cutter automatically determines all needed process parameters. A cutting path is seen on the screen of the control panel during the whole work process. The operator can distinguish the components that have already been cut (these are marked in different colors from those as yet uncut) and the location of the cutting device at any given moment of the cutting process.

A full-size preprinted marker, an offload screen, an overhead projector, marking on the cut component surface, and adhesive labels can be used to identify cut components, simplify their bundling, and reduce the risk of errors in the work process.

After the components are cut fully and placed in the storage area of the cutting table, the operator off-loads and groups the components and includes the bundle tickets.

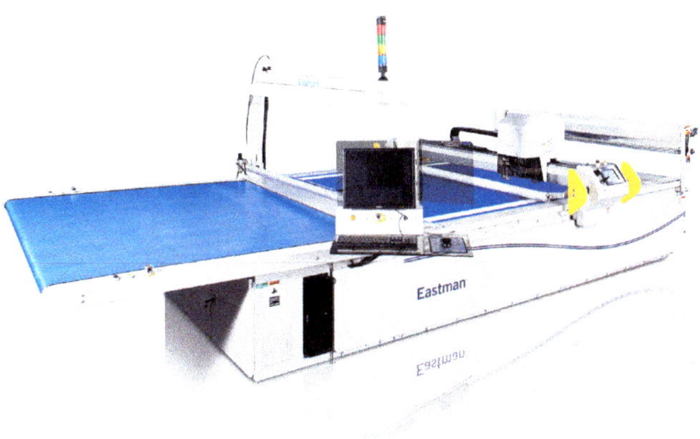

Figure 6.4 The automated multi-ply cutting system Raptor by Eastman.

6.6.2 Automated knife cutters

Knife cutting is the most widely used automated method for processing textile materials. The cutting is done by multi-tool cutting head, which can carry out a variety of cutting operations (see Fig. 6.5).

The choice of each tool for the cutting process depends on the material to be cut and the configuration of the required contours and cutting operations. The profiles of the components are cut by rotary blade knives, drag knives, and oscillating knives; and punches are used to make holes.

- *Rotary blade knife*—during the work process, the rotary blade knife rolls over the material (see Fig. 6.6). The knife is used to cut contours of the components. The diameter of the knife depends on the properties of the material and the complexity of the shape to be cut.
- *Drag knife*—has a sharp angled blade (see Fig. 6.7). The knife is used to cut detailed contours, sharp corners, small circles, and notches. During the cutting process, the knife is dragged along the profile of the cut component. The angle of the blade depends on the properties of the material.

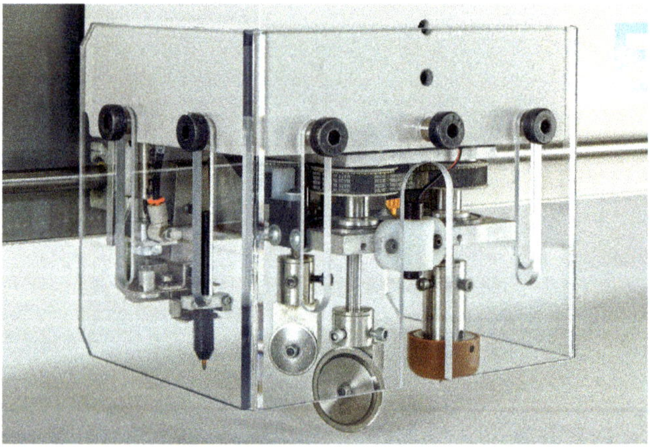

Figure 6.5 The cutting head with tools of a single-low ply cutter by Eastman.

Figure 6.6 A rotary blade knife.

- *Oscillating/reciprocating knife*—performs vertical movement up and down in regular rhythm. The knife is used to cut multi-ply spreads, thick materials, and complex shapes.
- *Punch*—creates small round holes in the cut components, where they cannot be easily or quickly cut by other blades (see Fig. 6.8). Different diameter punches are used to make different size holes.

Other operations during cutting process can be performed using, notch tools (to cut V or slit notches), and different marking tools (to identify cut components).

Each job may be cut in several ways. The markings are normally done first for the entire work zone. This is followed by any necessary punch or drill marks. Next, all the contours of the components are cut using the chosen cutting tools. The cutting process is carried out in separate work zones. All the cutting operations (cutting of contours, notches, drills, etc.) are performed in a specified work zone. The material is then moved and the cutting job is started in the next work zone, continuing the cutting of the components. Cutting speed is dependent on many different factors, the most important of them are as follows: the properties of the material being cut, the number of material plies being cut, the cutting tool used, and the shapes being cut.

Figure 6.7 A drag knife.

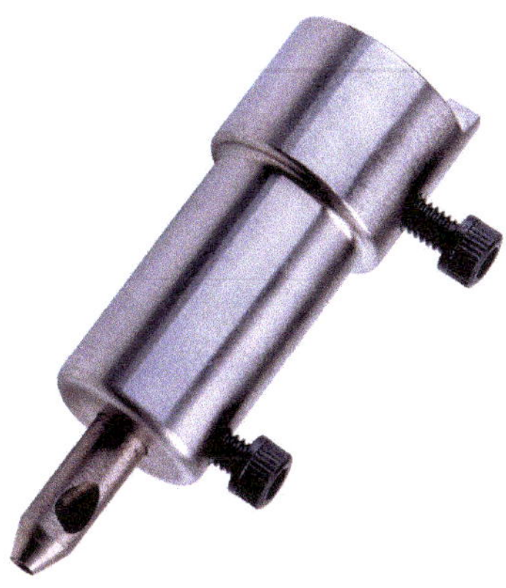

Figure 6.8 A punch.

6.6.2.1 The single and low-ply automated cutting

Single and low-ply cutting is performed by automated cutters, which can work with material heights up to 30mm (maximum thickness of the compressed material). The cutting is performed by rotary ("pizza wheel") and drag knives.

The rotary blades (18 and 28mm diameter) are efficient in cutting contours of the components. The rotary blade rolls over the fabric and, because of it, does not cause its displacement during cutting process. As the rotary knives cannot ensure very high cutting precision, drag knives are used to cut detailed contours, sharp corners, small circles, and notches.

If only one ply is cut, the material quality control, the matching of patterns and the marking of cut components can be done during the cutting process. Advanced cutters also ensure printing certain patterns on the components during the cutting process. Single-ply cutting is used for sample making, in made-to-measure manufacturing and cutting of styles from intricate pattern fabrics.

Low-ply cutting is used to process small orders and preproduction runs. They can be cut also by high-ply cutters (using oscillating knives), however high-ply cutters are more expensive and their use in low-ply cutting is not efficient.

6.6.2.2 The high-ply automated cutting

Multi-ply cutting is used for textile materials which are porous and air permeable. The multi-ply spread, prepared on a spreading table, is moved to the cutting table and covered with a lay of air impenetrable polyethylene film cover (see Fig. 6.9). Using a vacuum system, all the fabric plies are fixed to the cutting table and pressed tightly together to form an airtight seal, which ensures the material is held together during the entire cutting process.

As the cut material is thicker and not homogenous, cutting tools used for single-ply cutting cannot be used and the normal processing time will be longer than that of

Figure 6.9 The fabric spread covered with polyethylene film during the cutting process.

single-ply cutting. The cutting process is carried out by a vertical high oscillation knife and a drill or a punch. Because of the vertical movement of the knife and the necessity of ensuring accurate cutting (also for the lowest plies), the cutting surface is covered with bristles that allow the knife to pierce all the fabric plies easily, not displacing them or causing any damage (see Fig. 6.10).

A large variety of high-ply knife cutting systems are available. Their power level is determined by the maximum thickness of the compressed material to be cut: 5, 7, and up to 10mm. High-ply cutters are working in lower speed and accuracy than the single-ply cutters. They are used for high volume, comparatively low cost goods such as garments and upholstery, where a very high degree of accuracy is not required.

High-ply cutters have several important advantages: they ensure very high productivity, high effectiveness cutting large orders, enough high quality of cut components, and a high degree of flexibility. However, they are more expensive than single-ply cutters, not efficient to cut small orders, and cannot be used to process the styles from intricate pattern fabrics where pattern matching is required. For several decades the high-ply cutters have been the most efficient cutting equipment used in garment industry. However, during the last decade small orders are replacing large ones and the high-ply cutters slowly losing their importance.

6.6.3 Automated laser cutters

Because of very high accuracy, high processing speed, flexibility, simple operation, and high degree of automation, laser cutting has become the second most widely use automated cutting method in the garment industry. Textile materials are treated

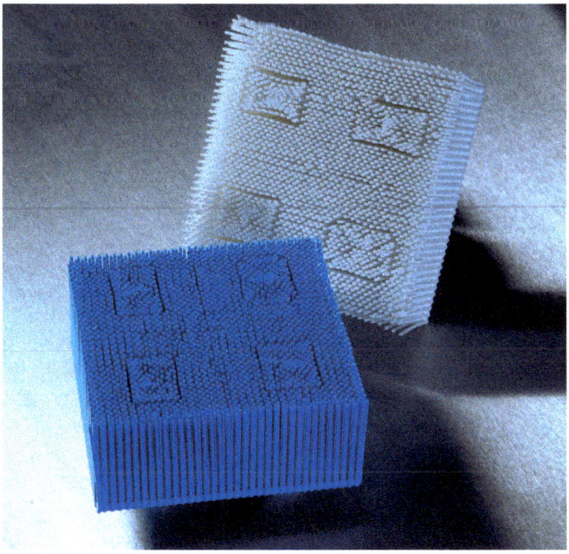

Figure 6.10 Bristle blocks covering cutting surface.

by carbon dioxide (CO_2) laser equipment to perform through cutting, kiss cutting, engraving, and marking. The laser cutting is tool free, no contact surface treatment method that has also several additional advantages—treated material is not distorted during the processing, there is no cutting tools wearing out, and there is no need for sharpening and replacement of the cutting tools.

Several companies produce laser equipment to treat different textile materials such as: Jeanologia (Spain), Macsa ID (Spain), Iberlaser (Spain), Sei Laser (Italy), Proel TSI (Italy), Puntoart (Italy), GMI (Italy), Eurolaser (Germany), Rofin (Germany), Trotec (Austria), BITO (United States), Golden laser (China), GBOS laser (China), Ant Laser (China), and Perfect Laser (China).

6.6.3.1 General characteristics of a laser cutter

To perform the cutting process, the light from a laser source is focused, intensified, and transferred by a lens or mirror to create a laser beam on the cut surface. As the beam of light strikes the material, fabric absorbs great part of the energy it receives, and rapidly heats up. To perform laser cutting, the laser beam is steered onto the material surface by a cutting head, following a computer-driven path.

The main parts of the laser cutting head are: a *focusing lens* and a *cutting nozzle*. By the help of the lens the laser beam is focused onto the material surface. The cutting nozzle guides the stream of compressed air into the cutting gap to drive out of it the debris and molten material and protect optical lenses from flying sparks, vapors, and particles generated on the cut material surface. The gas also cools the heat-affected material zone and thus increases cutting quality.

During the work process, the cut material is kept flat and fixed on a *cutting surface*. As the cutting surface also has to support the work of an extraction system to evacuate smoke particles and soiling from the cut material, the metal grid is most often used as a table surface cutting textiles. While performing engraving, marking or kiss cutting, cutting surface can be a metal plate.

By the help of special *extraction systems*, the smoke particles and soiling created in the cutting process are drawn away above and below the cut material surface. A suction device located beneath a grid or the perforated cutting surface collects cutting emissions under the material and drives them away. The suction of air under the cut material also helps to fix it on the work surface and prevents already cut components from slipping and lifting. Above the cut material the cutting emissions use to be collected in two different ways: laterally or upward. In the case of lateral extraction, the emissions are evacuated via lateral exhaust slots that are usually located at the rear. In case of upper extraction, cutting emissions are collected around the laser beam directly upward. Upper/lateral extraction is very important when the functioning of the lower extraction unit is restricted (e.g., performing engraving or marking).

6.6.3.2 Laser treatment methods

During the last decade, laser through cutting, kiss cutting, engraving, and marking are successfully replacing many traditional textile finishing methods, such as: cutting, printing, discoloration, embroidering, pleating, and wide range of denim finishing methods.

Through cutting

During laser through cutting, laser beam fully melts or burns one or more material plies. The through cutting is used to cut out separate material components and to create different decoration (wholes) in the material, manufacturing garments, finishing textile materials, and creating embroideries.

Kiss cutting

Laser kiss cutting is used to cut the top layer of a textile material without cutting another attached material. Cutting the fabric on the fabric and stitching them together, different kind of multiple-layer embroideries, appliqués, and labels are created.

Engraving and marking

Engraving and marking are two more laser treatment methods that can be performed by laser cutting equipment by changing processing parameters. During laser engraving, the laser beam physically removes only the part of material surface and creates on it a noticeable visually and in touch cavity. Engraving is widely used in carpet and rug manufacturing, producing garments from textiles with low cut pile and fleece. During laser marking the laser beam slightly alters its properties or appearance. The color change of the treated material is the most typical physical effect of this process. This method is used widely in denim, as well as, cotton and linen fabric. Laser marking is also used to create codes, brand names, and engrave other info on finished garments, or other textile and leather labels.

Using different process parameters, both the processes, surface engraving and marking, can be done by the same equipment.

6.6.4 Laser cutting of textiles of different origin

Textiles of different origin differently react on laser treatment creating, different kind of desirable and undesirable changes in the material structure.

6.6.4.1 Textiles of natural origin (cotton, linen, wool, silk)

Natural organic materials vaporize quickly, when cut with a laser resulting in smooth edges. Annealing marks in light brown/orange tinge can appear on the cut edges. The thicker is the fabric, the more apparent are the marks. After the cutting, the edges of the fabric can still be frayed.

While performing the laser marking, most part of natural fabrics is marked in a brown/orange tinge. Indigo color denim fabrics treated by laser lose their color, and the marks on the fabric surface appear in white. The dense fabrics are more suitable for marking than the light ones. While marking very light fabrics, the laser beam could cut the material through. One or few natural origin material plies can be cut by laser.

6.6.4.2 Textiles of synthetic origin (polyester, nylon, polyamide)

Cutting synthetic materials with laser beam, melts material edges and, protects them from fraying. This is an advantage in many applications. In garments manufacturing hard edges resulting from melting can irritate skin. Therefore, laser cutting cannot be used to process garments that have direct contact with the body. Engraved marks in the fabric are usually slightly glossy and darker in contrast to the material surface.

The fiber melting complicates multi-ply cutting process if the cut edges of several plies fuse together. In order to avoid this problem, textiles of synthetic origin are cut only by laser in one ply.

6.6.4.3 Cutting intricate textiles

There are several kinds of fabrics whose processing by manual or automated knife cutting methods are problematic. Their cutting by laser is more effective as they ensure higher quality and cutting speed. The laser cutting of various intricate textiles has been discussed in the following section.

Artificial fur

Artificial fur is usually cut manually, separating its substrate material by scissors because of pile, which must not be compressed and cut/damaged during cutting process. Contactless laser cutting is performed from the left side of the material keeping it in uncompressed state. Individual strands creating pile of the fur remain intact, while laser cuts only the substrate material in high speed and quality.

Velcro

Knife cutting is difficult and cannot give high precision because of Velcro material's dual structure. Similar to artificial fur, Velcro is cut easily by laser from its left side separating only its base material. Laser processing ensures much higher cutting quality and speed.

Very light materials

Lightweight fabrics may be distorted due to moving or dragging by bladed tools during the processing. Using laser, high precision and speed cutting are performed in contact-free way without clamping or fixing the light material.

Fabrics with intricate patterns

The laser cutting is more effective while cutting fabrics with intricate patterns than the knife cutting. This is because the automated pattern matching methods developed till now ensure a simple fabric ply effective processing (see Section 6.4). It can be performed in higher speed, keeping the fabric ply in uncompressed state.

6.6.5 The use of water-jet systems for cutting textiles

Water-jet cutting is a universal separation method that is increasingly used in garment industries. "Pure" water-jet cutters are used to cut soft materials such as paper, foam, plastic, rubber, leather, composites, and textiles. Ordinary tap water is pressurized to ultrahigh levels (up to 6000 bar) and forced through a small orifice in an industrial jewel to form an intense cutting stream. The jet stream moves at a speed of up to 2.5 times the speed of sound.

Water-jet cutting does not generate levels of heat that would harm textile fibers or fabrics. It is possible to process very narrow and sharp corners and difficult contours. There is no deformation as the cutting forces are very small. Clamping devices are rarely needed. As only water is used, the cutting device does not need to be sharpened, changed, or cooled during the cutting process. In addition, only a short cutting program needs to be created and no time loss occurs in setting the machinery and equipment for subsequent process. The jet is so thin that it does not even wet the material. The water consumption is low and it produces a low wear and tear of the components. The maintenance costs of the water-jet system are low comparing with other kind of cutting systems. Although there is no danger of fusing as in the laser cutting methods, there is a tendency for these edges to mesh together, making it difficult to separate the plies.

6.6.6 Multi-purpose cutters

Multi-purpose cutting systems are developed to use more than one cutting method to process a wider range of materials than the traditional cutting systems.

6.6.6.1 Blade cutters plus lasers

These dual purpose cutting systems combine a different type of blade cutter with a high or low power CO_2 laser. A rotary blade cutter works with materials that are not suited to laser cutting, while laser cutting can be combined with creasing tool operations. Vacuum tables can be built with quick change cutting surfaces or with dual zones suitable for either laser or blade cutting.

6.6.6.2 Blade cutters plus ultrasonic tools

Cutting systems can work with thick and difficult materials (corrugated cardboards, composites, rubber, cork, and filter materials) using blade cutters. When fitted with ultrasonic tools, the cutting system can also cut technical textiles that cannot be cut by conventional methods, and has the capacity to cut and seal the edges of synthetic fabrics. The table top is zoned to allow for both blade and ultrasonic cutting.

6.7 Fusing of cut components

Today about 80% of all garments require the use of interlining to increase their strength and stability and to improve the shape and crease resistance. Fusing is a process in which cut components are fused with interlinings that are coated with thermoplastic resin by the application of heat and pressure over a controlled period.

The fusing is carried out by special fusing presses. They are divided into two groups according to their work processes such as: discontinuous work process (flat) fusing presses and continuous work process fusing presses. Discontinuous fusing presses work sequentially and separately from other fusing processes. They are less productive and are suitable for small and medium production units. Continuous fusing presses enable an ongoing process by moving the components on a conveyor belt. They offer a higher level of productivity, are more energy efficient, and can ensure fusing without material shrinkage and fading. Because of these advantages, continuous presses are designed for a variety of production units.

Several well known companies produce fusing presses which include: Meyer (Germany), VEIT-Group (Germany), Macpi (Italy), Konsan (United Kingdom), Hashima (Japan), and Oshima (Taiwan).

6.7.1 Continuous fusing press and its main parts

The operation of the press depends on a continuously moving conveyor belt, which moves the face fabric and its interlining components in and out of the heating chamber. The press consists of the following main parts: a work surface to move the components; a heating chamber that heats the components; pressure rolls to press the components; and a cooling system to cool down the fused components.

6.7.1.1 Work surface

The work surface of the press is a heatproof, seamless conveyor belt. The width of the belt determines the power of the press. The work surface can be made up of several conveyor belts (see Fig. 6.11):

- front loading conveyor belt: positioned at the front of the fusing machine to load components (1);
- feed (lower) conveyor belt (1): maintains contact with the upper conveyor belt during the heating (by the upper (3) and lower (4) heating zone) and pressing (by pressure rolls (5 and 6)) process;
- upper conveyor belt (2): maintains contact with the feed (lower) conveyor belt during the heating and pressing process; and
- exit conveyor belt (7): cools the fused components and moves them to an unloading operator.

Belts are manufactured from adhesive resistant materials and have a teflon nonstick surface, which can withstand temperatures up to 230°C and which assure an accurate temperature during the heating process. The presses have special belt cleaning systems that provide both cleaning and lubrication to keep the belts in good condition during the fusing process. They can also be equipped with a special return belt system that returns the fused components to the front of the machine.

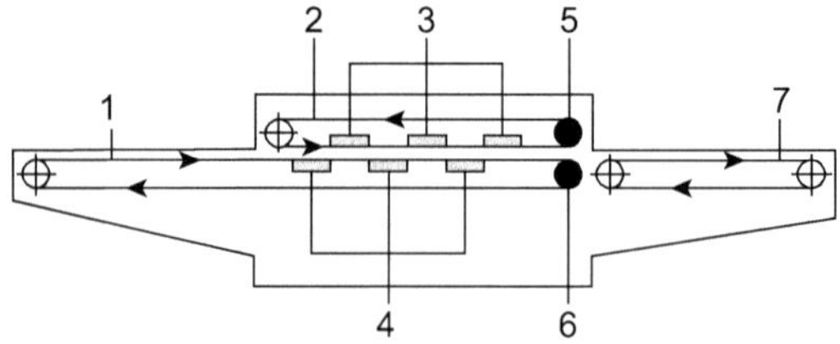

Figure 6.11 The schema of a continuous fusing press: (1) loading and feed (lower) conveyer belt, (2) upper conveyer belt, (3) upper heating zone, (4) lower heating zone, (5, 6) pressure rolls, and (7) exit conveyer belt.

6.7.1.2 Heating chamber

The heating chamber of a press consists of several individually controlled separate heating zones for the even distribution of heat. By the use of an electronic thermostat, the range of temperature difference can be kept to ±3°C. The temperatures for the upper heating zones (3, which heat the interlining) and the lower heating zones (4, which heat the face fabric) can be adjusted separately and precisely using a special heat control system. A high-quality fused assembly is thus created, maintaining the face fabric structure (avoiding shrinkage or fading) and controlling the adhesive resins on the type of interlining that has been selected.

6.7.1.3 Pressure rolls

After the face fabric and interlining have been heated in the upper and lower heating zones, the components are moved between the upper (5) and lower (6) pressure rolls to press them together. One or two pairs of large diameter pressure rollers assure stable and equal pressure. As the pressure rollers are fixed on the back of the upper and lower belts, the face fabrics and interlining cannot become entangled around them. The pressure is usually generated pneumatically. Most fusing presses have silicon rubber coated pressure rollers to ensure a more uniform and larger area of pressure than hard pressure rollers. The silicone coating also can enable the simultaneous pressing of thick and thin fabric components.

6.7.1.4 Cooling system

After fusing, the components must be cooled to stabilize the fusing effect, to maintain their permanent shape, and to prevent deformation during the unloading process. The cooling process can be performed on the exit conveyor belt (7), which moves the components to an unloading point, or by using special cooling conveyors with air-blowing or vacuum systems or water-cooled pressing plates.

6.7.1.5 Additional equipment

Fusing presses can be enhanced by a variety of additional equipment, such as:

- feeding conveyors with several loading belts that are positioned on the front of a fusing press to create ergonomic workplaces and increase productivity;
- collection stackers that stack fused components and require fewer operators;
- return belts that return the fused components to the front of the press when only one operator is working with the machine;
- waistband fusing devices to fuse long and narrow waistband rolls.

6.7.2 Advanced fusing technologies to avoid fabric shrinkage

The most serious problem that often arises during the fusing process is the material shrinkage. It complicates fabric cutting process, reduces fabric utilization efficiency, and leads to fabric loss. The reason for shrinkage is the thermal shock of the fabric, which occurs when the material is exposed to a sudden and rapid change in temperature. It causes structural stress and irreversibly changes the properties of the material.

To avoid the fabric damage the heating temperature should be raised gradually by extending its processing time. However, conventional fusing presses cannot ensure gradual heating of the components and different extent of fabric shrinkage occurs very often. As a result, the dimensions of the fused components are reduced and their joining with other parts of the garment can become embarrassed or even impossible.

To avoid the above described complication in the component joining, the extent of shrinkage for each fabric should be predetermined and added to dimensions of the components in their markers. However, because of the wide variety of textile materials, their tests and corrections of pattern pieces are seldom performed. Most often, the occurrence of further problems are prevented by adding safety buffers around the fusible components or their blocks in markers and performing two-step cutting. In the first step, the fusible components or their blocks are cut slightly bigger. After fusing, the fine cutting is completed and the extra material remaining after fabric shrinkage is removed and material is wasted.

Advanced fusing technologies offer new generation fusing presses, which can ensure more qualitative fusing process and avoid material shrinkage. The continuous work process fusing press has a long heating chamber and very sensitive heating system. The heating chamber of the press consists of several, up to 5, 7, even 9, or 12, individually controlled separate heating zones. Every heating zone has different temperature and can heat up components step by step in longer heating time. Temperatures for the upper heating zones, which heat the interlining, and the lower heating zones, which heat the face fabric, can be adjusted separately taking into account qualities of the materials.

The long heating chamber with several heating zones extends the heating time and ensures gradual temperature rise so that the fabric can be fused perfectly avoiding shrinkage even under the lower temperature than in a traditional short heating chamber press.

The heating capacity and the geometry of the heating chamber of the press determine the material to be fused. Lighter fabrics generally require lower heating capacity and the use of a press with shorter heating chamber with less heating zones can be much beneficial. The main heat should come from the face fabric side, to have the melting resins flow toward the face fabric. For heavier fabrics and sandwich or multilayer fusing, the presses with higher heating capacity and a longer heating chamber with several heating zones are needed. Besides, only heating from both top and bottom sides with separate thermocontrol ensures the correct temperature balance.

Continuous fusing presses with extended heating chambers and separate heat control systems are developed by companies such as: Meyer (Germany), VEIT-Group (Germany), Reliant Machinery (United Kingdom), The Martin Group (Italy), and Oshima (Taiwan).

6.8 Future trends in automation of textile material spreading and cutting

In the nearest future, because of very high competition in apparel industry, manual work in a cutting room will have to be replaced fully with highly automated equipment. The general tasks of automation in the cutting room are as follows: to increase work productivity, to increase cut components quality, to ensure maximally efficient use of raw materials, to ensure maximal flexibility of the work process, and to minimize work process dependence of human recourses. Introduction of automation should improve the following:

- production planning and control systems to plan and schedule work process more effectively and reduce fabric loss,
- fabric defect elimination to reduce spreading time and fabric loss,
- automated pattern matching methods for single- and multi-ply spreading of fabrics with intricate patterns,
- universal spreaders and cutters to process the most diverse materials with the same machinery,
- multi-purpose cutters to cut wider range of materials and perform additional operations during fabrics cutting,
- simplify operating of automated systems and improve work process monitoring and control systems to reduce the role of the operator in the work process,
- online support to an operator, remote technical assistance, predictive maintenance, and anticipation of breakdowns.

6.9 Conclusion

By the help of automation, the most drastic changes in the garment manufacturing has happened in the cutting room. From manual work process, fully dependent on human recourses, fabric cutting has become highly organized, productive, and very flexible process. The significance of the work, results of the cutting department, is the whole production process, determines necessity to continue advancement in the fabric cutting processes to reach the maximal automation level with the minimal human intervention. The newer technologies such as laser cutting and water-jet cutting are increasingly used in fabric cutting because of the increased productivity, accuracy, and quality of cutting. Because of increased demand for high-quality garments and rising labor costs, garment industries are adopting automated cutting tools into their production lines. In near future, many of the garment industries will be almost replacing the manual cutting methods with automatic cutting tools and equipment.

Further reading

Bowers, M.R., Agarwal, A., 1993. Hierarchical production planning: scheduling the apparel industry. International Journal of Clothing Science and Technology 5 (3/4), 36–43.

Chow, Y.F., Chan, A., Kan, C.-W., 2012. Effect of CO_2 laser irradiation on the properties of cotton fabric. Textile Research Journal 82, 1220–1234.

Dascalu, T., Acosta-Ortiz, S.E., Ortiz-Morales, M., Compean, I., 2000. Removal of the indigo color by laser beam–denim interaction. Optics and Lasers in Engineering 34 (3), 179–189.

Degraeve, Z., Gochet, W., Jans, R., 2002. Alternative formulations for a layout problem in the fashion industry. European Journal of Operational Research 143, 80–93.

Gersak, J., 2013. Design of Clothing Manufacturing Processes: A Systematic Approach to Planning, Scheduling and Control. Woodhead, Cambridge.

Gutauskas, M., Masteikaite, V., 1997. Mechanical stability of fused textile systems. International Journal of Clothing Science and Technology 9 (5), 360–366.

Gutauskas, M., Masteikaite, V., Kolomejec, L., 2000. Estimation of fused textile systems shrinkage. International Journal of Clothing Science and Technology 12 (1), 63–72.

Hui, C.L., Ng, S.F., Chan, C.C., 2000. A study of the roll planning of fabric spreading using genetic algorithms. International Journal of Clothing Science and Technology 12 (1), 50–62.

Jevšnik, S., Jelka Geršak, J., 1998. Objective evaluation and prediction of properties of a fused panel. International Journal of Clothing Science and Technology 10 (3/4), 252–262.

Jevšnik, S., Jelka Geršak, J., 2001. Use of a knowledge base for studying the correlation between the constructional parameters of fabrics and properties of a fused panel. International Journal of Clothing Science and Technology 13 (3/4), 186–197.

Juciene, M., Urbelis, V., Juchnevičienė, Ž., Čepukonė, L., 2013. The effect of laser technological parameters on the color and structure of denim fabric. Textile Research Journal 84 (6), 662–670.

Kan, C., Yuen, C., Cheng, C., 2010. Technical study of the effect of CO_2 laser surface engraving on the colour properties of denim fabric. Coloration Technology 126, 365–371.

Kan, C.-W., 2014a. CO_2 laser treatment as a clean process for treating denim fabric. Journal of Cleaner Production 66, 624–631.

Kan, C., 2014b. Colour fading effect of indigo-dyed cotton denim fabric by CO_2 laser. Fibers and Polymers 15, 426–429.

Kim, S.J., Kim, K.H., Lee, D.H., Bae, G.H., 1998. Suitability of nonwoven fusible interlining to the thin worsted fabrics. International Journal of Clothing Science and Technology 10 (3/4), 273–282.

Lai, S.-S., 2001. Optimal combinations of face and fusible interlining fabrics. International Journal of Clothing Science and Technology 13 (5), 322–338.

Mahrle, A., Beyer, E., 2009. Theoretical aspects of fibre laser cutting. Journal of Physics D: Applied Physics 42, 175507.

Nayak, R., Khandual, A., 2010. Application of laser in apparel industry. Colourage 57, 85–90.

Nayak, R., Padhye, R., 2015. Garment Manufacturing Technology. Elsevier, Amsterdam.

Nayak, R., Padhye, R., 2016. The use of laser in garment manufacturing: an overview. Fashion and Textiles 3 (1), 1–16.

Ng, S.F., Hui, C.L., Leaf, G.A.V., 1998. Fabric loss during spreading: a theoretical analysis and its implications. Journal of the Textile Institute 89 (1), 686–695.

Ng, S.F., Hui, C.L., Leaf, G.A.V., 1999. A mathematical model for predicting fabric loss during spreading. International Journal of Clothing Science and Technology 11 (2/3), 76–83.

Ortiz-Morales, M.N., Poterasu, M., Acosta-Ortiz, S.E., Compean, I., Hernandez-Alvarado, M.R., 2003. A comparison between characteristics of various laser-based denim fading processes. Optics and Lasers in Engineering 39, 15–24.

Tyler, D.J., 2003. Carr and Latham's Technology of Clothing Manufacture. Blackwell Publishing, Oxford.

Viļumsone–Nemes, I., 2012. Industrial Cutting of Textile Materials. Woodhead, Cambridge.

Vilumsone-Nemes, I., 2015. Fabric spreading and cutting. In: Nayak, R., Padhye, R. (Eds.), Garment Manufacturing Technology. Woodhead-Elsevier, Cambridge, UK, pp. 221–248.

Walter, L., Kartsounis, G., Carosio, S., 2009. Transforming Clothing Production into a Demand-Driven, Knowledge-Based, High-Tech Industry. Springer-Verlag, London.

Wong, W.K., Chan, C.K., Ip, W.H., 2000. Optimization of spreading and cutting sequencing model in garment manufacturing. Computers in Industry 43 (1), 1–10.

Wong, W.K., Guo, Z.X., Leung, S.Y.S., 2011. Applications of artificial intelligence in the apparel industry: a review. Textile Research Journal 81 (18), 1871–1892.

Wong, W.K., Guo, Z.X., Leung, S.Y.S., 2013. Optimizing Decision Making in the Apparel Supply Chain Using Artificial Intelligence (AI): From Production to Retail. Woodhead, Cambridge.

Yoon, S.Y., Park, C.K., Kim, H.-S., Kim, S., July 2010. Optimization of fusing process conditions using the Taguchi method. Textile Research Journal 80, 1016–1026.

Yuan, G.X., Jiang, S.X., Newton, E., Fan, J.T., Au, W.M., 2012. Application of laser treatment for fashion design. Journal of the Textile Institute 103, 48–54.

Automation in material handling

7

Volker Lutz, Hans-Christian Früh, Thomas Gries, Josef Klingele
Institut fuer Textiltechnik of RWTH Aachen University, Aachen, Germany

7.1 Introduction

The manufacturing of clothing can be divided into two categories such as: joining of fabric components or pattern pieces and material handling.

Till today, both in the clothing industry and technical textiles production a significant portion of the works of joining and handling work is being performed manually (Nayak and Padhye, 2015). In high-wage countries the material handling plays a significant role, which leads to high cost and offers a potential for automation. Material handling plays an important role in textile as well as clothing industries to increase the productivity and save the labor cost. In garment manufacturing, majority of the material handling involves the transportation of fabrics, cut components, trims, and finished garments from one work station to the other. Automatic material handling involves suitable tools or equipment or even robots to transport the materials.

In garment manufacturing, significant time is spent in material especially fabric handling, cut or semifinished components handling, and storage. The manufacturing costs of technical textiles constitute of 72% compared with the costs for fiber (23%) and matrix material (5%). Besides the demand for economic and efficient technologies there is a claim to the handling in terms of precision and a gentle material treatment. The most important reason for that is the limp material behavior of textile semifinished parts (i.e., fabric).

From semifinished to ready-made products the processes of separation, joining, and handling have to be undertaken by the material-handling equipment. Furthermore, superordinated quality assurance systems are of great relevance for automated production processes. For the selection of methods for automated production of textile products, a holistic view of the textile mounting process is necessary. Fig. 7.1 shows the overview of the basic process steps in textile forming.

Studies show that the handling and the mass transport are still predominantly undertaken manually. For example, in the manufacture of a garment about 80% of the total labor cost is due to handling. In other words the amount of the transport and handling operations around the sewing machine measured on the total assembly time is about 80%. In the field of seat cushion production, floor mats and vehicle cradles for automobile, a handling time ratio compared with the joining process of ~2/3 in the total production time (Koch, 1992; Egbers and Bühler, 1993; Szimmat, 2007; Wulfhorst, 1998; Hanisch and Krockenberger, 1993; Gutsche, 1992; Gebauer et al., 2004; Stephan, 2001).

Automated handling processes are an important aspect of efficient concepts for garment manufacturing in the future. However, there is still a gab of mature automated handling solutions for garment-related textiles in the production chain. As the knowledge of automated handling of textiles for reinforcement materials is further

Automation in Garment Manufacturing. http://dx.doi.org/10.1016/B978-0-08-101211-6.00007-0

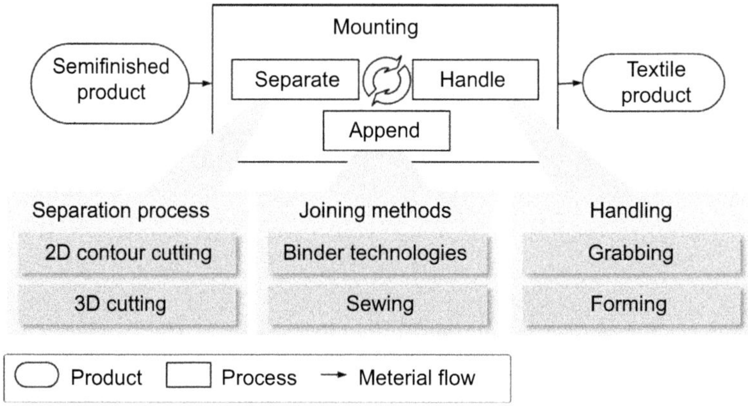

Figure 7.1 Overview of the basic process steps in textile forming.

developed compared with the handling of garment textiles, the transfer of results and solutions seems to be applicable. Hence the aspects of automation in composite materials can be used to identify the concepts for automation in garment handling.

To illustrate the potential for automation, 415 companies from the German automotive, shoe, and protective clothing industry were examined in 2004. The results proved that 85% of the breezy body parts are delivered as bales, and 77% are automatically placed and trimmed. The layers are cut with 79% as multilayers. None of the companies then automatically pick up the blanks. The handling in the joining process takes place with 79% manually, 21% semiautomatically; and 72% without the use grippers, whereas the remaining 28% use almost exclusively needle or scrap grippers. In total, 59% of the companies run a singling process, of which none is automated, although in 51% of the cases the stacks are even ordered (Szimmat, 2007). These results strongly prove the gab of automation for handling processes in textile manufacturing. It is expected that the results are lower in particular for garment manufacturing because of the higher degree of flexible textiles and huge amount of different variants.

The handling of flexible products, compared to bend-resistant materials, e.g., metals, is significantly more challenging. As clothing production involves handling of fabrics in many operations, it is extremely difficult to design tools and equipment for accurate handling of fabrics. In this chapter, first the concept of handling, subsequently the handling-relevant material properties are described. Various types of grippers for handling textile materials also described in this chapter. The use of radio-frequency identification (RFID) in material handling has made the process much easier compared with the traditional ticket systems, which is also a part of this chapter.

7.1.1 Definition of material handling

The Verein Deutscher Ingenieure (VDI)-Directive 2860 defines the handling as well as transport and storage as a partial function of the material flow. Handling means "creating, defined change or temporary maintenance of a predetermined, spatial arrangement of geometrically determined bodies in a reference coordinate system" (VDI, 1990). Compared to the transport and storing of items, handling is defined by the

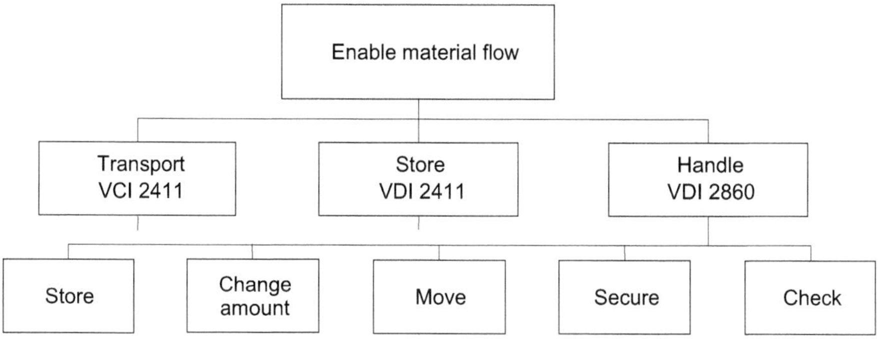

Figure 7.2 Definition of handling corresponding to the VDI standards. *VCI*, Verband der Chemischen Industrie e.V.; *VDI*, Verein Deutscher Ingenieure.

position and orientation of an object. It can be divided into five subfunctions: save, change quantities, move, secure, and check. In Fig. 7.2, the partial functions of the handling according to the VDI-Directive 2860 are shown in an organizational chart (VDI, 1990; Gutsche, 1992).

The individual subfunctions can be divided up further to elementary functions, combined functions, and additional functions, which are described in the VDI-Directive. The definitions are held very general to ensure a broad applicability. The strong, general breakdown serves the structured finding of solutions for specific applications. For each handling task the corresponding partial functions are differently pronounced (VDI, 1990; Stephan, 2001).

7.1.2 Properties of material and processes

For optimum handling, the knowledge about the specific properties of textiles is essential. Textiles are considered to be particularly difficult to handle because they have a component behavior, which is difficult to predict. Moreover, the measurement of the characteristic properties is often difficult to perform. There are a large number of different fabric qualities, which often have inhomogeneous and anisotropic even within the same material. These can, in turn, vary from one production batch to another production batch. Even "pure materials" are made of several components with variability in their properties. This further increases the variance of the properties. Many of the textile properties are also climate-dependent, e.g., shape, and dimensions, depend on the temperature and humidity of the work place.

In any case, textiles have greater tolerances and deviating behavior compared to semifinished products made from other materials. Curling edges are a good example of such behavior. Important for handling is the fact that many textiles are highly sensitive to dirt and mechanical damage due to forces acting on the surface. Coarseness, elongation, drapability, surface texture (roughness and hairiness), air permeability, adhesiveness, frictional behavior, surface weight, sliding resistance, wettability, and thickness are some of the properties (Koch, 1992; Stephan and Jensen, 2004; Gutsche, 1992; Stephan, 2001; Hanisch and Krockenberger, 1993; Hou, 1993). Various properties are explained in the table (Table 7.1), and reference is made to possible measuring methods. In his work, Gütsche developed an assessment of the importance of the component

Table 7.1 Importance of specimen properties on partial steps of textile handling

● Strong ◑ Average ○ Low	Partial steps of handling					
Specimen properties	**Supply**	**Shaping**	**Gripper principle**	**Size of gripping area**	**Gripper kinematics**	**Separation**
Type of material	◑	●	●	●	○	◑
Formstability	●	●	●	◑	●	●
Elasticity	○	●	●	○	◑	●
Surface properties	●	○	●	◑	○	◑
Permeability	○	○	●	●	○	◑
Adhesion of the surface	●	●	●	●	●	●
Friction	◑	○	◑	○	●	●
Mass per area	○	◑	○	●	●	◑
Dimension of specimen	●	●	○	◑	●	●

Based on Gutsche, C., 1992. Beitrag zur automatisierten Montage technischer Textilien Berlin. Techn. Univ., Diss. (Zugl. München, Wien: Hanser, 1993).

properties within the handling process (Koch, 1992; Hou, 1993; Monkman et al., 2007; Gutsche, 1992; Stephan, 2001; Gebauer et al., 2004; Stephan and Jensen, 2004). For various process steps, Table 7.1 gives an example of a qualitative evaluation.

The results by Gütsche (Gut92) showed that the type of material has a strong importance on nearly all partial steps of handling. This is obvious because of the fact that the type of material is directly influencing all major material and mechanical properties. The permeability of a specimen will in general influence the decision of the gripping technology.

7.2 Gripping technologies for textile handling

The handling of amplification semifinished products is for various reasons technically challenging. The softness of the textiles can even under small pressure (dead weight, air resistance) lead to impermissible deformations. Furthermore, the low sliding strength of many reinforcing textiles in aspect of a distortion-free handling is a major challenge; especially the automated handling is made difficult by these factors (Greb, 2013). According to the state of the art the handling and the reshaping of the textiles are still dependent on a large manual operation.

In automated processes, there is often the need for the textile to not only being transported in the handling but also being shaped. Many handling systems are therefore designed as portal- or robot-guided end effectors. They have a gripper guide system with gripping elements adapted to the handling task. The gripper guide can be rigid (e.g., rigid frame construction), actively driven (e.g., pneumatic actuators), or passively movable (e.g., flexible molding foam). In the following section the functional principles of some gripping elements are presented. Afterward technical approaches for automated handling systems are described.

Vacuum grippers are suitable for the handling of closed or dense surface structures such as reinforcing materials. In the case of vacuum grippers, the gripping element rests on the gripping material and is connected to a vacuum generator. The vacuum is generated by electric (e.g., pumps, blowers) or pneumatic (e.g., ejectors) means. Because of the pressure difference, the gripping material is pressed onto the gripper. Due to the high porosity and the associated air permeability of the textiles, high-volume flows are usually required. Regarding the designs of the vacuum grippers, a distinction is made between bellows and flat suction pads. Bellows grippers are distinguished by the fact that they adapt themselves to the surface contour of the gripping material and during the suction process they show a lifting effect. This can be used for singling.

Bernoulli grippers are also established for the handling of the textile semifinished products. With these grippers, an air stream is passed over the gripping material. Because of the effect of Bernoulli the gripping material is sucked. Gripper types are shown schematically in Table 7.2.

A reliable form-fitting handle of the textiles allows needle grippers. Pneumatic cylinders are used to drive pointed gripper needles into the material to be gripped. By angularly interlocking the needles, the textiles can be gripped positively. The penetration depth of the needles can be adjusted for some commercially available models (J. Schmalz GmbH, 2011). This can be used in a targeted manner to grab a defined number of textile blanks from a stack (Greb, 2013). Fig. 7.3 indicates the subprocesses of automated handling tasks and relevant technical solutions (Table 7.3).

Rollers are suitable for handling textiles with large dimensions. In the case of freezing grippers, an active medium between the gripper and the gripping material is collected by the Peltier elements below the freezing point. The ice crystals provide a temporary adhesion and allow handling of the material (Jensen and Stephan, 2002). Furthermore, various special methods exist for gripping textile semireinforcements. Special composite grippers have an integrated vacuum generator and a defined discharge of the sucked air. High volumetric flows allow multiaxial non crimp fabrics to be safely lifted (Ehinger, 2012). Surface grippers, which ensure the holding force via electrostatic effects, are described in the literature (Brecher et al., 2012a). Adhesive foils can also be used for handling (Prust, 2011). High-frequency ultrasonic waves can be used to build up a pressure field that can be used for handling. In the so-called ultrasonic levitator with suction edge, the gripping material is held in suspension above a sound transducer, at the same time sucked on the edge and thereby fixed in its position (Prust, 2011). Handling systems are used for specific handling tasks, which can be equipped with various gripping elements. A commercial handling system is the composite handling system from Brötje Automation GmbH, Wiefelstede. It allows the handling of textile semireinforcements for composite material. The blanks are gripped by vacuum grippers. With the combined

Table 7.2 Schematic representation as well as advantages and disadvantages of gripping elements for preform handling (Part 1)

Gripping element		
Bellow suction pads 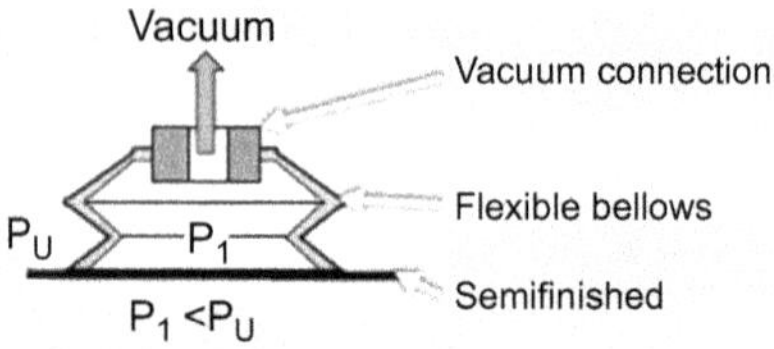	Advantages	• Low unit costs • Small construction space • Separating possibility
	Disadvantages	• Handled material must be air-tight • Vacuum generation required • Leaves minute deformations
Flat suction pads 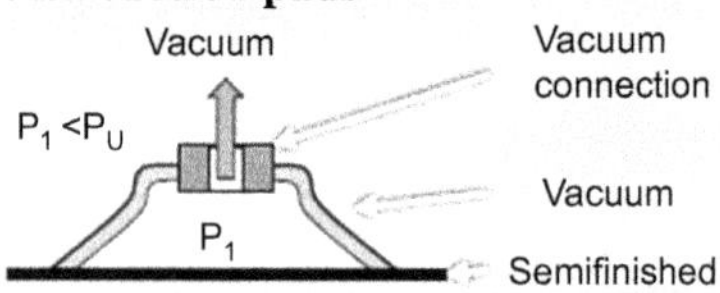	Advantages	• Low unit cost • Small construction space • Textile is not lifted
	Disadvantages	• Handled material must be air-tight • Vacuum generation required • Leaves minute deformations
Bernoullie gripper 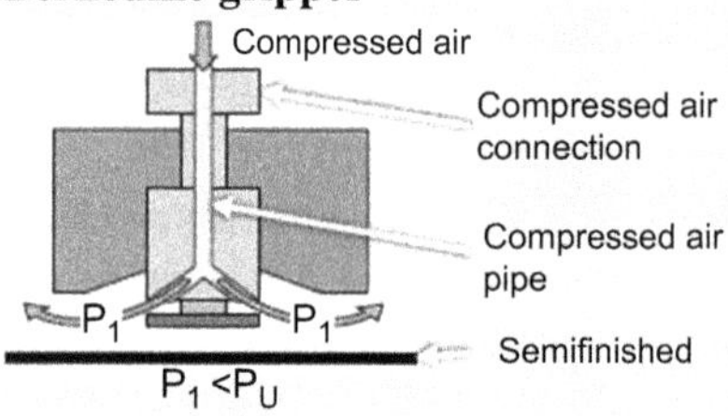	Advantages	• Contactless gripping • Direct use of compressed air • Small deformations
	Disadvantages	• Pollution through compression • No lateral fixation • Air consumption

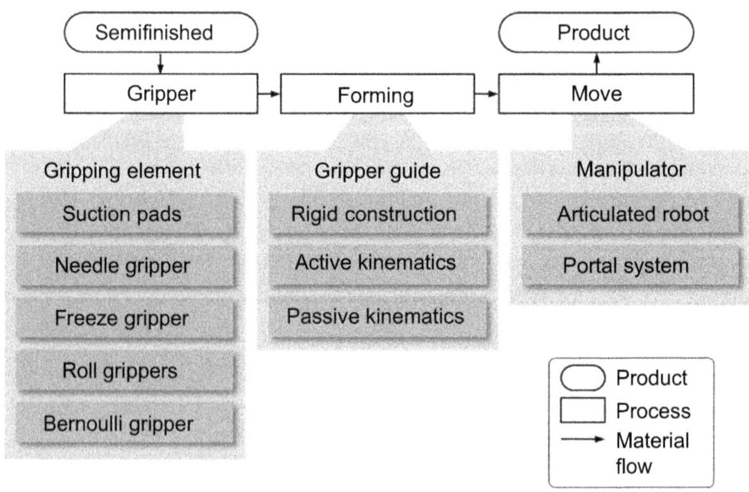

Figure 7.3 Subprocesses of automated handling tasks (top) and relevant technical solutions (below).

Table 7.3 **Schematic representation as well as advantages and disadvantages of gripping elements for preform handling (Part 2)**

Needle gripper 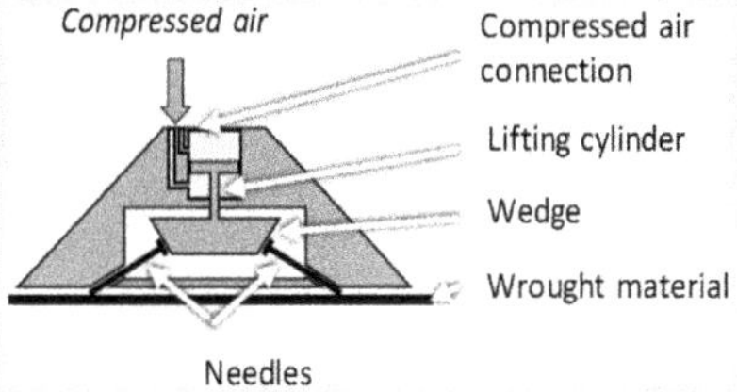	Advantages	• High holding force • Adjustable needle stroke • Only compressed air pulse required
	Disadvantages	• High per unit cost • Damage to the textile • Large installation space
Freeze gripper 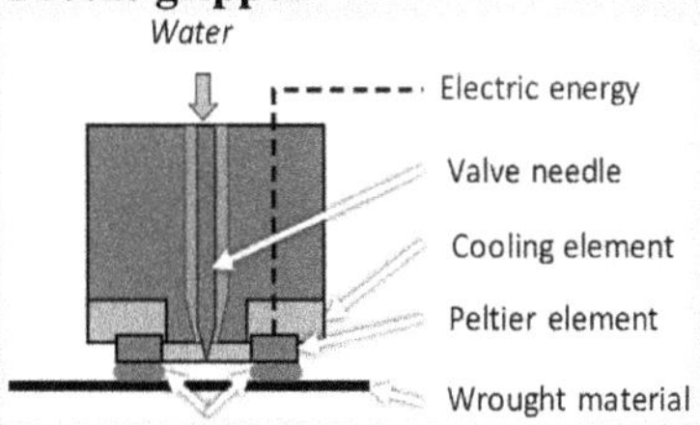	Advantages	• Low textile damage • Dividing well possible • High holding force can be achieved
	Disadvantages	• High per unit cost • Separate control required • Low process stability
Roll gripper 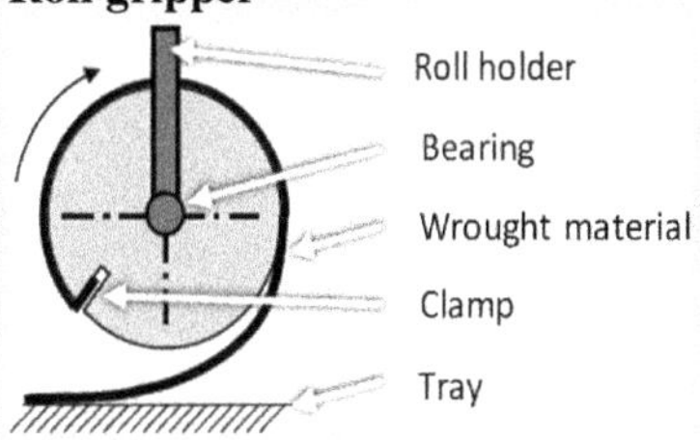	Advantages	• High holding force • High gripping and depositing rates • Low-impact to the textile
	Disadvantages	• Only simple geometries • Dividing is limited possible • Gripping material must be deformable

gripping and draping system (composite handling and draping system), the textiles can also formed by means of a flexible membrane film (Brö13). For handling of fabrics with low transversal strengths, simple suction lifting devices can be used.

According to the principle of a surface gripper, textile blanks are vacuumed over the entire surface. The gripping element is moved manually. The warp-free handling allows the production of high-quality sample plates in laboratory scale (Meyer, 2008).

The aspects of gripping technologies in composite material handling are also applicable on garment manufacturing environments considering different textile properties through analogy observations.

7.3 Automation in material handling related to high-performance textiles

For the gripping and forming of glass fiber noncrimped fabrics a robot-guided gripper was introduced at Institute für Textiltechnik der RWTH Aachen University. About an unequal transmitting gear, the textile can be gripped on a flat surface by pneumatic

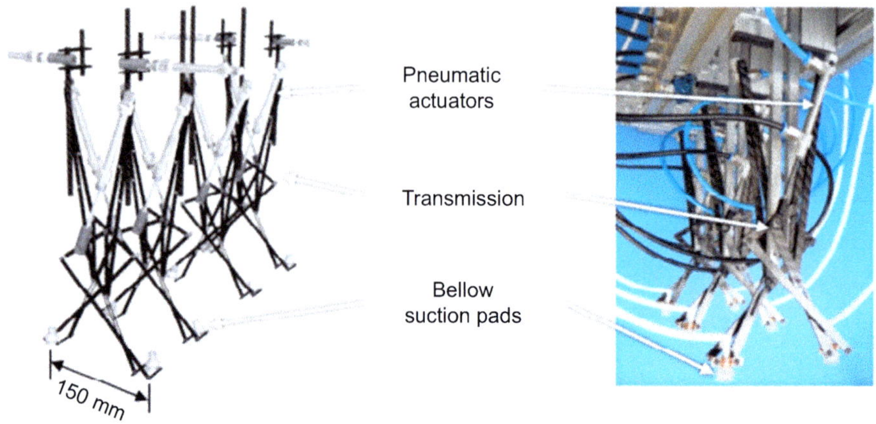

Figure 7.4 Handling system for automated gripping and draping textile materials: CAD model (left) and technical implementation at Institute für Textiltechnik der RWTH Aachen University-Preformcenter (right).

actuators and can be positioned in a shell mold. As gripping elements bellow suction pads are used, this allows the placement of several gripping elements in the narrowest space (Klingele and Devaux, 2013).

A form-flexible pad filled with granules is used in the gripping system "FormHand" for gripping and forming textiles. By evacuating the pad, the system flexibly adapts to different surface geometries (Raatz and Löchte, 2013). In the "AutoRTM" project an end effector has been developed with which various semicut blanks can be handled. The textiles are gripped with vacuum thereby. In the end effector, different areas of the gripper surface are activated via a chamber system (Eberth et al., 2006).

Ehinger describes a handling system based on the principle of a vacuum surface gripper (Ehinger, 2012).

Individual areas of the assembly tools can be switched on and off thus different cutting geometries can be gripped and deposed. The flexible surface of the gripper made of foam material allows the draping of the textiles in the preforming process (Ehinger, 2012).

In the "AutoPreforms" project an active handling end effector for curved preforms was developed. The system is robot guided and has a spring-mounted freeze gripper (Henning and Gries, 2008). In project "FOR860" an adaptive gripping system according to the fin-jet principle was developed at the Fraunhofer Institute for Production Engineering (IPT, Aachen). With the gripping system, textiles can be placed on curved surfaces (Brecher et al., 2012b). Fig. 7.4 indicates the handling system for automated gripping and draping of textile materials.

7.3.1 New conveyer systems

A new approach to transport single textiles between the process steps has been developed by SoftWear Automation Inc., Atlanta, US. The system is designed as a conveyer based on vacuum ball systems. Each ball can be driven in two directions. Each ball with vacuum and drives is called a budger. The conveyance system consists of distributed budgers in a transport table along the way of transportation. With this technology the stiffness or flexibility of flat textiles is not relevant to the defined movement over various distances. Even wrinkles could be eliminated by pulling with at least two budgers in the opposite direction. In combination with modern vision systems to detect the contour of the textile a precise placement of textiles becomes possible. Compared to gripping technologies, the conveyer system using clusters of budgers will allow an always flat and defined transportation (http://www.softwearautomation.com/#!technology/ch7k).

7.4 Digital tracking with radio-frequency identification

In garment manufacturing for mass production cloth hangers have established to transport the semifinished or finished garments. The use of automated cloth hangers with a rail system guiding clamps or hangers to a dedicated destination in the production has improved the material flow in terms of time efficiency. Automated cloth hangers conduct the product until delivery. Today the use of RFID technology makes a 100% localization of goods possible during the production. The RFID can be implemented in the hangers or in the cloth as a label for identification (Nayak et al., 2007, 2015).

The use of RFID was a major key to improve the efficiency, the customer service, and the availability of products on the sales floor. The data, which RFID transmits, can help to build a better overview on products on the sales floor (Tellkamp and Quiede, 2005). The RFID technology can be used as an identification of more than one product at a time and without intervisibility between the reader and the writer (IWL09). RFID tags or chips are easy to reuse because they are rewritable. The tracking with RFID is easier than the one by hand. It helps to cut down the time of the shipment and the tracking of the shipped products because a 100% control of the products is worthwhile. When using the RFID, the collector does not need to scan a single barcode, which has been in use for the last 20 years. The information of the product is directly sent to the computer when the RFID reader captures the product (Tellkamp and Quiede, 2005).

During a project cooperation between Kaufhof AG and Gerry Weber, they tested the use of the RFID technology within the supply chain. They used this project to test if the technology of RFID is viable for clothing. As readers for the RFID tags they used movable and fixed readers, which are attached to shelves, as well as readers, which were fixed under the table next to the cashier. The readers, which were installed on the sales floor used their RFID range to identify the clothes next to them.

Other companies tested the technology in the United Kingdom and the United States. The projects all resulted to be successful. Nevertheless, there is still work to be done according the technology and the business cases. The implementation costs were also a milestone to reach. To sum up, all projects were successful. It could be shown that the efficiency within the supply chain could be improved. The customer service was improved as well as the product availability on the sales floor. Furthermore, it was shown that the sales rate has improved while using the RFID technology on the shop floor (Tellkamp and Quiede, 2005).

With the help of RFID, products can be detected very easily in the goods receipt. There is less effort for the workers and customers can save time at the cashier (Nayak and Padhye, 2014). Within the project of Kaufhof and Gerry Weber, some problems had been detected. When the RFID was used with suspended products, the RFID tags could not be detected, as the clothes sometimes were to near to each other and the reader could just detect one tag. Furthermore, metal pieces at the shelves on the shop floor, sometimes limited the range of the RFID. Therefore, more readers are needed, which means additional cost of installation. RFID tags or chips are relatively costly and they are cheaper when used more than one time (Tellkamp and Quiede, 2005).

As a next step, the coexistence of barcode and RFID has to be a possibility. The use of the RFID can help for shop-in-shop solutions. Another application of RFID in the future could be a self-checkout system for customers. Currently, the prices of RFID tags are still several times higher than the bar codes (Sarma, 2001). Based on the experience of existing projects to implement RFID in garment business the costs of installation should not exceed EUR 0.10–0.12. If the RFID tag is only used once, the tag cost is a more important cost driver. If, on the other hand, the RFID tag is used more than once, it decreases the importance of procurement costs (Tellkamp and Quiede, 2005).

7.5 Conclusion

In this chapter various aspects of automated gripping technologies to handle textiles have been discussed using the parallel aspects of automated preforming process chains. Depending on the process steps and the textiles to be handled, different gripper technologies can be chosen. Permeability and the ease of undesired textile deformations during handling operations are the main challenges to be addressed. Various grippers are often attached to multifunctional gripper guides enabling active or passive forming of textiles. The handling process itself is often carried out by robots or portal systems. In the future, gripper textile interaction needs to be investigated to reduce gripping-induced defects while using needle grippers or reduction of energy consumption using vacuum gripper systems.

However, new transportation systems such as conveyers using vacuum rollers and vision systems may close the gap of manual transportation between process steps. The reduction of manual handling quantities will result in a strong increase of production efficiency. Apart from the mechanical transportation the possibility to track every piece of a garment at any time will allow 100% knowledge of material flow and added value even for the customer.

The garment industry could be one of the first industries to use RFID at the product level. At present, most applications are still in pilot-experimental stage. Much will depend on how many companies are actively driving the introduction of RFID. It is not yet final how the actual introduction of RFID might be implemented and adopted by the garment industries. In the foreseeable future there will be a coexistence of barcode and RFID. In this case, corresponding transitional scenarios must be established. It is not yet clear how the RFID tags will be installed on the individual product level. Thus, it is possible, e.g., in hanging labels, in ease tapes or directly into the clothing to integrate. All methods have certain advantages and disadvantages and implications (e.g., in the set of reusability or usability of the RFID tag over the logistics chain) (Tellkamp and Quiede, 2005).

However, both approaches to improve handling technology and digital tracking technology will enable garment production to increase efficiency by the reduction of handling costs.

References

Brecher, C., Emonts, M., Ozolin, B., Haubert, K., February 27, 2012a. Elektrostatischer Greifer kann textile Halbzeuge handhaben Maschinenmarkt. Available from: http://files.vogel.de/vogelonline/pdfarticles/mm/themenkanaele/automatisierung/montagetechnik_handhabungstechnik/articles/354748/354748.pdf.

Brecher, C., Emonts, M., Ozolin, B., Schnabel, A., Greb, C., November 29–30, 2012b. PreformherstellungundHandhabungfürdieSerienfertigungvonFaserverbundstrukturbauteilen. In: Dörfel, A. (Ed.), Proceedings of the 6th Aachen-Dresden International Textile Conference, Dresden. Institute of Textile Machinery and High Performance Material Technology (ITM), Dresden. TU Dresden, 2012, Datei: lecture_ozolin.pdf.

Brö13, Brötje Automation GmbH, October 10, 2013. Device for Use in the Manufacture of Fiber-Reinforced Components. Patent US8556617 B2.

Eberth, U., Fastert, M., Friedrich, M., Klein-Lassek, M., Krafft, H.-M., 2006. Neue Technologien für die automatisierte RTM-Fertigung (AUTO-RTM) : CFK-Bauteile in Stückzahlen bis 20.000 pro Jahr Abschlussbericht. Composite Technology Center Stade, Stade.

Egbers, G., Bühler, G., 1993. Grundlagenuntersuchung zur automatischen Handhabung großflächiger Werkstücke in der Maschenkonfektion: Schlussbericht zum Forschungsvorhaben AiF-Nr. 8449 Denkendorf: o.V. .

Ehinger, C.A., 2012. Automatisierte Montage von Faserverbund-Vorformlingen München. Utz, 2013 Zugl. München, Techn. Univ., Diss.. Available from: http://nbn-resolving.de/urn/resolver.pl?urn:nbn:de:bvb:91- Diss-20121004-1107974-0-5.

Gann, T., 2009. Prozessoptimierung mit RFID in der Bekleidungsindustrie, Prozessoptimierung mit RFID in der Bekleidungsindustrie, IWL AG https://www.iwl.de/images/stories/events-presse/logistik-tage-2009/vortraege/09-2-iwl_prozessoptimierung_mit_rfid_in_der_bekleidungsindustrie.pdf.

Gebauer, I., Dörsch, C., Müller, D.-H., September 28–29, 2004. Handhabung biegeschlaffer Materialien – Zentrale Herausforderung in der automatisierten Produktion von Composite-Bauteilen AVK-TV:7. In: Internationale AVK-TV Tagung für verstärkte Kunststoffe und duroplastische Formmassen Baden-Baden.

Greb, C.W., 2013. Systematische Gestaltung von Fertigungsprozessen textiler Verstärkungsstrukturen für Hochleistungs-Faserverbundkunststoffe. Shaker, Aachen. Zugl. Aachen, Techn. Hochsch., Diss., 2013.

Gutsche, C., 1992. Beitrag zur automatisierten Montage technischer Textilien. Techn. Univ., Diss., Berlin. Zugl. München, Wien: Hanser, 1993.

Hanisch, G., Krockenberger, O., 1993. Flexibles Greifen und Vereinzeln biegeschlaffer Flächengebilde. Eine technische Herausforderung mit hohem Rationalisierungspotential Produktion und Management. wt 83 (1993), H. 9, S. 138–140.

Henning, K., Gries, T., Flachskampf, P., 2008. Wirtschaftliche Herstellung von Faserverbundbauteilen mit Hilfe automatisiert hergestellter Preforms. Shaker, Aachen.

Hou, M., 1993. Verfahren zum automatischen Handhaben und Positionieren labiler Stoffzuschnitte in der Bekleidungsfertigung. Univ., Diss., Stuttgart.

http://www.softwearautomation.com/#!technology/ch7k.

J. Schmalz GmbH, 2011. In: (Hrsg.): Vakuum Komponenten Katalog 2012/2013. Schmalz, Glatten. Available from: http://katalog.schmalz.com/.

Jensen, L., Stephan, J., 2002. Gefriergreifer : neue Anwendung der Peltier-Technik KI Luft- und Kältetechnik. 38, H. 12, S. 572–576.

Klingele, J., Devaux, D., Gries, T., May 6–9, 2013. Process Chains for the Production of Novel Binder Preforms. In: Education and Green Sky - Materials Technology for a Better World : SAMPE 2013 Conference and Exhibition. SAMPE, Long Beach, California. - Covina, pp. 1531–1544. Datei: 186-142-1-PB.pdf.

Koch, W., 1992. Untersuchungen zum Greifen und Vereinzeln von Zuschnitteilen. Univ., Diss., Stuttgart.

Meyer, 2008. O.:Kurzfaser-Preform-Technologie zur kraftflussgerechten Herstellung von Faserverbundbauteilen. Univ., Diss., Stuttgart. URN: urn:nbn:de:bsz:93-opus-34329.

Monkman, G.J., Hesse, S., Steinmann, R., Schunk, H., 2007. Robot Grippers. WILEY-VCH, Weinheim.

Nayak, R., Padhye, R., 2014. Introduction: the apparel industry. In: Nayak, R., Padhye, R. (Eds.), Garment Manufacturing Technology. Elsevier.

Nayak, R., Padhye, R., 2015. Garment Manufacturing Technology. Elsevier.

Nayak, R., Chatterjee, K., Khurana, G., Khandual, A., 2007. RFID: Tagging the New Era. Man-made Textiles in India. 50, pp. 174–177.

Nayak, R., Singh, A., Padhye, R., Wang, L., 2015. RFID in textile and clothing manufacturing: technology and challenges. Fashion and Textiles 2, 1–16.

Prust, D., 2011. Entwicklung einer auf trockenen Adhäsion basierenden Greifvorrichtung. Shaker Verlag, Aachen. Zugl. Aachen: Techn. Hochsch.,Diss., 2011.

Raatz, A., Löchte, C., June 11–12, 2013. Form-flexible Handling Technology (FormHand) for Automation in RTM Preforming 7. CFK-Valley Stade Convention, Stade. Available from: http://www.cfk-convention.com/fileadmin/Convention_2013/Referenten/Vortraege/CFK_Conv2013_RAATZ.pdf.

Sarma, S., 2001. Towards the 5c Tag. Auto-ID Center. archive.epcglobalinc.org/publishedresearch/MIT-AUTOID-WH-006.pdf.

Stephan, J., Jensen, L., 2004. Gefriergreifer – die neue Greiftechnologie für textile Materialien Mittex. 111, H. 1, S. 13–14.

Stephan, J., 2001. Beitrag zum Greifen von Textilien. Techn. Univ., Diss., Berlin. Zugl. Berlin: IPK, 2001.

Szimmat, F., 2007. Beitrag Zum Vereinzeln Flächiger Biegeschlaffer Bauteile. Univ., Diss., Berlin. Zugl. Stuttgart: Fraunhofer IRB, 2007.

Tellkamp, C., Quiede, U., 2005. Das Internet der DingeBuchuntertitel Ubiquitous Computing und RFID in der Praxis: Visionen, Technologien, Anwendungen. Handlungsanleitungen Herausgeber Elgar Fleisch. Friedemann Mattern.

VDI 2860 05.90 Montage- und Handhabungstechnik – Handhabungsfunktionen, Handhabungseinrichtungen; Begriffe, Definitionen, Symbole, 1990. .

Wulfhorst, B., 1998. Textile Fertigungsverfahren: Eine Einführung München. Hanser, Wien.

Further reading

Grundmann, T.C., 2009. Automatisiertes Preforming für schalenförmige komplexe Faserverbundbauteile. Shaker, Aachen. Zugl. Aachen: Techn. Hochsch., Diss., 2009.

Reimerdes, H.-G., Göttner, W., Gries, T., Henkel, F., Schmachtenberg, E., Meyer-Noack, S., 2005. Industrielle Herstellung von FVK-Strukturbauteilen Kunststoffe, vol. 95,H. 3, S. 72–75.

Rödel, H., 2011. Konfektionstechnik für Faserverbundwerkstoffe. In: Cherif, C. (Ed.), Textile Werkstoffe für den Leichtbau : Techniken, Verfahren, Materialien, Eigenschaften. Springer, Berlin, pp. 403–452.

Schnabel, A., Greb, C., Gries, T., November 27, 2012. Innovative Preformingverfahren Abschlusskolloquium der Forschergruppe, vol. 860, Aachen.

Steinhilber, M., Greiner, J., Mann, B., Eberhardt, C.-P., Wechs, T., Bürkner, S., Ischtschuk, L., Jaeger, G., Hermes, C., Staffenberger, T., Kronreif, G., Klautzsch, M., Fragner, S., Richtsfeld, M., 2009. REDUX – Realisierung einer durchgängigen Prozesskette zur effizienten Produktion von CFK-Strukturen in textiler Preform-/RTM-Technik Ottobrunn [u.a.]. EADS Deutschland. Available from: http://edok01.tib.uni-hannover.de/edoks/e01fb09/612828719.pdf.

Wulfhorst, B., Moll, K.-U., 1998. Aufbau, Prüfung und Aufmachung der Harze und Fasern Aachen: RWTH Aachen, Forschungsbericht Teilprojekt 1. In: SFB 332 (Hrsg.): Arbeits- und Ergebnisbericht 1996/1997/1998 des Sonderforschungsbereich 332 "Produktionstechnik für Bauteile Aus nichtmetallischen Faserverbundwerkstoffen", vol. 332. SFB, Aachen, pp. 1–40.

Application of robotics in garment manufacturing

8

Thomas Gries, Volker Lutz
Institut für Textiltechnik of RWTH Aachen University, Aachen, Germany

8.1 Introduction

The art of sewing has not fundamentally changed since the first seamstress put needle and thread to fabric thousands of years ago. Even with great engineering advances including mechanized looms and sewing machines, the way sewn goods are produced is just as labor-intensive today as it was 100 years ago. To further complicate matters, today's consumers want inexpensive, high-quality goods delivered to their doorstep within days, pushing the limits of the traditional manufacturing business model to its breaking point.

Over the past few decades, sewn goods manufacturers lowered overhead by relocating operations to the developing countries paying the lowest wages. However, this business strategy is becoming increasingly difficult to maintain because of rising labor costs in the developing countries, a global shortage of skilled labor, and a change in consumer behavior pushed by fast fashion brands and social media platforms (Nayak and Padhye, 2015a). New market conditions have made the sewn garment industries mature enough for a new age of automation (Reddy, 2016).

Sewing, with a percentage of 85% of all joining methods, represents the most important textile joining technology. As an essential process step of the manufacturing of clothes and technical textiles, sewing contributes with an approximately percentage of 35%–40% of the total costs (e.g., male outerwear) to a considerable part to the added value of textile products (Textil und Mode Jahrbuch, 2011). Due to the industrialization clothing manufactures originated and the work organization and the operational procedure of individual workstations changed significantly toward sewing lines and sewing cabins. A fundamental characteristic of the sewing industry is the high personnel commitment and the low level of automation (Moll, 1999a). This personnel commitment led to a drift of 90% of the sewing industry toward low-wage countries, especially to southeast Asia.

Survey results revealed that the material handling and the converting of fabric into garment are still realized manually (Fig. 8.1). For example, nearly 80% of the factory cost of the manufacture of a piece of clothing is related to the handling cost and the handling time is also about 80% of the overall production time. Hence, it is essential to use automation in material handling to achieve cost benefits.

The percentage of the transportation and treatment processes for apparel clothing represents 80% measured at the overall manufacture time. In the production of other textile components such as seat cushion, doormats, and roof of vehicles for

Automation in Garment Manufacturing. https://doi.org/10.1016/B978-0-08-101211-6.00008-2

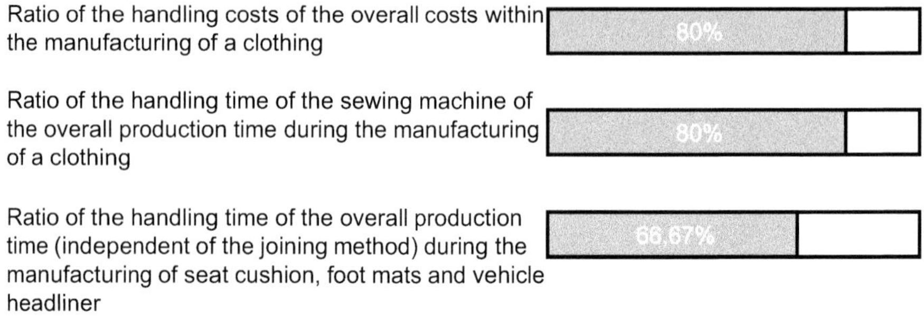

Figure 8.1 Industry figures relating to the material handling of the clothing and automotive industry.

automobiles, the handling time regardless of the joining technique represents 2/3 of the overall manufacture time. In the field of technical textiles, several operations are also performed manually. The difficulty in achieving automation levels can be ascribed primarily to the inhomogeneous characteristics of textiles. Fig. 8.1 illustrates the industry figures relating to the material handling of the clothing and automotive industry (Egbers and Bühler, 1993; Szimmat, 2007; Hanisch and Krockenberger, 1993; Gutsche, 1993; Gebauer et al., 2004; Stephan, 2001).

8.2 Automation and robotics for sewing

Given that clothing needs to be modified according to season, purpose and consumer demand, large-scale production is especially labor-intensive (Nayak and Padhye, 2015b). Partial automation has aided in faster production, improved quality, and increased quantity but it's still not a viable option for several high-cost labor countries. As a result, these countries are heavily dependent on imports still today (The Fung Group, 2015). There are various principles in the literature for the automation of the sewing process, which must fulfill the following two challenges for the realization of an automated manufacture of clothing: (1) *Quality*: In cooperation with the three-dimensional (3D) sewing, the concept of a moving tool allows a very high quality, which manifests itself in the repeatability and minor manufacturing errors. (2) *Flexibility*: Within a product, e.g., shape of a skirt, it is possible, due to the fast-adjustable flexible shaped body, the individual-layer cutting and the sorting and buffering capacities of the transport system, to produce different fabrication sizes, material qualities, and patterns quickly and in frequent alternation (Brozio, 2001).

During sewing, as in many high-cost industries, a higher automation level or robotics offers the potential to replace manual work steps, e.g., by robots and automated handling devices. Robots do not tire and can, except for maintenance and repair times, carry out endless repetitions with constant precision 24 h/day. This improves the performance and quality of production. Furthermore, they can eliminate the inefficiencies of humanization relating to the working environment because they are resistant to environmental pollution (such as dust, noise, and heat) and can take over monotonous,

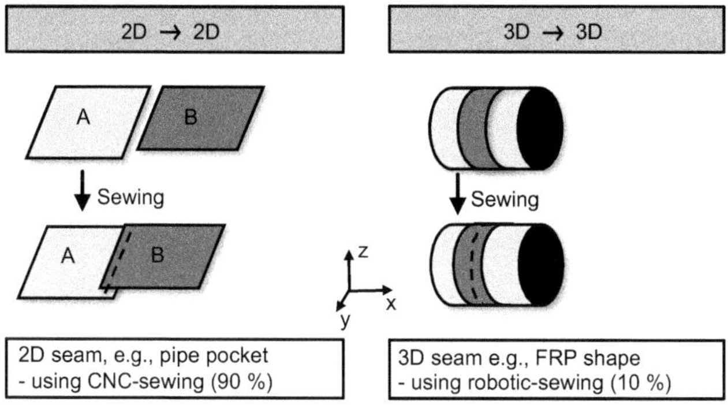

Figure 8.2 Principle and example of a 2D→2D seam (left), of a 3D→3D seam (right) *CNN*, computer numerical control; *FRP*, fiber reinforced plastics.

as well as difficult physical work from humans. In addition, the removal of human beings from the machine cycle eliminates a bottleneck, which otherwise slows the overall process (Hanisch and Krockenberger, 1993; Jauch, 1992; Kondratas, 2005).

Fig. 8.2 describes the formation of two-dimensional (2D) and 3D seams. In the case of a 2D→2D seam, the textiles are submitted to the sewing process laminary and are also removed laminary. In the case of a 3D→3D seam, the textiles are already submitted to the sewing process spatially. With both principles, the seam does not provide additional spatial transformation of the sewn textiles.

For the creation of a 2D→2D seam, especially the computer numerical control (CNC) sewing technology is widely used among all automation solutions. With the advancement of the sewing heads toward mobile single and double sewing heads, they can be guided 3D in space by means of robots. In the case of a 3D→3D seam, the sewing head is guided by the robot over the textile along the previously programmed seam path during the sewing process. When the sewing process is completed, the 3D textile component is removed from the shape. When sewing clothes, the sewing is mainly due to the 2D→3D seam principle (Fig. 8.3). The textiles, which are flat before the sewing process in the form of fabric, become a spatial 3D structure through the seam. The 3D form of the stitched textiles is necessary, for example, in the case of clothing, to allow a wear comfort adapted to the body. To be able to produce a 2D→3D seam, either both fabrics must be sewn with different contours or a curvature along the seam must be produced by means of different tension in the textile, the so-called "seam width." A combination of seam contours and tension in the textile is also possible to realize a 3D positioning of the sewn textiles.

Following the established technologies and automation approaches in sewing, technologies are described and compared with each other. A substantial distinction of the sewing result is the dimensionality. Since the 1980s many research projects have been realized worldwide for the automation of the sewing process. The projects focused on the sewing automation of the clothing industry. In the period from 1982 to 1991, the Japanese Ministry of Trade and Industry (MTI), provided US $ 55 million for the development of new textile manufacturing processes. Under the name TRAASS, 27 companies merged to develop a fully automatic production system for clothing.

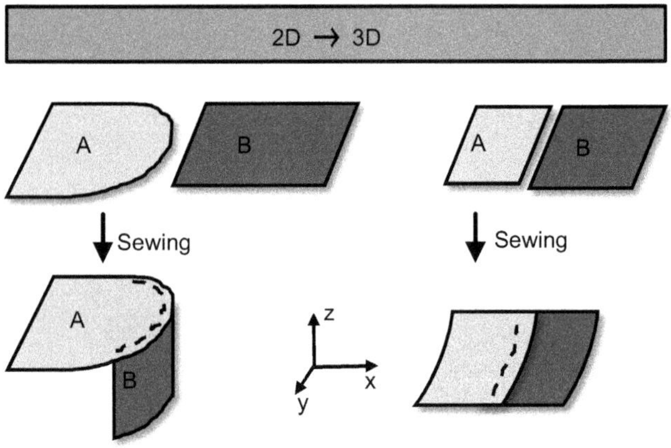

Figure 8.3 Principle of a 2D→3D seam. Contour dependent 3D-shape creation (left), dependent on the length of the edge 3D-shape (right).

The pilot system was planned to produce a ladies' blazer. The project resulted in solutions for various subproblems, such as automatic functions for a needle threading device or automatic needle replacement. However, the planned complete system could not be implemented due to the lack of technologies, high investment costs, and the lack of tolerance requirements (Jones and Stylios, 2013; Anon., 1993a,b, 1995, 1998; Winck, 2009; Zöll, 2002).

In the United States, the Textile/Clothing Technology Corporation $(TC)^2$ organization was setup in 1980, with the aim of developing an automated production line to significantly reduce costs. Apart from American institutes, the project involved more than 20 garment and textile companies as well as fiber manufacturers. The planned automated production lines were specifically designed for one component. The finished piece of clothing should then be produced from these various automatically manufactured components with manual control. However, solutions suitable for the serial production for the production line were not presented (Zöll, 2002).

In Europe, three sewing automation projects were carried out between 1986 and 1997 under the name BRITE/EURAM. In contrast to the TRAASS and $(TC)^2$ projects, these ones were not targeted to create a complete automation but rather for specific partial automations. From these projects, technologies for the handling (e.g., freezer grippers) and for positioning of workpieces have emerged. However, none of these technologies could be converted into a product, which is ready to go into mass production (Anon., 1993a,b; 1995; 1998; Wulfhorst et al., 2006; Zöll, 2002).

8.3 Computer numerical control technologies for sewing process

The CNC technology is the most important 2D-sewing operation for small- and large-area flat sewing applications. Due to the use of fixed sewing frames, the flexibility and scalability of this technology are limited. The CNC sewing technology allows the

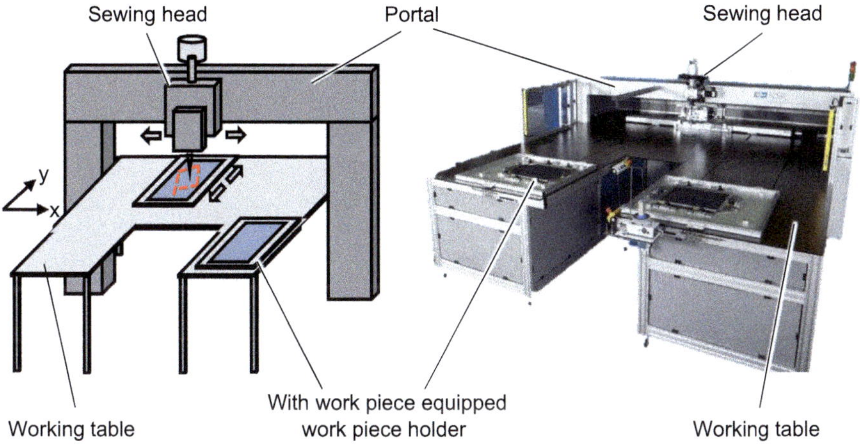

Figure 8.4 Computer numerical control sewing factory with two workpiece holder for a fast change of the workpiece.
Source: http://www.ksl-lorsch.de/produkte/automotive/interieur/.

sewing of one or more layers of textile, which must be clamped laminar. With the aid of a CNC, it is possible to place a free seam path on the workpiece. By clamping the workpiece into a workpiece holder during CNC sewing, the complex handling of flexible textiles is avoided. The sewing material, which is usually to be finished with an ornamental seam in the CNC sewing process, is clamped in a rigid workpiece holder and afterward guided under the sewing head in an X- and Y-movement. Depending on the size of the overall equipment, there are machine models in which the sewing head is guided in an X- and Y-movement and the workpiece is fixed. Fig. 8.4 shows the principle of a combined sewing material and sewing head guide. 3D seams cannot be realized with this method. The sewing process is used for ornamental and design seams (Fig. 8.5). Beyond that, several 2D-textile layers can be connected to each other by means of the programmed seam contours.

The development of CNC sewing has already succeeded in automating the creation of the seam pattern. However, a human operator is still required to program the sewing process and monitor the seam quality, there is still no fully automated seam pattern generation (Jones and Stylios, 2013). In addition, the fixed dimensions of the workpiece holders severely limit the use of variable sewing product sizes.

In the work process using CNC sewing technology, it is particularly efficient to work with at least two clamping frames. While one clamping frame is clamped in the machine, the second one can already be equipped with new workpieces. With the help of quick-change systems, e.g., based on magnetic closures, the operator can place the machine's clamping frame without additional locking and automatically lock it to the X-Y system of the machine.

The size of the CNC field is basically limited by the dimensions of the linear axes. Large machine types can cover a sewing area up to 3 m × 3 m. Machines with small sewing field sizes of less than 10 cm × 10 cm are generally referred to as small-field CNC machines. A typical application for small-field CNC machines is, for example,

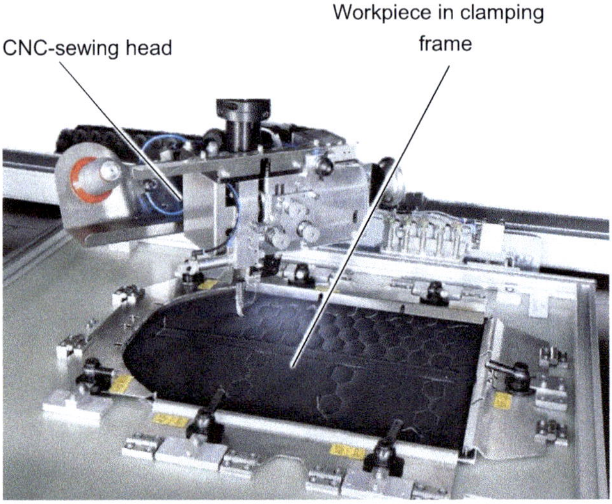

Figure 8.5 Computer numerical control-sewing machine for ornamental seams for car seats. *Source:* http://www.ksl-lorsch.de/produkte/automotive/interieur/.

the automatic sewing of labels into clothing. If, in addition to the small-field CNC machine, further aggregates are available, e.g., for the transportation of the machine, the complete system is referred to as an automatic sewing machine.

8.4 Sewing automats and sewing units

Sewing automats and units have been developed to offer semi or fully automated solution for defined production steps in garment production, e.g., buttonholes or slit pockets. In general, this sewing process is limited to one specific 2D-seam application. There have been attempts in the past to fully automate the sewing process. Most of the previous systems relied on clamps to hold the fabric taught, making it more rigid and less susceptible to distortion. This system limited automation to certain operations during the sewing process such as when sewing on buttons or pockets. Aligning two pieces of fabric correctly and feeding them through the sewing head without slippage or buckling—while maintaining the correct tension levels—has proven to be a process better managed by human hands (http://www.textileworld.com/textile-world/knitting-apparel/2016/05/the-rise-of-robotic-automation-in-the-sewing-industry/).

Semiautomatic sewing machine systems are understood to be systems that allow sewing of at least two laminar textiles based on a fixed seam path. Thus, 3D semifinished products are produced from laminar textiles. In addition to a CNC-controlled seam path, the automatic sewing machines have feeding systems that transport and position the workpiece to a clamping position. After the sewing process, the stitched textile is automatically released from the clamping and transported to a defined storage area.

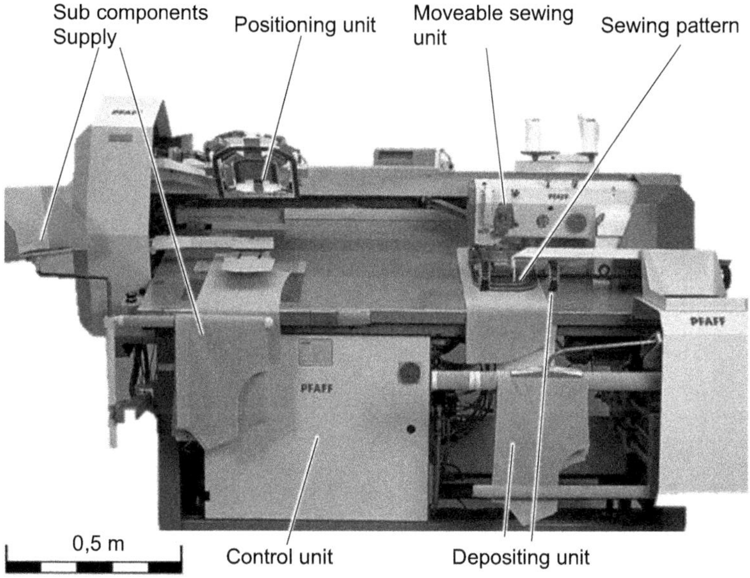

Figure 8.6 Pfaff 3588: programmable pocket sewing machine for shirts, blouses, jeans, or working clothes (Dürkopp Adler AG, 2009).
Source: https://www.pfaff-industrial.de/de/service-support/downloads.

For the garment industry, automatic feeding systems of textile semifinished products are already commercially available. For the mass production of clothing elements, e.g., of men's clothing various systems have been developed. These handling systems are designed for individual seams. The operator positions the textile at a starting point. The machine then pulls the material into the sewing process (Durkopp Adler AG, 2009).

The machine system is based on a standard sewing machine for the required sewing stitch. Handling and positioning components are applied to and around the sewing table. The fabric is fixed and transported via belts, rollers, suction pads, or vacuum pumps. Furthermore, there are clamping frames which are pressed onto the textile. Various seam shapes can be realized by moving the frame. This technique is mainly used for decorating textiles (Dürkopp Adler AG, 2009). Fig. 8.6 shows a pocket sewing machine from PFAFF Industry systems and Machine GmbH, Kaiserslautern. With this system, an operator has to insert the cut components, the positioning, transport, sewing, and dropping, are performed fully automatically (http://www.vetron-europe.com/; Dürkopp Adler AG, 2009; Gottschalk, 1996; Anon., 1998).

For many 2D-sewing tasks, nowadays automation solutions are offered in the form of semiautomatic machines. However, these automatic processing systems offer only partial automations for the manufacturing of individual seams. A submission by an operator is furthermore required. To create a complex 3D textile, many process steps of these semiautomatic machines are necessary, whereby an economical and flexible production is also difficult to achieve. These solutions are limited to specific

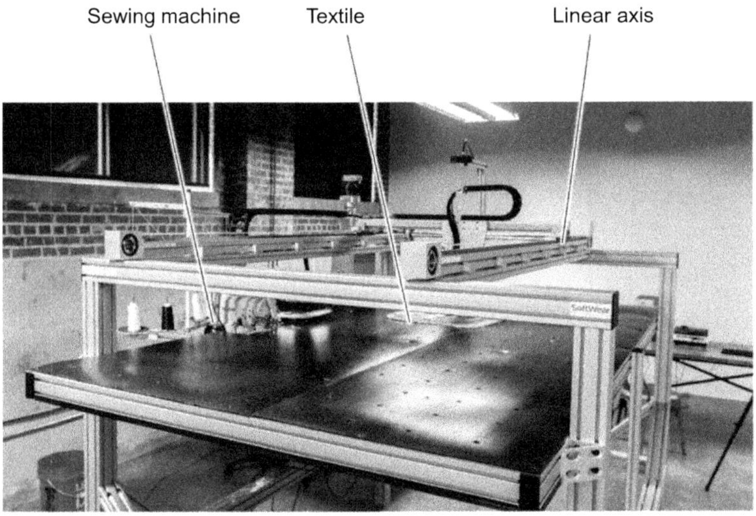

Figure 8.7 Sewing unit by SoftWear Automation Inc. for flat sewing goods.
Source: http://www.textileworld.com/textile-world/knitting-apparel/2016/05/the-rise-of-robotic-automation-in-the-sewing-industry/.

applications. Sewing systems for the sewing of pockets in trousers, shirts, or jackets, sewing systems for sewing seams (trouser step seam, trouser seam), as well as CNC small- and large-area sewing machines, are state of the art. The installed workpiece carriers or the control unit must be individually adapted for each size. This results in financial limitations due to high tool costs with different workpiece dimensions.

Atlanta-based SoftWear Automation Inc. recently introduced a radical new approach to sewing automation (Fig. 8.7). The company has developed a system that eliminates fabric distortion issues by relying on an advanced computer vision system. The camera tracks stitching at the needle and coordinates the precise movement of the fabric using lightweight robots (http://www.textileworld.com/textile-world/knitting-apparel/2016/05/the-rise-of-robotic-automation-in-the-sewing-industry/).

By the integration of vision systems, the automated process is not depending on further programming skills in production. Changing the product from small to large scale will be possible without programming a new seam path. This technology will become the reality in future world of garment manufacturing.

8.5 Robotics for three-dimensional sewing operations

In many approaches and research projects robots have been used to either guide the textile through the sewing process or movement of a sewing head mounted to a robot. Compared to the production of composite materials, robots have not yet been established in garment production, for example, due to high investment costs.

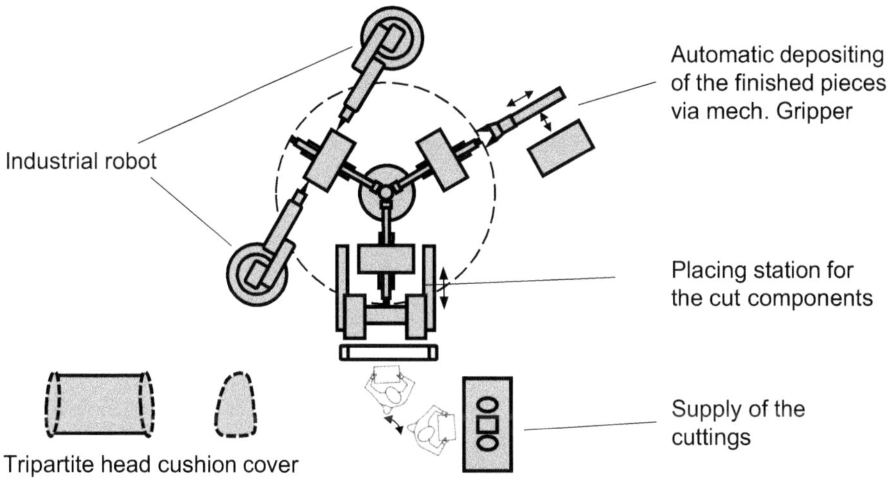

Figure 8.8 Integrated three-dimensional sewing system for the automatic sewing of a head cushion cover (Winck, 2009).

8.5.1 Integrated three-dimensional sewing system

The integrated 3D-sewing system refers to a system in which the cut components are provided tensioned in a room in a 3D manner, are sewn. The system was developed for the sewing of a head cushion cover of a car seat head cushion. The structure is shown in Fig. 8.8.

An operator fixes the cut components into a workpiece carrier, following the automatic process is started. First, the cut components are clamped and thus tensioned around a shaped body. The parts are then positioned to each other and one step further navigated into the sewing station by rotating the rotary table. In the sewing station the two circumferential seams are sewn by a robot. The rotary table is rotated again and the sewn material is moved into the unloading station. There, the clamps are released and the resulting 3D shell is peeled off with a mechanical gripper. After a repeated rotation, the shaped body again reaches the feeding station. The processes of the stations run parallel. Thus, the cycle time is calculated from the longest processing time of a station plus the transport time (Gries et al., 2014; Winck, 2009; Anon., 1998).

A major disadvantage of the integrated 3D-sewing system is the high investment costs due to the two industrial robots and the automated clamping and turning devices. Furthermore, the system proves to be inflexible with regard to different geometries of the cuttings because new moldings are produced for other product designs and the industrial robots have to be manually reprogrammed.

8.5.2 Three-dimensional sewing with robots for preforms

A particular advancement of the last decade is the one-sided sewing process, in which the textile is held stationary and the sewing head is guided over the textile. These sewing methods are primarily interesting for composite applications, especially due to

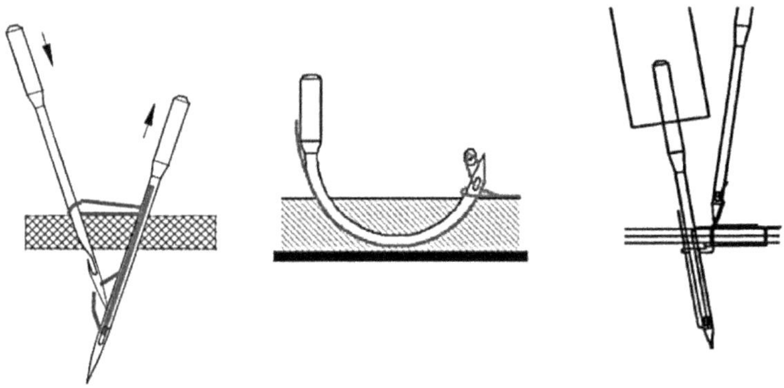

Figure 8.9 Schematic description of the three most important one-sided sewing technologies (Laourine, 2005).

the possibility to sew complex component geometries, for example, with 3D reinforcing ribs. In contrast to clothing, the textiles used for composite must not be folded or pleated to avoid damage to the reinforced fibers.

In case of large components, sewn with conventional sewing methods, correspondingly wide sewing machines must be used and the suitable large sewing material must be guided. Because the movement is carried out by the sewing head, in the case of one-sided sewing techniques, very large components can be sewn and a synchronously moved gripper or lower thread system, underneath the workpiece, can be dispensed.

For the guidance of the sewing heads, the use of robots has been established. The four most important one-sided sewing methods used for composite today are all guided by a robot or a sewing portal. These sewing processes are shown in the following in Fig. 8.9.

At the Institute for Textile Technology Aachen (ITA), Aachen, Germany, the "ITA-Preformcenter" makes it possible to map and evaluate the process chain from the textile reinforcement semifinished product to the finished preform. The robot cell of the "ITA-Preformcenter" has been manufactured by Keilmann Sondermaschinenbau GmbH, Lorsch. A sketch of the individual components can be found in Fig. 8.10.

It consists of a 6-axis industrial robot type KR150 from KUKA Roboter GmbH, Augsburg, which is assembled on a portal made of steel girders and has an additional linear axis to increase the working space. The industrial robot can access a wide variety of effectors, as listed below:

- Various gripper systems (needle, freezer, vacuum, etc.)
- Blind stitch sewing head (Keilmann Sondermaschinenbau GmbH, Lorsch)
- Tufting sewing head (Keilmann Sondermaschinenbau GmbH, Lorsch)
- ITA one-sided sewing head
- Binder application system
- Camera and laser sensors for quality control

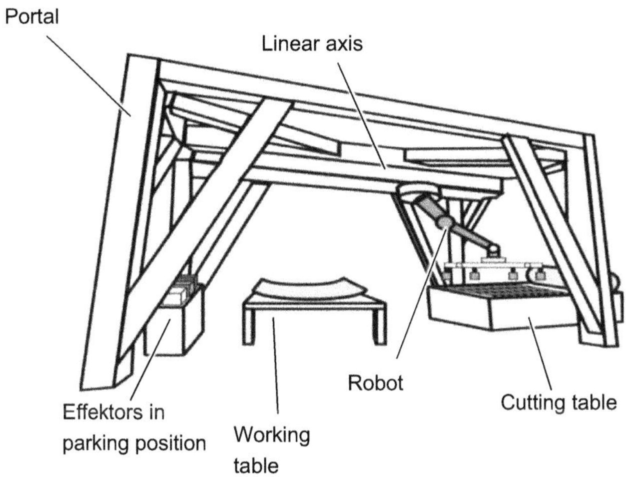

Figure 8.10 Schematic presentation of the ITA-Preformcenter (Grundmann, 2013).

8.5.3 Three-dimensional sewing operations with automated sewing units

In addition to robotics further principles to flexibly guide textiles through the sewing process are developed in research projects worldwide. In this chapter all known approaches are discussed and compared. An automatic sewing system that sews up slightly curved cuttings is presented within the dissertation of Thomas Gottschalk. Within tests the construction is appraised by using body and arm cut components of men's jackets. The transportation and the feed movement are realized with the help of three rubber coated roller pairs.

The two roller pairs in the front of the sewing machine can thereby be adjusted transversely with respect to the sewing direction. A laser distance sensor monitors the seam allowance for each layer. The rollers and linear units are controlled by a programmable logic controller. The seam lines are generated in the AutoCAD software of Autodesk Inc., San Rafael, United States via a Lisp macro. The adjustment of the curvature is effected by oblique warping of the respective position by means of the transverse feed units (Gottschalk, 1996).

The prerequisite for the automated sewing system are the cut components with large radii of curvature of the contour, which allow a certain degree of elongation and oblique warping. These characteristics are not guaranteed in the materials of the cuttings for car seat covers.

8.6 Real-time sewing cell with two lightweight industrial robots

During the robot-guided sewing process, unlike the integrated 3D-sewing, the workpiece is guided by an industrial robot rather than the sewing head. Within the robot-guided sewing process, a sewing cell consists of an industrial C-frame sewing machine and two robot arms, each of which guides the upper or the lower workpiece. Finally, the material is gripped by the robots and fed to the joint patch. At the same time, the sewn edges are monitored by sensors and the robot arms are controlled in real-time via an analogue board. The sewing cell is shown in Fig. 8.11.

The disadvantages of the robot-guided sewing are a very limited possible radius of curvature of the seam sections and the restricted working areas of the industrial robots. An automated sewing cell was developed in a financial funded project (budget about €6.4 million (http://sfweb2.sintef.no/SINTEF-Raufoss-Manufacturing/Prosjekter/3D-sewingjoining/)), in which the Norwegian research community, the Scandinavian research organization SINTEF, the TU Trondheim, and several Norwegian companies participated. The sewing cell consists of two robots with mechanical grippers equipped with force sensors. With these sensors, the tension of the textile is constantly monitored and regulated during the sewing process. The system is connected to a commercially available industrial sewing machine. A one-dimensional diode array is used to monitor the edge guidance. The control of the industrial robots is carried by a piece of special software consisting of a Linux computer, the Robot Operation System

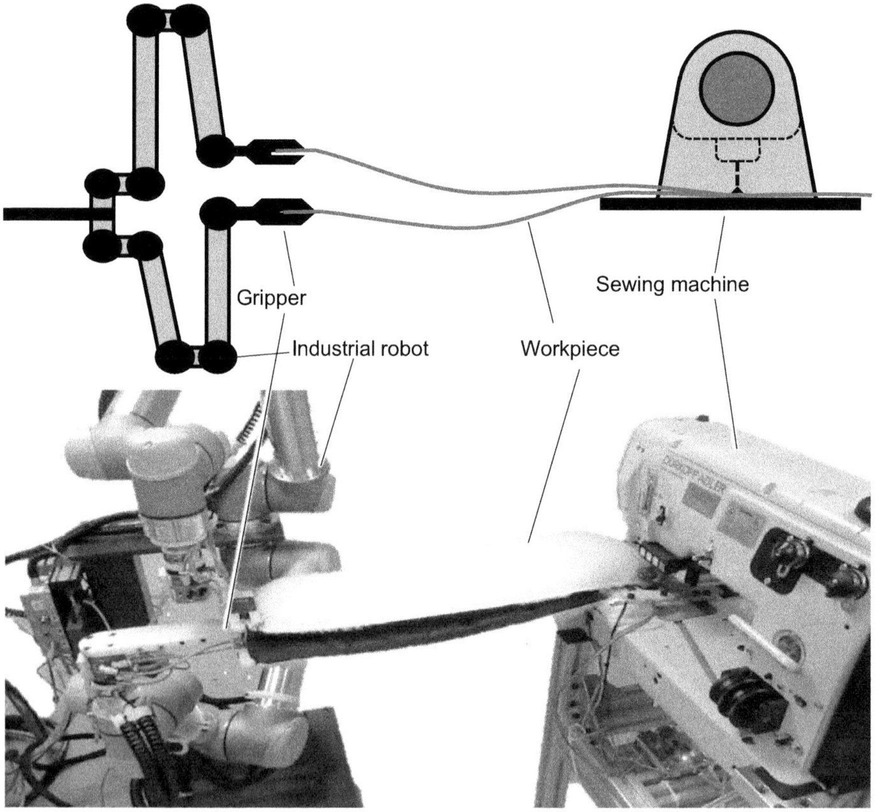

Figure 8.11 Sewing cell with industrial robots for workpiece feeding (Schrimpf et al., 2013).

and several low-level controllers. This system allows a real-time control of the sewing process. The movements of the robots are controlled in real-time via the edge sensor (Schrimpf et al., 2013; Schrimpf and Wetterwald, 2012).

In this sewing cell, with restrictions, cut components for armchair covers can be sewn in two layers regarding the tolerances. The restrictions relate, on the one hand, to the radius of curvature of the seam section, which must be small. On the other hand, the working space and the movements of the links of the industrial robots are restricted; therefore no continuous process is possible with certain geometries of the seam section. Furthermore, the problem of the positioning by mechanical grippers and the initial positioning under the needle were not considered.

In the United States of America, the DARPA (Defense Advanced Research Projects Agency) offered a project for automated sewing of clothing fabrics with 1.2 million US dollars (https://www.fbo.gov/index?s=opportunity&mode=form&tab=core&id=54151eccf003c66ee3598b28 fc092fe). The consent was received by the SoftWear Automation Inc., Atlanta, Georgia, United States (spin-off from the Georgia Institute of Technology in Atlanta, Georgia, United States). In addition to the development of a textile transport system made of vacuum balls, a system for automatic sewing material guidance was developed using an alternative presser foot transport (Blombach, 2010). This sewing foot moves up and down by 180 degrees shifted to the needle movement. While the foot rests on the textile, it moves translative in the sewing direction and rotative to the orientation of the next point of the seam. The axes are driven by coil motors because these allow very high accelerations with simultaneously high forces.

An optical camera system, which detects the orientation of the warp and weft threads of a woven fabric, serves to monitor the position of the cut components. Thus, the orientation and position of a cut component can be determined from image to image. For this purpose, the textile must be appropriately marked during the cutting process (Winck et al., 2009; Blombach, 2010; Winck, 2009; http://www.softwearautomation.com/#!technology/ch7k).

The alternative sewing foot transport is not suitable due to the comparatively high forces required for the sewing of foam-laminated cover materials. In addition, the placement of a pretension in the textile, comparable to the holding by an operator, cannot be implemented with temporary pinching.

In 2015, Xi'an TYPICAL Europe GmbH, Kaiserslautern introduced a machine concept "Autoseam" under its brand name VETRON, which can also be automated by means of an additional alternating sewing foot transport in the form of a position ring that leads flat textiles through the sewing process. The machine recognizes the edges of two superposed fabrics; the system scans the edges stitch by stitch, corrects itself, and sews the seam exactly at the same distance from the edge. With the system, it is possible to sew two textiles with different edge contours along their edges and create a 3D seam. The principle and the introduced machine are shown in Fig. 8.12.

In 2016 a robot-based sewing system with the name SEWBO was introduced (Fig. 8.13). The system was developed by Sewbo, Inc., Seattle, United States. With the help of an industrial robot, all necessary seams of a T-shirt were automatically sewn for the first time. This was possible by the fact that the textiles were previously treated with a water-soluble polymer and become rigid by this preliminary process.

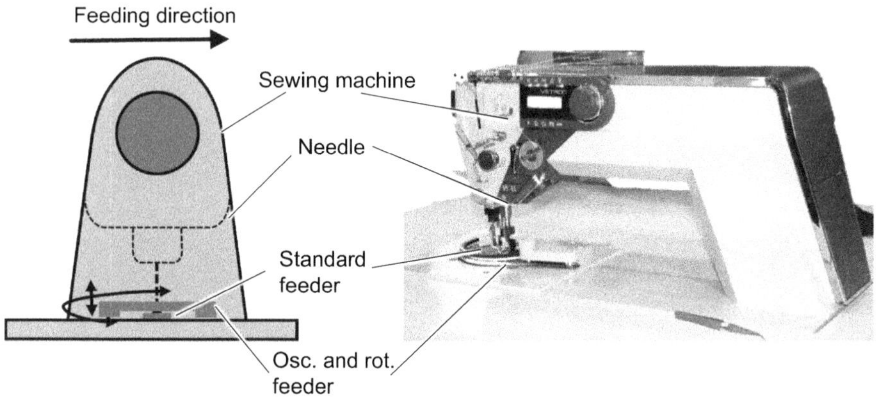

Figure 8.12 Vetron Autoseam machine for the automatic manufacturing of three-dimensional seams.
Source: http://www.vetron-europe.com/.

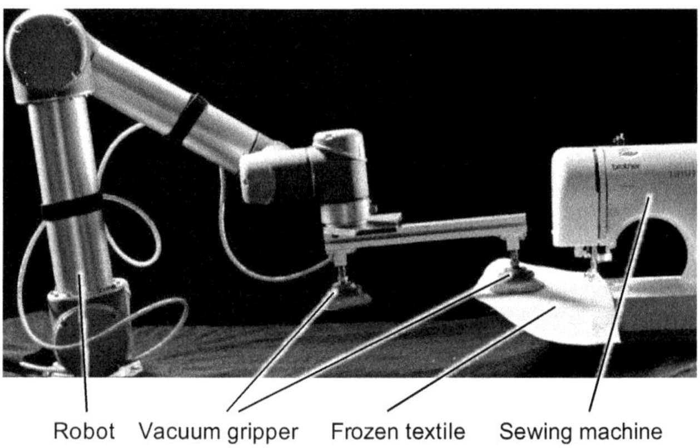

Figure 8.13 SEWBO robot sewing system guides a frozen textile.
Source: http://www.sewbo.com/author/jonzornow/.

The stiffness of the textile makes it possible for the robot to guide the textiles through the process such as a solid material. Following the sewing process, a washing process is performed to completely remove the polymer from the textile. The additional use of the polymer as an auxiliary agent and further washing processes increase the production costs so that the cost-effectiveness of the overall process is questionable.

Within the project speedfactory, funded by the Federal Ministry for Economic Affairs and Energy, a new sewing system to automatically sew two layers of textiles has been developed. The seam is placed according to a seam path coming from design data and is monitored by a vision system to ensure seam position quality. At the Institut für Textiltechnik of the RWTH University (ITA) the patented system to guide textiles with different contour through the sewing process has been developed and validated. The system consists of two transport rollers, which are placed in front of the

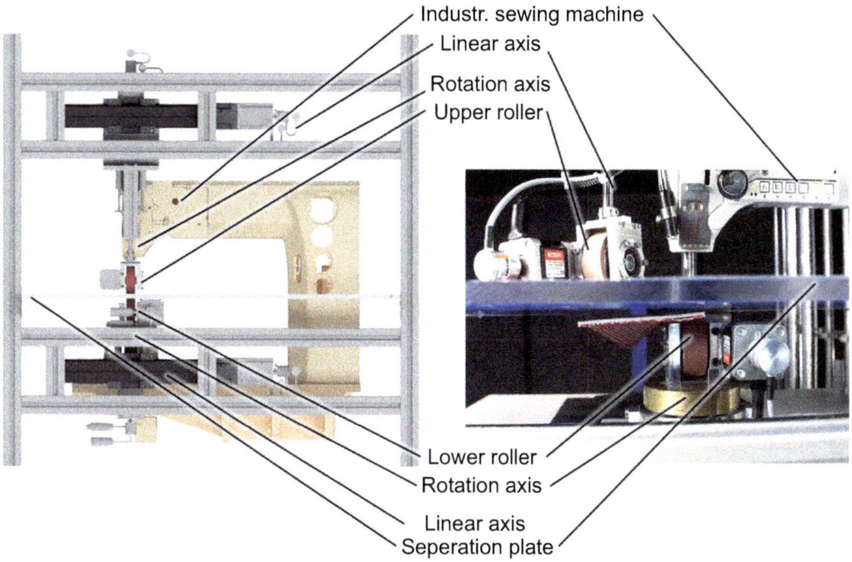

Figure 8.14 Automated sewing system developed by ITA (Lutz, 2017).

needle, one roller for each of two individually guided textile layers. The principle and the system are shown in Fig. 8.14.

The textiles are separated into a top and bottom area by a separation plate. Both top and bottom area use the same principle. The transport roller can be placed with a certain angle and position toward the needle to achieve total flexible seam geometries. In addition each roller can be precisely driven to individually exert tension to the textile layers. The result is a new system with the capability to produce seams with defined fullness and flexible seam paths for a high variety of materials. Using a modular concept the system could be mounted to already existing machines. With this system sewing of 2D- and 3D-seam dimensions is possible with existing sewing machines and the same amount of workspace needed compared with a manual workspace (Simonis and Lutz, 2017).

As with the 2D workflows, 3D-sewing requires greater personal independence to ensure productivity, quality, and profitability. It is only when most of the process operations in the clothing industry are mechanized, especially during sewing, which several production stages can be linked and automated (Moll, 1999b). The intensive use of robotics and intelligent manufacturing are important in competing successfully in today's competitive international economy (Michelini and Razzoli, 2013). Some apparel companies are much ahead of the competition on automation. Later this year, Germany sportswear giant "Adidas" will open "Speedfactory," its first shoe manufacturing plant controlled largely by robots. The company, which employs over a million workers in contract factories, and sources most of its production in Asia, is tapping automated production as a way to quickly produce and deliver goods in response to consumer demand in major markets closer to home. The first Speedfactory will be in Ansbach in Southern Germany (Abnett, 2016).

The plant is expected to reach a mature production level of 500,000 pairs of shoes annually. This will help to improve supply chain issues that have traditionally plagued the industry. In the case of footwear, consumers receive products many months after a retailer has placed an order. That can result in reduced sales, quality, and customer experience (Abnett, 2016). As the pace of automation continues—the IFR (International Federation of Robotics) predicts double-digit growth in global sales of industrial robots between 2015 and 2018—Ford predicts that, in the fashion industry as well as the broader economy, "robotic technologies and machine learning technology are likely to eventually automate nearly any job that is on some level routine and predictable" (Abnett, 2016).

8.7 Advantages and disadvantages of automation in sewing

The use of automation and robotics helps in improving the quality and efficiency but they are expensive and need program changes when the style of the garment is different. Table 8.1 describes the advantages and disadvantages of application of automation and robotics in sewing.

8.8 Conclusion

This chapter presented automation solutions for 2D- and 3D-sewing operations, that are established in varied intensity in industrial applications. The technical solutions for 2D-sewing are already in use in the production of clothing, e.g., for longitudinal seams of trousers or piped pockets. In addition, the CNC sewing technique is intensified used to produce decorative seams, e.g., in the automotive sector. On the other hand, the much more complex 3D-sewing is known to very few established applications, such as the sewing of 3D fiber composite preforms. However, the operations requiring 3D shaping and assembly of the workpiece presuppose expansion and further development of the existing mechanical approaches.

Robot-based sewing systems have already been developed but have only been used in special applications in small series production so far. Therefore, on the one hand, systems to produce clothing and on the other hand to produce mass products (such as headrests, airbags, etc.) have been developed. The limitation in various sizes of the textiles has severely limited its suitability for use. *Moll* has clearly described this problematic aspect of automatic sewing machines as: "An automatic sewing machine, for example, cannot sew parts with arbitrarily different seam lengths to one another" (Moll and Zöll, 1998). Further problems for the automation of the manufacture are the choice of the sewing technique (e.g., double locked stitch, double chain stitch), the limited design possibilities, the preliminary and subsequent process steps or the handling times (Moll and Zöll, 1998; Stephan, 2001). Although there are several difficulties ahead of the automation and use of robotics, the days are not far when the garments will be fully manufactured by industrial robots.

Table 8.1 Advantages and disadvantages of automatic sewing

<table>
<tr>
<td>CNC-sewing (2D-sewing)
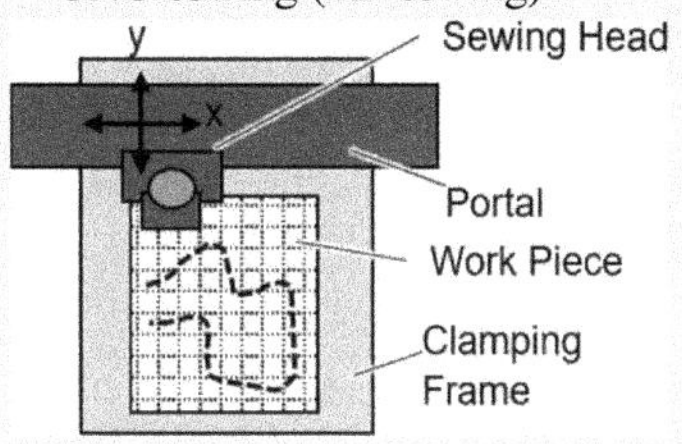
</td>
<td>Advantages

Disadvantages</td>
<td>• Free seam design
• High productivity
• Mass production capable
• Manual preparation
• High setup times
• High investment</td>
</tr>
<tr>
<td>Sewing units (2D-sewing)
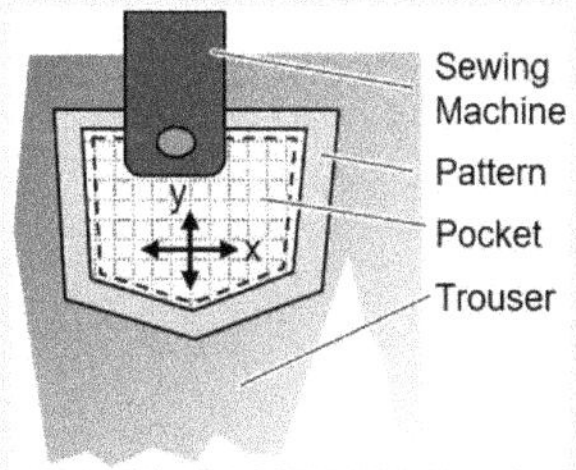
</td>
<td>Advantages

Disadvantages</td>
<td>• Low investment
• High productivity
• Semiautomated preparation
• Fixed seam operation
• Low flexibility (seam design)
• Highly dependent on material</td>
</tr>
<tr>
<td>Integrated sewing (3D-sewing)
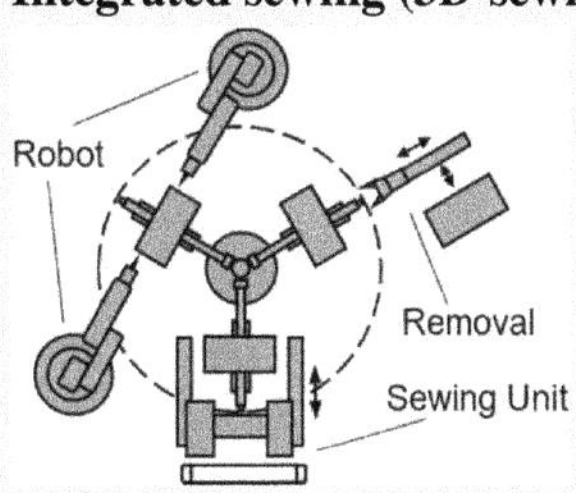
</td>
<td>Advantages

Disadvantages</td>
<td>• Automated preparation
• High productivity
• High Seam quality
• Need professional staff
• High investment
• Large workspace</td>
</tr>
<tr>
<td>Robot sewing (3D-sewing)
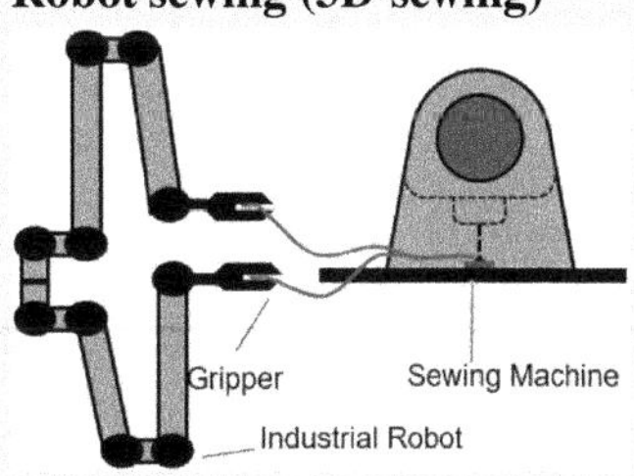
</td>
<td>Advantages

Disadvantages</td>
<td>• Free seam designs
• Allows seam fullness
• Seam quality
• High investment
• Programming necessary
• Slow</td>
</tr>
<tr>
<td>Alternative foot transport (3D-sewing)
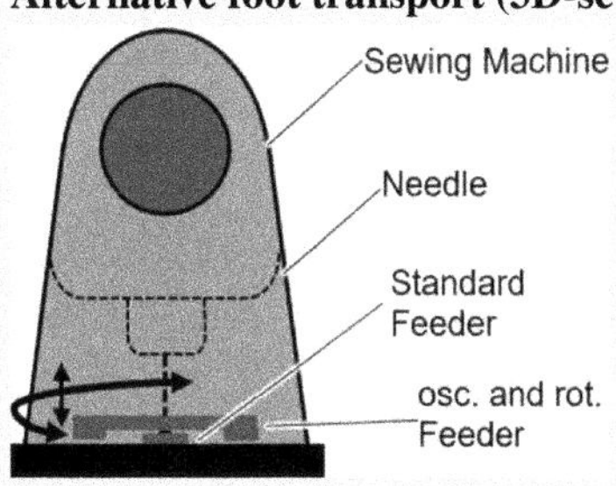
</td>
<td>Advantages

Disadvantages</td>
<td>• Low investment
• Low workspace
• Easy operation
• Fullness not possible
• Seam dependent on textile edge
• Manual placement of textile parts</td>
</tr>
</table>

2D, two-dimensional; *3D*, three-dimensional; *CNC*, computer numerical control.

References

Abnett, K., May 2016. The Robot Opportunity. Fashion-Tech BoF.

Anon., 1993a. BRITE/EURAM-Projekt P 2242. Flexible Montagezellen BW Fashion Technics. H. 3, S. 20–22.

Anon., 1993b. BRITE-Projekt P 1535. Sequenzautomatisierung BW Fashion Technics. H. 4, S. 12–14.

Anon., 1995. Robotex Project from Brite/EuRam. Automated Manipulation of Flexible Materials during Sewing BW Fashion Technics. H. 17, S. 23.

Anon., 1998. Robotex Projekt von BRITE/Euram Bekleidung/Wear Fashion technics. Nr. 9.

Blombach, A., 2010. Richtig & Falsch München. Sanssouci Verlag.

Brozio, M., 2001. Anwendung des Dualen Entwurfs auf die Entwicklung eines robotergesteuerten 3D-Nähsystems. Wissenschaftsverlag Main in Aachen, Aachen.

Dürkopp Adler AG, 2009. World of Sewing Technology – Garment 2009. Dürkopp Adler AG, Bielefeld.

Egbers, G., Bühler, G., 1993. Grundlagenuntersuchung zur automatischen Handhabung großflächiger Werkstücke in der Maschenkonfektion: Schlussbericht zum Forschungsvorhaben AiF-Nr. 8449 Denkendorf: o.V.

https://www.fbo.gov/index?s=opportunity&mode=form&tab=core&id=54151eccf003c66ee3598b28 fc092fe.

Gebauer, I., Dörsch, C., Müller, D.-H., 28.-29. September 2004. Handhabung biegeschlaffer Materialien – Zentrale. Herausforderung in der automatisierten Produktion von Composite-Bauteilen. In: 7. Internationale AVK-TV Tagung für verstärkte Kunststoffe und duroplastische Formmassen Baden-Baden.

Gottschalk, Th., 1996. Automatisiertes Nähen von Zuschnitten ungleicher Kontur. Techn. Univ., Diss., Berlin.

Gries, Th., Veit, D., Wulfhorst, B., 2014. Textile Fertigungsverfahren, Aufl. 2. Hanser, München.

Grundmann, T.C., 2013. Automatisiertes Preforming für schalenförmige komplexe Faserverbundbauteile. RWTH University, Aachen.

Gutsche, C., 1993. Beitrag zur automatisierten Montage technischer Textilien. Techn. Univ., Diss., Berlin. 1992; Zugl. München, Wien: Hanser.

Hanisch, G., Krockenberger, O., 1993. Flexibles Greifen und Vereinzeln biegeschlaffer Flächengebilde. Eine technische Herausforderung mit hohem Rationalisierungspotential wt 83. H. 9, S. 138–140.

Jauch, T., 1992. Automatisierte Montage nicht formstabiler Bauteile mit Industrierobotern – Konzepte für die Montage von Dichtungselementen. Univ., Diss., Kaiserslautern. Zugl. Kaiserslautern: Univ., 1993.

Jones, I., Stylios, G.K., 2013. Joining Textiles: Principles and Applications. Woodhead Publishing Limited.

Kondratas, A., 2005. Robotic gripping device for garment handling operations and its adaptive control. Fibres and Textiles in Eastern Europe 13 (4), 84–89.

http://www.ksl-lorsch.de/produkte/automotive/interieur/.

Laourine, E., 1. Mai 2005. Einseitige Nähtechnik für die Herstellung von dreidimensionalen Faserverbundbauteilen, Auflage: 1. Shaker.

Lutz, V., 2017. Arbeitsablaufbasierte Verteilung automatisierter Teilarbeitsschritte am Beispiel von Näharbeitsschritten – Workflow-Based Distribution of Automated Work Steps Regarding the Textile Sewing Process (Dissertation). RWTH Aachen, Shaker Verlag.

Michelini, R.C., Razzoli, R.P., 2013. Robotics in clothes manufacture. International Journal of Mechanical Engineering and Applications. 1 (1), 17–27. http://www.sciencepublishing-group.com/j/ijmea.

Moll, P., November 1999a. 50 Jahre (Industrie?)nähmaschinen. Bekleidung wear.
Moll, K.-U., 1999b. Nähverfahren zur Herstellung von belastungsgerechten Fügezonen in Faserverbundwerkstoffen Dissertation. RWTH Aachen, Shaker Verlag.
Moll, P., Zöll, K., 1998. Prozeßvoraussetzungen für eine Automatisierung in der Näherei. Jahrbuch der Bekleidungsindustrie, p. 1.
Nayak, R., Padhye, R., 2015a. Introduction: the apparel industry. In: Nayak, R., Padhye, R. (Eds.), Garment Manufacturing Technology. Elsevier.
Nayak, R., Padhye, R., 2015b. Garment Manufacturing Technology. Elsevier.
Reddy, K.P., June 2016. The rise of robotic automation in the sewing industry article based on presentation given at the 2015. In: Textile World Innovation Forum.
Schrimpf, J., Wetterwald, L.E., May 2012. Experiments towards automated sewing with a multi robot system. In: IEEE International Conference on Robotics and Automation Saint Paul, Minnesota, USA.
Schrimpf, J., Lind, M., Mathisen, G., 2013. Real-time analysis of a multi-robot sewing cell. In: IEEE International Conference on Industrial Technology (ICIT) Kapstadt.
http://www.sewbo.com/author/jonzornow/.
http://sfweb2.sintef.no/SINTEF-Raufoss-Manufacturing/Prosjekter/3D-sewingjoining/.
Simonis, K., Lutz, V., 2017. Abschlussbericht Zum Verbundvorhaben Speedfactory. Institut für Textiltechnik der RWTH Aachen, Bundesministerium für Wirtschaft, Berlin.
http://www.softwearautomation.com/#!technology/ch7k.
Stephan, J., 2001. Beitrag zum Greifen von Textilien. Techn. Univ., Diss., 2001, Berlin. Zugl. Berlin: IPK.
Szimmat, F., 2007. Beitrag zum Vereinzeln flächiger biegeschlaffer Bauteile. Univ., Diss., Berlin. Zugl. Stuttgart: Fraunhofer IRB, 2007.
Textil und Mode Jahrbuch 2009 + 2010, 2011. Gesamtverband der deutschen Textil- und Modeindustrie e. V. Berlin.
http://www.textileworld.com/textile-world/knitting-apparel/2016/05/the-rise-of-robotic-automation-in-the-sewing-industry/.
The Fung Group, July 8, 2015. Fung Business Intelligence Centre Global Retail & Technology Flash Report: Robotic Sewing Machine 2015. The Fung Group, New York/Hogkong
http://www.vetron-europe.com/.
Winck, C.R., 2009. Fabric Control for Feeding into an Automated Sewing Machine (Master thesis). Georgia Institute of Technology.
Winck, R.C., Dickerson, S., Book, W.J., Huggins, J.D., 2009. A novel approach to fabric control for automated sewing: IEEE/ASME. In: International Conference on Advanced Intelligent Mechatronics Singapur.
Wulfhorst, B., Gries, T., Veit, D., 2006. Textile Technology. Hanser Verlag, München.
Zöll, K., 2002. Nähtechnik zur Fertigung textiler Hüllen Aachen. Techn. Hochsch, Diss.

Further reading

https://www.businessoffashion.com/articles/fashion-tech/the-robotics-opportunity-manufacturing-efficiencies.
Monkman, G.J., Hesse, S., Steinmann, R., Schunk, H., 2007. Robot Grippers. WILEY-VCH, Weinheim.
Koch, W., 1992. Untersuchungen zum Greifen und Vereinzeln von Zuschnitteilen. Univ., Diss., Stuttgart.
Krockenberger, O., 1995. Entwicklung einer flexibel automatisierten. Nähanlage Stuttgart Universität, Diss.

Automation in sewing technology

9

Prabir Jana
National Institute of Fashion Technology, New Delhi, India

9.1 Introduction

The word automation is claimed by some to be a contraction of the term "automatic operation." The roots of automation can be traced to mechanization, which can be defined as the transfer of skills and manual activities to machine operation (Liberatore, 1996). Automation has been defined by manufacturing engineers as a technology concerned with the application of mechanical-, electronic-, and computer-based systems to operate and control production (Groover, 1987). Glock has defined three stages in the advancement of sewing technology; mechanization, automation, and robotics (Glock and Kunz, 2000). Robotics are the most advanced form of automation, which are computerized, reprogrammable, and multifunctional manipulators designed to move materials, parts, tools, or specialized devices through variable programmed motions for the execution of a variety of tasks (Rosenberg, 1983) and is discussed in a separate chapter.

The early 1970s and 1980s saw extensive research and development activities in the United States, Europe, and Japan, aiming toward a totally human less sewing factory that is flexible as well as productive. The important research and development initiatives in sewing area included pretreatment technology for stiffening and bending fabrics, temporary joining of pieces for efficient sewing assembly, sewing head movable-type-automated sleeve mounting, work movable-type skirt waist belting, spatial clamp system shoulder pad wrap sewing, a mechanism that can grip flexible fabrics such as a worker and technology for transporting fabric items being processed between different workstations (Jana, 2003a). Always looking for ways to reduce operating costs, although the apparel manufacturers became increasingly comfortable with automation (Bobbin Show International, 1995), the primary hurdle toward automation of sewing was to handle the dimensionally unstable fabric. Despite the appearance of some early prototypes (of automated sewing systems) at machine fairs, none of these initiatives have produced commercially exploitable outcomes (Totterdil, 1995). Such workstations improve the ergonomics, efficiency, and safety of operations and may be purchased off-the-shelf or custom developed by plant engineers or research and development (R&D) departments (Glock and Kunz, 2000; Jana, 1999).

This chapter "Automation in sewing technology" is broken down into several sections. The section basic sewing kinematics explains the mechanism of stitch formation and continuous and cyclic feeding of fabric to complete sewing task. The next section discusses the evolution and basic principles of automation, the hardware and software

Automation in Garment Manufacturing. http://dx.doi.org/10.1016/B978-0-08-101211-6.00009-4

systems typically used in sewing automation. The sewing machine with under bed trimmer (UBT) and with automatic bobbin changer describes the use of simple hardware and software to achieve the basic level of automation in sewing. The next sections give examples of automated workstations for different apparel and nonapparel merchandise categories. The final section discusses the future trends and researches in sewing automation and important sources for further information.

9.2 Basic kinematics for continuous and cyclic sewing machines

Industrial sewing machines are of two types; continuous and cyclic. Kinematics for both continuous and cyclic sewing machines are made up of four mechanisms, the take up mechanism, the needle piercing mechanism, the fabric feeding mechanism, and the hook rotating mechanism. Sewing is a complicated process involving synchronized motion of the four parts. In continuous machine the fabric feeding mechanism is linear, whereas in cyclic it is nonlinear.

9.2.1 *Kinematics for continuous sewing machines*

The first functional sewing machine was invented by the French tailor, Barthélemy Thimonnier in 1830. The invention sparked fear among other tailors for their livelihood, and Barthélemy Thimonnier's workshop was burnt down. After 16 years, Elias Howe, the American inventor, patented an improved sewing machine in 1846. Earlier sewing machines used only one thread and a chainstitch that could unravel. Howe's revolutionary machine was the first that used two separate threads, one threaded through the needle and one in a shuttle; it was powered by a hand crank. A sideways-moving needle with its eye at one end would pierce the fabric, creating a loop of thread on the other side; a shuttle would then push thread through the loop, creating a tight lockstitch. Howe's business did not thrive. Others, such as Isaac M. Singer made slight modifications in the machine and built successful businesses. Howe sued those who had infringed on his patent and won royalties on all machines sold. Howe died the year his patent expired.

The sewing machine has anatomical parts such as machine *head*, machine *arm*, *throat* plate, *teeth* in feed dog, and presser *foot*. All four mechanisms of sewing machine are synchronized through a single source of rotatory motion in the main horizontal shaft, which runs through the machine arm from flywheel/handwheel at right to machine head at left. The shaft gets its rotatory motion through manual rotating of handwheel or through a belt drive that is either driven by pedal crankshaft or electric motor. All industrial sewing machines are fitted with electric motor, which is housed below the machine bed. The motor pulley drives the flywheel/handwheel shaft by a belt through a slot in the machine bed. The important mechanical motion principles used are cam-follower, crankshaft, and gear drive. Fig. 9.1 shows cam-follower, Fig. 9.2 shows gear drive, and Fig. 9.3 shows crankshaft mechanism.

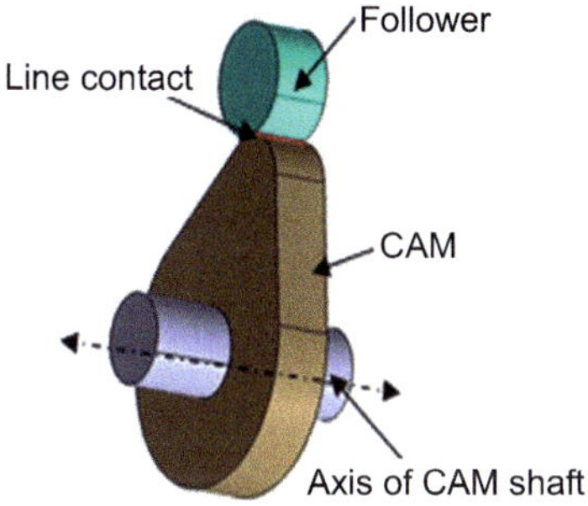

Figure 9.1 Cam-follower mechanism.

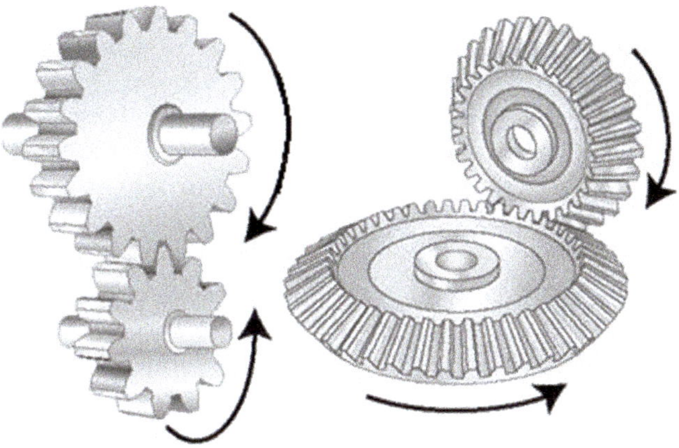

Figure 9.2 Gear drive mechanism.

Figure 9.3 Crankshaft mechanism.

From flywheel shaft, the motions are transmitted through two different systems; belt driven and cam-follower driven. The belt driven kinematics is used by brands that used sealed lubrication-type machine head (German origin brands). The cam-follower driven kinematics is used by brands that used open tub lubrication-type machine head (Japanese origin brands).

9.2.1.1 The cam-follower driven kinematics

The right corner of the main shaft (arm shaft in Fig. 9.4) has two cam-follower mechanism; rotation of main shaft transmits up–down motion to two connecting rods, which are connected to two rocking shafts (front and rear bed shafts in Fig. 9.4). The rocking shafts are so-called because these shafts get small amount (usually lesser than 30degrees) of rotatory motion from the two connecting rods. The other ends of the front and rear rocking shafts are connected to a specially shaped bar where feed dog is mounted. While the rear rocking shaft is connected to the feed dog bar by a vertical connector, the front rocking shaft is connected to the feed dog bar by a horizontal connector. The rear end of the feed dog bar gets to and fro motion and front end of the feed dog bar gets up–down motion (refer Fig. 9.5). Fig. 9.4 shows the sewing machine head with cam-follower kinematics.

The motions are so synchronized that the feed dog bar gets "to" motion (away from operator) from rear rocker shaft, followed by "down" motion from front rocker shaft, followed by "fro" motion (coming back toward operator) from rear rocker shaft and lastly úp' motion from front rocker shaft (refer Fig. 9.5). The resultant theoretical motion is a rectangular motion in a vertical plane. In reality the vertical and horizontal motions are slightly overlapped at the transfer point (means, while the feed dog bar is nearing completion of it's forward motion, the down motion also starts simultaneously), resulting smoothening of the corners of rectangle. The actual motion of the

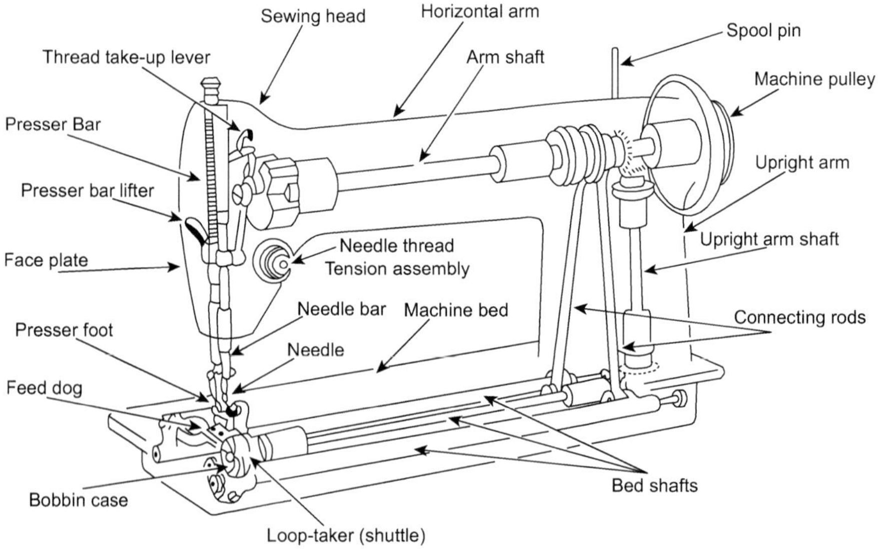

Figure 9.4 Machine head with cam-follower driven kinematics.

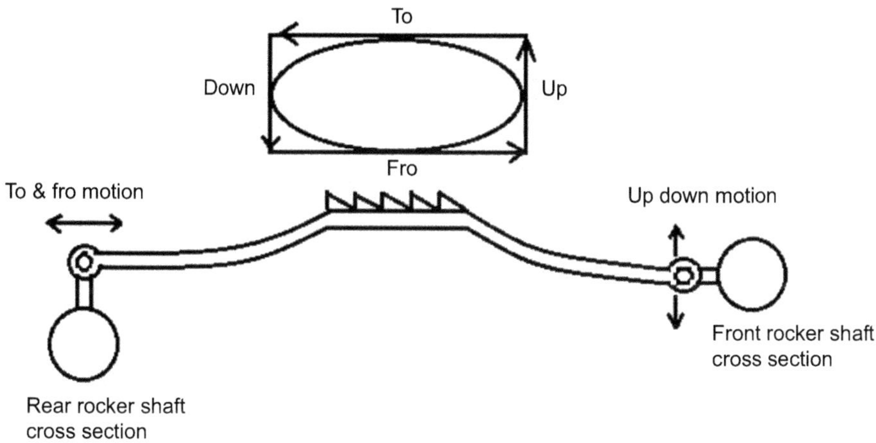

Figure 9.5 Feed bar connection with rocker shaft and elliptical motion of feed dog.

feed dog is therefore elliptical. One 360 degrees rotation of the main shaft will result one complete elliptical cycle of feed dog.

After the cam-follower, there is a bevel gear in the main shaft, which transfers rotatory motion to one vertical shaft that runs through the bottom of the machine. The vertical shaft is further connected to horizontal hook shaft by another bevel gear. The hook shaft is housed below the machine bed, positioned centrally between the two rocker shafts and hook is mounted on the left end of the hook shaft. The number of gears is so numbered that one 360 degrees rotation of the main shaft will result two 360 degrees rotation of the hook shaft.

There is a cam mounted in the left end of the main shaft (inside the machine head), which is connected to the take up lever moving is a vertical plane. The shape of the cam is such that one 360 degrees rotation of the main shaft will result one complete skewed elliptical motion of the take up lever.

The main shaft is also attached to the needle bar through a crank; resulting, the needle bar moves vertically giving up–down movement to the needle. The fabric is kept in horizontal plane flat on machine bed and the needle moves vertically to penetrate the fabric. One 360 degrees rotation of the main shaft will result one complete up–down cycle of the needle.

9.2.1.2 Belt driven kinematics

While the above kinematics are present in most of the industrial sewing machine sold today, earlier German origin brands (such as Pfaff) used to follow slightly different kinematics, where a toothed belt was used to drive another shaft at the bottom rear of the machine. The bottom shaft further drives the hook shaft by gear mechanism. The bottom rear shaft is connected to front rocker shaft by cam-follower mechanism and the front rocker shaft is attached with one end of the feed bar by vertical connection. The bottom rear shaft is connected with the other end of the feed bar by horizontal connection. Quite similar to cam-follower driven kinematics, the vertical connection

Figure 9.6 Belt-driven kinematics.

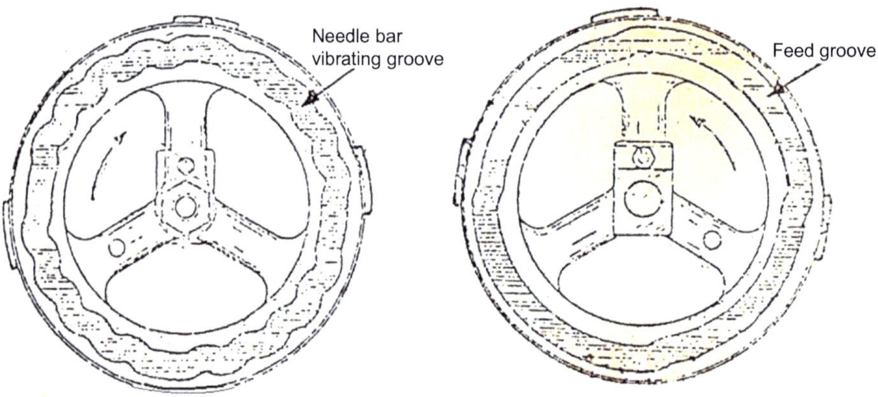

Figure 9.7 Front and back view of two cycle face computer-aided manufacturing used in buttonhole and button sew machines.

gives to and fro motion to feed dog and horizontal connection gives up–down movement to the feed dog. Fig. 9.6 shows the belt driven kinematics.

9.2.2 *Kinematics for cyclic sewing machines*

These sewing machines complete the sewing work in a short automatic cycle (Carr and Latham, 1999), thus the name cycle or cyclic (Jana, 2015) or automatic (Kilgus, 1996). Glock categorized these machines as semiautomatic special purpose machine where the operator places the materials or garment part, activates the machine, and the machine completes the cycle of preprogrammed number of stitches (Glock and Kunz, 2000). While the fabric is held by clamps, sewing guides or templates, the movement of clamp is controlled by cam-follower mechanism or electronics. Typical applications of this type of machines are button sewing, buttonhole sewing, and pattern tacking machines. Fig. 9.7 shows the front and back view of two faced cam; one groove is vibrating the needle bar and another groove driving the feed clamp.

In these machines needle moves only up–down to form the stitch and the clamp feed move the fabric horizontally (x-axis) or vertically (y-axis) creating the required shape. The movement of clamp is controlled by cam-follower mechanism or electronics. In a cam-follower mechanism the specific pattern of button sew (N or X for example), specific length and width of buttonhole or specific shape of tack (I or D for example) can be made by one specific cam-type and for any change of pattern and/or shape of tack, the cam need to be replaced and involves machine downtime. In electronic machine, the movement of cam is controlled by microprocessor, thus different pattern and/or shape of tacks can be sewn easily by changing the program in the microprocessor.

Button sewing machines are either single chainstitch (107-type) or lockstitch (304-type) machine head where the fabric placed over the cloth clamp and button is placed in button clamp. Once the operator starts the machine, the button clamp comes down and touches the cloth clamp; button and cloth clamp together moves horizontally and vertically as per cam-follower or microprocessor movement and at the end of the cycle the thread is cut and empty button clamp move up to original position for sewing the next button. During 1990s there were chainstitch button sewing machines (Pfaff 3303), where needle bar used to oscillate horizontally between two holes, in case of four hole button, the button clamp used to move at appropriate moment to bring the second pair of holes in line of sewing (Kilgus, 1996). These oscillating bar button sewing machines were preferred for sewing buttons with reinforced washer at the back of the fabric (for button down collar shirt).

Button hole sewing machines are either lockstitch zigzag (304-type) or double chainstitch zigzag (404-type) machine head, where the fabric placed over the cloth clamp. Once the operator starts the machine, the fabric holding frame comes down, grips the fabric against the clamp and movement of the clamp starts. Buttonhole is formed by a series of zigzag stitch with varying degree of needle throw in synchronization with X-Y axis clamp movement. The hole in the buttonhole is created by slashing/cutting fabric either after sewing or before sewing. In ordinary straight buttonhole (shirts and blouses) once the sewing cycle is complete the knife comes down to slash the center of buttonhole, called cut-after mechanism. In eyelet buttonhole machine, button space is created by die cutting and both cut-before or cut-after mechanism possible. Traditionally, knife size depends on the length of buttonhole and knife needs to be replaced while changing buttonhole length; however, nowadays small knifes are mounted and based on the buttonhole length the knife calculates and makes multiple cutting strokes as required. No need to change knife saving time and increasing flexibility in handling multistyle. Cutting is generally done by slashing the fabric by a diagonal shaped knife against a slit; however, for some loosely woven fabrics with stronger yarns, the flat-shaped knife with chopping action against a Teflon base is appropriate to get a clean cut.

Pattern tacking machines are lockstitch zigzag (304-type) machine head where the fabric placed over the cloth clamp. The machine sews a predetermined number of stitches in a particular shape of pattern using a combination of lockstitch and zigzag stitch. The most common shape for cam-follower driven machine was a small "bar" or "I" that was used to reinforce the stress points such as pocket opening, zip opening, and thus the name of these machines was "bar-tacking machines." The microprocessor controlled machine can create any shape, and therefore the name of the machine changed to more generic "pattern tacking machine." The cost of these machines depends on the maximum size of the tack and corresponding clamp size. Machines are available from approximately 6 cm^2 to 4 m^2 stitch area.

9.3 The building blocks of automation

Automation is a technology that minimizes various costs incurred by both routine labor and routine human intellectual reasoning, by designing and building machines that perform human-desired operations with minimal or no human intervention. These machines are self-activated, self-acting, self-determining, self-regulating, and self-reliant (Liberatore, 1996). Sewing automats are engineered workstation, which is defined as separate category (Glock and Kunz, 2000) the combination of equipment and work aids that are needed to perform a designated operation and include the machine table, power supply, compressed air supply, clamp stands, mobile work racks, pneumatic cylinders, and programmable logic controllers (PLCs). While automating any operation, there is logic/decision (brainpower) and movement (muscle power); whereas logic and decision making is achieved by electronics, the movement is achieved by pneumatics, electropneumatics, and mechatronics (Jana, 1999). Automation includes feedback for controlling automated systems (Odrey, 1992). The important hardware and software components that are commonly used in sewing automation can be categorized as mechatronics, pneumatics, sensors, and vision systems. Another important category of hardware used in sewing automation is laser pointers, which act as a visual guide for manual positioning of components.

9.3.1 Mechatronics

The first type of "mechatronic" control or microprocessor control is the PLC. A PLC is a simpler, more rugged microcontroller designed for environments such as a factory floor. Input is usually given from switches such as push buttons controlled by machine operators or position sensors. Timers can also be programmed in the PLC to run a particular process for a set amount of time. Outputs include lamps, solenoid valves, and motors, with the input–output interfacing done within the controller (Noval and Dolezal, 2002).

9.3.2 Pneumatics

Pneumatics word is originated from the Greek πνευματικός pneumatikos, which means "coming from the wind." It is the use of pressurized gas to do work in science and technology (Anon., n.d.-a). Pneumatics is also defined as engineering science pertaining to gaseous pressure and flow (Anon., n.d.-b), or the branch of mechanics that deals with the mechanical properties of gases (Anon., n.d.-c). Most sewing machinery mechanization, that uses pneumatic applications, uses pressures of about 80–100 pounds per square inch (psi) (500–700 kPa). Air is freely available in the pneumatics; most factories are preplumbed for compressed air distribution, which makes it very easy to set up a manufacturing process. Pneumatic systems tend to have long operating lives and require very little maintenance. The basic components in a pneumatic circuit are variable restrictors, directional control valves, actuators, filters, regulators, and lubricators and air cylinders.

9.3.2.1 Variable restrictors

Variable restrictors are used to control the volume of airflow discharged through a needle cooler tube, thread blower tube, or some type of material guidance blower. It is important to remember that a variable restrictor will restrict the airflow equally in both directions. Variable restrictors with a check valve, also called a "speed control," are sometimes used to control the speed of an air cylinder. A speed control will restrict the airflow in one direction but allow the air to flow freely in the other direction.

9.3.2.2 Directional control valves and actuators

A directional control valve is a device, which connects, disconnects, or changes the direction of airflow in a circuit. The first thing that needs to be determined is the number of positions the valve has. Most valves have two positions, but some valves do have three positions. The number of positions a valve have is represented in its symbol by a series of squares. Fig. 9.8 shows the symbol for a 2-position, 2-way directional control valve. It is a 2-position valve because it consists of two squares. It is a 2-way valve because if you look at any one square it has two ports labeled 1 and 2. Generally speaking, if each square has two ports it's a 2-way valve, three ports is a 3-way valve, and four ports is a 4-way valve. Fig. 9.9 shows the symbol for a normally open 2-position, 3-way, solenoid-activated, spring-return directional control valve. The actuator symbol for a solenoid is attached to the top square. The top square shows us how the three ports are connected when the valve is activated, and bottom square shows the deactivated state. In its activated state, port 1 is blocked and port 2 is connected to port 3. When the valve is deactivated and the spring has returned it to its normal condition, the port 1 connects to port 2, and port 3 is blocked. In its normal or deactivated state,

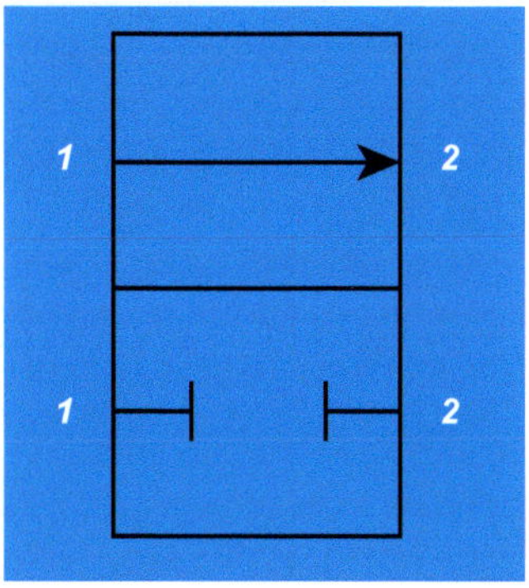

Figure 9.8 Symbol for a 2-position, 2-way directional control valve.

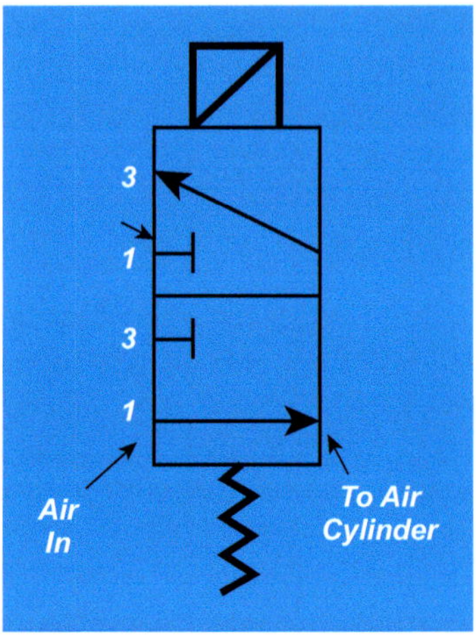

Figure 9.9 Symbol for a normally open 2-position, 3-way, solenoid-activated, spring-return directional control valve.

this valve allows air to go through it to an air cylinder, we say this valve is normally open.

Examples of use of directional control valves are as follows:

- 2-position, 2-way—Used primarily to activate needle coolers, thread blowers, etc.
- 2-position, 3-way—Used to activate single-acting air cylinders or for any component with one air connection that must be allowed to exhaust the air so it can return to its original position.
- 2-position, 4-way—Activates double-acting air cylinders or any component with two air connections. A 4-way valve will connect pressurized air to one end and allow exhaust air to escape from the other end of a component in one position, but in the other position it will reverse the pressurized and exhaust connections.

9.3.2.3 Actuators

An actuator is the means by which the valve is energized and deenergized. Actuators are represented by symbols that are added to the ends of the directional control valve's symbol.

Fig. 9.10 shows some actuator symbols that represent manual operation. Fig. 9.10(a) shows the symbol that is used to indicate that a push button is depressed to energize the valve, Fig. 9.10(b) indicates lever operation and Fig. 9.10(c) is universal symbol. Similarly Fig. 9.11 shows different types of mechanical actuator symbols.

9.3.2.4 Filters, regulators, and lubricators

Filters, regulators, and lubricators (FRL) are generally used together on each machine to condition the air properly. When used together they are referred to as FRL's, which is shown in Fig. 9.12. In the pneumatic system, the airflows through the filter first, the regulator second, and, if needed, the lubricator third.

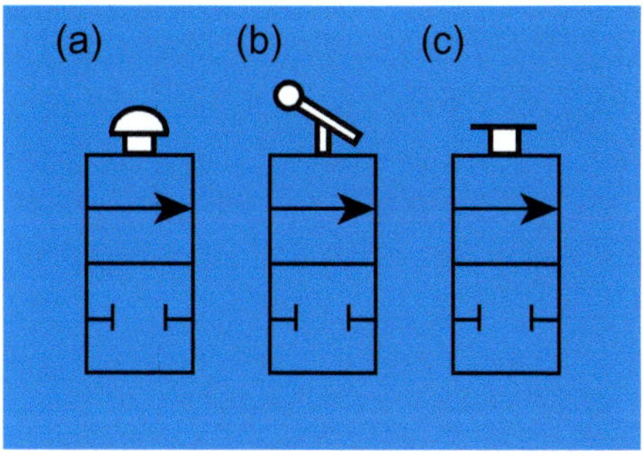

Figure 9.10 Manual actuators. (a) Push button, (b) lever, and (c) universal symbol.

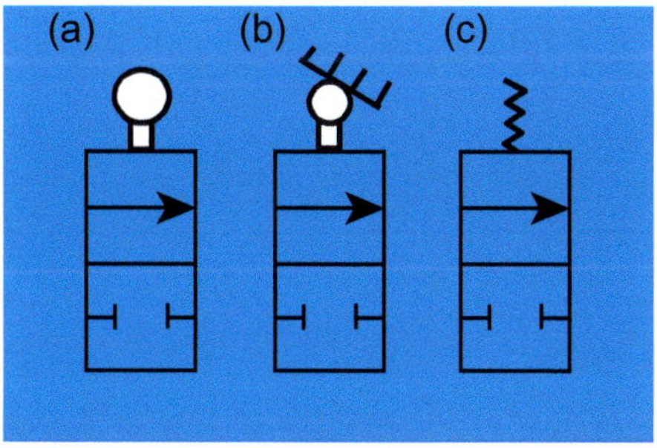

Figure 9.11 Mechanical actuators. (a) Roller, (b) cam-follower, and (c) spring.

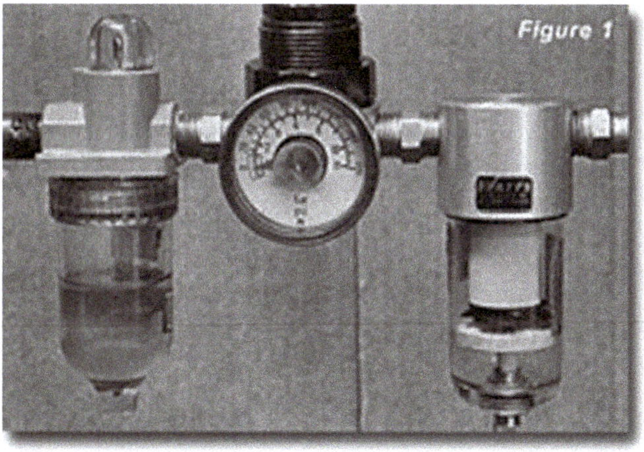

Figure 9.12 Filters, regulators, and lubricators.

9.3.2.5 Air cylinders

An air cylinder is a device that converts pneumatic energy into mechanical energy. At this time we are concerned with two basic types of air cylinders: single-acting and double-acting cylinders. Fig. 9.13 shows the symbol for a typical single-acting, single-rod spring-return air cylinder. It is single-acting because the air can only enter in one end; single-rod because the rod only extends out one end of the cylinder; and spring-return because a spring inside the cylinder will return the rod to its normal position when high pressure air to the cylinder is shut off. Fig. 9.14 shows the symbol for a double-acting, single-rod air cylinder. When high pressure air enters port B it will push the piston to the left, which will pull the cylinder rod into the cylinder. When the high pressure air is removed from port B and enters port A the piston will push the cylinder rod to the right, extending it.

Fig. 9.15 shows the symbol for a double-acting, double-rod air cylinder. It is double-acting because air can enter and be discharged from either end of the cylinder; and double-rod because the rod will extend out the left end of the cylinder when air enters port B, or out the right end of the cylinder when air enters port A.

There are several factors that determine the amount of force an air cylinder can produce. For example, a large diameter cylinder will produce a greater force than a smaller diameter cylinder. In addition, increasing the air pressure to the cylinder will increase the amount of force.

The speed of an air cylinder is determined by the flow of air in cubic feet per minute (CFM) to the air cylinder. If we restrict the flow to a cylinder, it will move slowly.

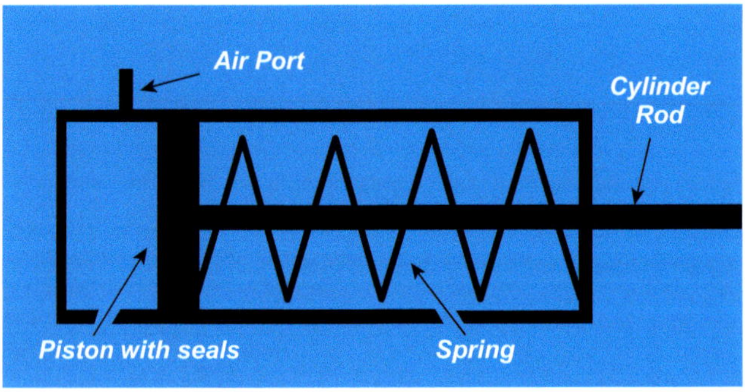

Figure 9.13 Symbol for a typical single-acting, single-rod spring-return air cylinder.

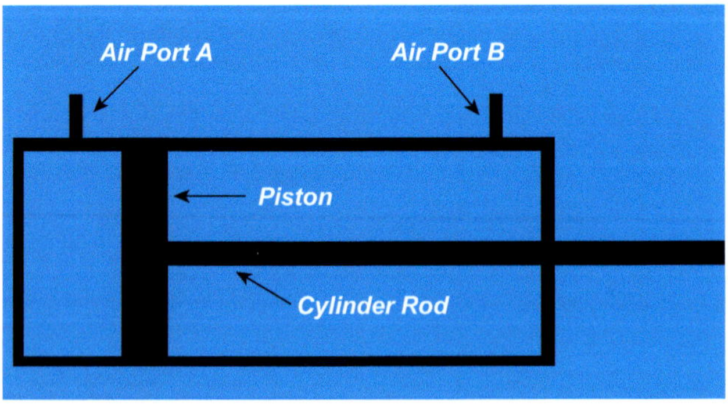

Figure 9.14 Symbol for a double-acting, single-rod air cylinder.

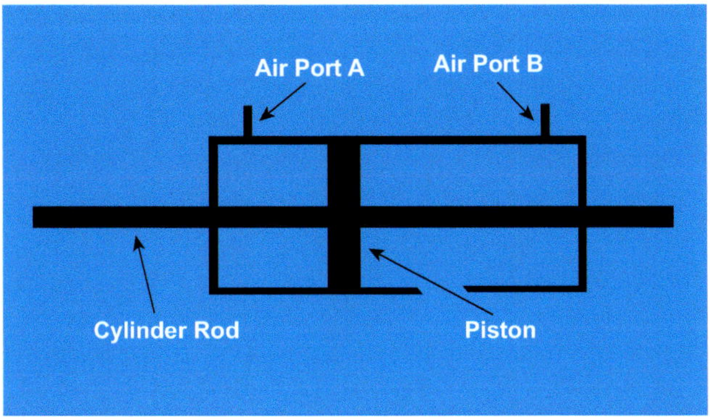

Figure 9.15 Symbol for a double-acting, double-rod air cylinder.

On the other hand, if we allow the air to flow to a cylinder unrestricted it will move with great speed, PRESSURE causes FORCE, FLOW causes SPEED.

9.3.2.6 Pneumatic circuits

The application of pneumatics to the sewn products industry is limited only by the imagination of mechanics and engineers. While pneumatics is sometimes used exclusively on a machine, many times electronics is used to control the operating sequence of the pneumatic components.

Figs. 9.16 and 9.17 show a circuit that could be used to control the speed of a single-acting air cylinder in one direction. This circuit consists of: A 2-position, 3-way solenoid-activated directional control valve, a speed control, and a single-acting, spring-return air cylinder. In Fig. 9.16, the solenoid valve has been activated. This allows the air to flow through the valve to speed control #1. Because of the orientation of the speed control the air entering will push the check ball off of its seat, allowing the air to flow through freely. The air then flows to the air cylinder causing the air cylinder to extend with full force.

In Fig. 9.17, the solenoid valve has been deactivated. This removes the high pressure air from the circuit and allows the spring in the air cylinder to retract the cylinder rod. As the spring retracts the cylinder, it pushes the air out of the cylinder and to speed control #1. Because the air is now entering the right side, it will push the check ball into its seat, blocking this passage for airflow.

Fig. 9.18 shows a circuit that could be used to raise a presser foot on a typical lockstitch machine. This circuit consists of:

1. A filter separator to remove contaminants and water from the air.
2. A regulator to reduce the air pressure to the correct pressure.
3. A lubricator could be used at this point, however, because the control valve is the knee operated-type, the exhaust air would contain some oil that could stain the operator's clothing. For this reason lubricators are sometimes left out of this type of circuit.

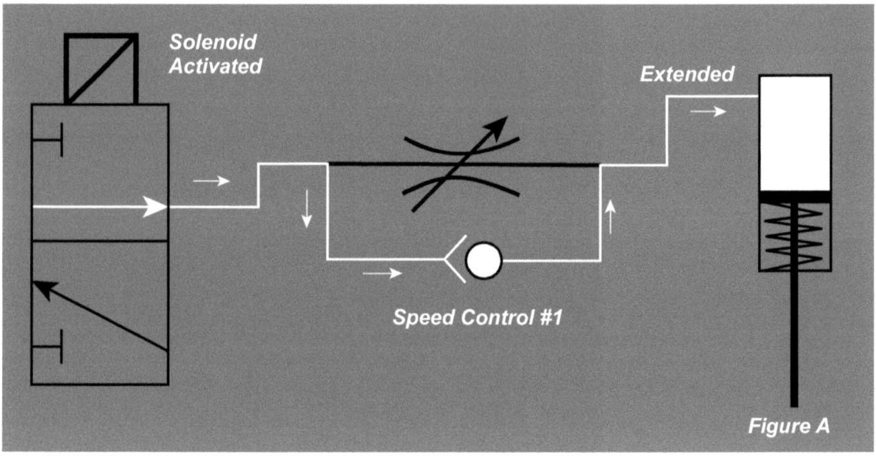

Figure 9.16 The solenoid valve has been activated.

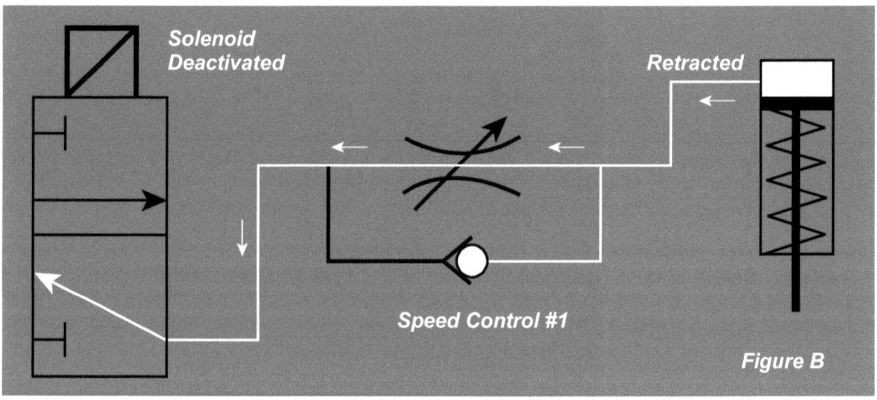

Figure 9.17 The solenoid valve has been deactivated.

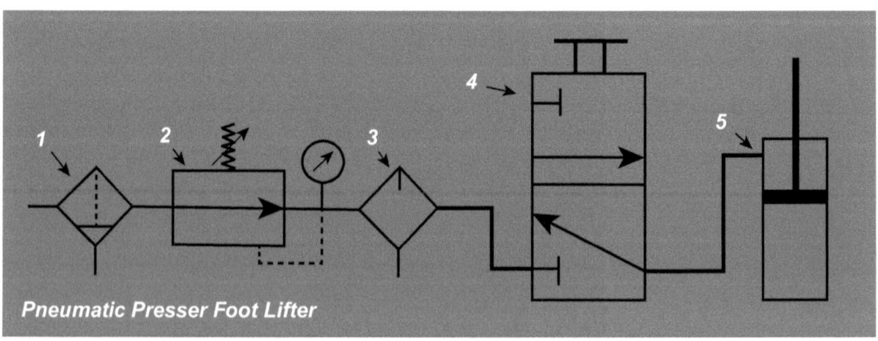

Figure 9.18 Circuit diagram for pneumatic presser foot lifter.

4. A 2-position, 3-way manually operated directional control valve. This must be a 3-way valve so that the exhaust air from the air cylinder can escape, allowing the presser foot to drop.
5. A single-acting air cylinder. Because this cylinder is lifting a presser foot, it is not necessary to have a spring-return cylinder as the presser foot spring will return the air cylinder as it pushes the foot back down.

9.3.3 Sensors

Sensor is a device that when exposed to a physical phenomenon (temperature, displacement, force, etc.) produces a proportional output signal (electrical, mechanical, magnetic, etc.) (Anjanappa et al., 2002). Out of a wide array of sensors, photoelectric proximity sensors are widely used in sewing automats. A photoelectric sensor consists primarily of an emitter for emitting light and a receiver for receiving light. When emitted light is interrupted or reflected by the sensing object, it changes the amount of light that arrives at the receiver. The receiver detects this change and converts it to an electrical output. Photoelectric sensors work in different modes; thru-beam, reflective, and retroreflective. These sensors use light sensitive elements to detect objects and are made up of an emitter (light source) and a receiver.

9.3.3.1 Thru-beam

In this case, the emitter and detector are two separate units. The emitter emits the light that is detected by the detector. A target is detected when it passes in-between the emitter and detector. Fig. 9.19 shows the working principle of thru-beam sensor.

The advantage of this mode is that the sensing range is more. The disadvantage is two parts need to be mounted separately and for long range the installation could be difficult because the emitter has to be put with the detection range of the detector. Such sensors are used in the sewing machine where the machine has to be stopped on detection of the end of a cloth overlap or a stepped portion formed at a terminus of the upper cloth on the lower cloth (Hagino and Ando, 1990). Pfaff 2481/2483 model offers such features called sensewmat.

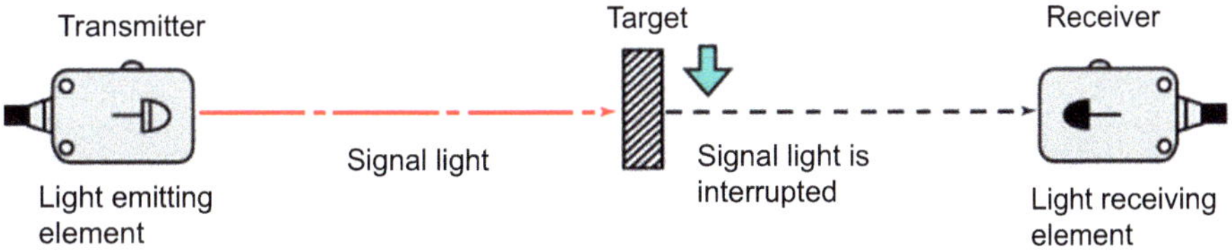

Figure 9.19 Thru-beam sensor.

9.3.3.2 Reflective

In this case the emitter and detector are put in the single package in such a way that their field of view crosses. Here the emitter continuously emits the light. When the target comes within the operating range of the sensor the light from the emitter is reflected off the target and detected by the detector. Fig. 9.20 shows the working principle of reflective sensor.

The advantages of this are low cost and easy installation. The disadvantages are short sensing distance. The sensing distance depends on target size, surface, and shape. These are very effective and appropriate for automating start–stop function of sewing machine based on presence of fabric ply. A drill small hole is created in the needle plate along the sewing line 1 cm before presser foot. The transmitter/receiver is installed beneath the hole so that the light is transmitting vertically upward through the hole, when the fabric is placed near needle point (means fabric is covering the hole), the light is reflected back from fabric, and the machine starts. As long as the fabric is covering the hole the machine keeps running, when last end point of fabric reaches the needle point (means the hole is left open), the light does not reflect back, and machine stops. Most serging automats are fitted with such sensors.

9.3.3.3 Retroreflective

The main components of this sensor are the emitter, detector, and the retroreflector. The emitter and the detector are in the same package. The retroreflector is placed little far from the sensor. The light from the emitter is reflected off the retroreflector and detected by the detector. When the target passes between the sensor and the retroreflector the beam is not reflected back to the detector. Here the problem can be that the beam could reflect from the target itself. For this the polarizing filter is used in the sensor. Hence only the light reflected by the retroreflector is detected by detector. Fig. 9.21 shows the working principle of retroreflective sensor.

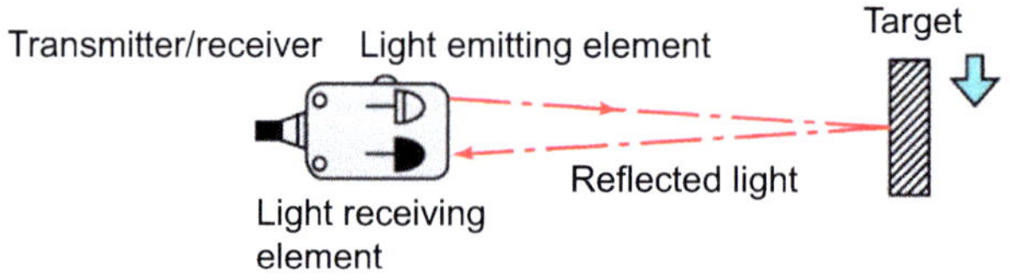

Figure 9.20 Reflective sensor.

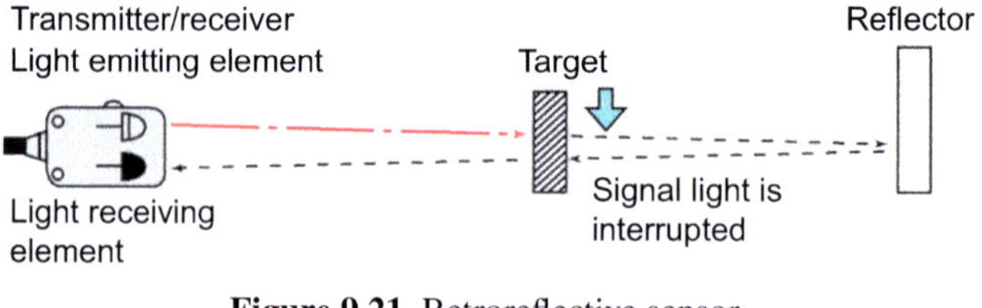

Figure 9.21 Retroreflective sensor.

The advantages of this are low cost and easy to install. The disadvantages are: separate retroreflector need to be used, it cannot be used to detect small objects as the target has to block the entire beam from emitter, performance could be affected in case of dirt on retroreflector, and problems in detecting clear targets.

9.3.4 Vision systems

Vision systems have the advantages derived from being noncontact sensors. Most vision systems are based on the detection and measure of some feature in objects such as perimeters, vertices, areas, holes, etc. (Casals, 1989). A vision system may consist of one or more cameras acting as sensors. The vision interface unit connects the cameras to the video display and the hardware to convert the image into a suitable digital form. The hardware includes a video buffering unit called a frame grabber and an image preprocessor. A computer using the image processing software extracts the information about the scene, interpretation of the image, and recognition of the objects for desired application (Mittal and Nagrath, 2003). The increased processing power of computing and miniaturization of chip size made things possible, which earlier were not thought to be so. The ability of a vision-based system to look (capture image), process the captured image at a split second (much like neurons in our brain), and direct mechanical actions is a reality. This is very similar to a sewing operator who is handling/guiding the creation of fabric during the sewing process (Henderson and Jana, 2017). Fig. 9.22 shows the working of Bother's vision sewing system.

9.3.5 Laser pointers

A laser is a device that emits light through a process of optical amplification based on the stimulated emission of electromagnetic radiation. The term "laser" originated as an acronym for "light amplification by stimulated emission of radiation." A laser pointer is a device with a power source and a laser diode emitting a very narrow

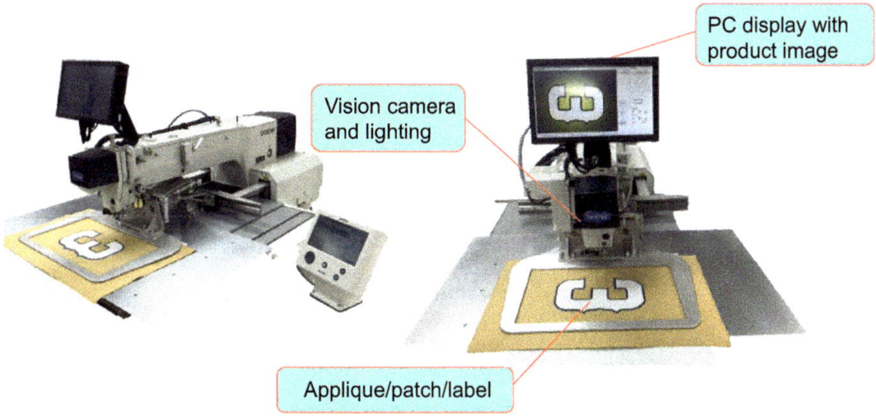

Figure 9.22 Vision sewing system from Brother.

coherent low-powered laser beam of visible light, intended to be used to highlight something of interest by illuminating it with a small bright spot or line of colored light. The small width of the beam and low power of typical laser pointers make the beam itself invisible in a reasonably clean atmosphere, only showing a point of light when striking an opaque surface (Anon., n.d.-d). These laser pointers are generally used in the loading section of sewing automats as a visual guide so that the fabric component can be placed matching with the laser line visible on the machine bed. The color of laser light used in sewing automat is usually red and safe to human eye.

9.4 Evolution of sewing automats

It is interesting to record that the need for automation in sewing was triggered by search for methods for rapidly manufacturing military apparel, should the need arise for quick and heavy mobilization of men for the armed forces (Casten, 1987). The Defense Logistics Agency (DLA), United States in collaboration with universities had spent $3.2 million during 1986–87 on advanced apparel manufacturing technology demonstration on a project to increase the use of advanced technology by apparel manufacturers (Casten, 1987). Some unclassified documents from DLA, United States (McPherson, 1987) shows that early automation initiatives was by North Carolina State University, that included operations such as combat trouser uniform front pocket, forming and sewing of waistband loops for pull on slacks, rail seamers (up to 28 ft) primarily for the home fashion, large area automatic sergers and binders, automatic small part sergers (single-ply guidance), sleeve seamers (double ply guidance), and buttonhole indexers for large parts as shower curtains to name some important operations. Some of the important researches in sewing automation can be summarized as:

- Unmanned (workerless) factory by Senkoken, Japan… year 1967–70
- Unmanned sewing system 2000 by Singer, United States… year 1967–69
- Automatic sleeve attaching in menswear by [TC]2 (Textile Clothing Technology Corporation), Draper Labs and Singer, United States… year 1986
- BRITE Project by Courtaulds PLC, United Kingdom and Pfaff, Germany… 1985–88
- Automated sewing system by Chalmers University and TEFO, Sweden
- Technology Research Association for Automated Sewing Systems (TRAASS) by Ministry of International Trade and Industry, Japan… year 1982–90

These researches definitely indicate that automation was aimed toward formal and commodity products and easy to handle fabrics (rather fashion products and dimensionally unstable fabrics).

Any sewing operation can be broken up into three steps; loading, sewing, and unloading (Jana, 1999). In commercially available workstations, loading is manual, while sewing and unloading steps are completely automated without any human intervention (Jana, 2014). Arthur D. Little divided automation into five consecutive elements; namely separation and pick up, orientation of piece in direction and plane, bringing two or more pieces or plies under proper orientation with each other in plane and rotation, control through the work point, and accurate stacking, so that subsequent operation can be performed on the combined piece again (Tyler, 2000).

9.4.1 Loading of fabric component

Loading involves picking up the fabric component/s by one/two hand/hands, placing/aligning/folding/matching and sliding below the presser foot. While mechanizing loading operation, identifying the right/wrong side of a fabric will require visual or touch sensor, separating plies and picking up a single ply from a stack of fabric will require specialized picker that will be able to separate top ply from a stack of fabrics and pick up. Although there are numerous patents for single ply picker from a fabric stack (Kamal, 1987), and even some of the patents are assigned to clothing brands such as Fruit of the Loom (Beasock et al., 1993) and Levi's (Blessing, 1977), only Cluett Peabody picker (Hughes et al., 1976) was commercially used in automats (Jet Sew Technologies, 1998). Placing the component at sewing table and sliding beneath needle point are achieved by conveyor belt.

9.4.2 Sewing of fabric component

Sewing involves sewing needle movement and occasional stopping for guiding/aligning/matching/measuring/pivoting till the sewing cycle is complete. While start of sewing may be mechanized by proximity or optical sensor, feed manipulation will require the tension adjusted belt-feed system, dynamic edge alignment of similar or dissimilar curve during sewing will require air-assisted multilayer edge guide, and pivoting the fabric keeping the dimensional stability intact will require spatial clamp feed or template sewing (Jana, 2015). Zippy edge guide from Profeel aligns edge of two plies together, while the plies are separated by a metal plate, where air jet continuously pushes the fabric edge to the guide. Air guide from Profeel can direct the movement of fabric during sewing by inbuilt air guides strategically located in the sewing table. Vetron AutoSeam—an automated sewing machine—that is capable of automated sewing of two opposite shape of fabric, by tracking the movement of fabric with intelligent cameras and controlling the movement of the fabric under the needle on a stitch by stitch basis (Frye and Jana, 2017). The patented ThreadVision system from Softwear Automation also tracks individual threads in each fabric, redefining the traditional coordinate system to thread-count, allowing for high accuracy panel tracking regardless of deformation. This system controls the feed and orientation of the fabric with multidegree freedom feed dogs to fully automate the sewing process (Frye and Jana, 2017).

9.4.3 Unloading of (disposing off) sewn component

Unloading- or disposing-off the sewn piece involves sliding/flipping/folding to arrange/stack/hang the piece in order once the thread is cut. Disposal involves severing the sewing thread as well as sliding/flipping the sewn component from table to stack/hang in a stacker/trey to make loading for next operation less cumbersome and less time-consuming (Carr and Latham, 1999; Jana, 2003a). Depending on the size of the sewed parts to be disposed off, the mechanism of disposal can be pick and place, slide and align, flipped to hang on fold, or concentrically arranged. There are two types of stackers; small part stacker and large part stacker.

Small part stackers are placed behind or on one side of the machine, as desired and controlled by thread trimmer, chain or tape cutter, and knuckle or knee switch. T-shirt

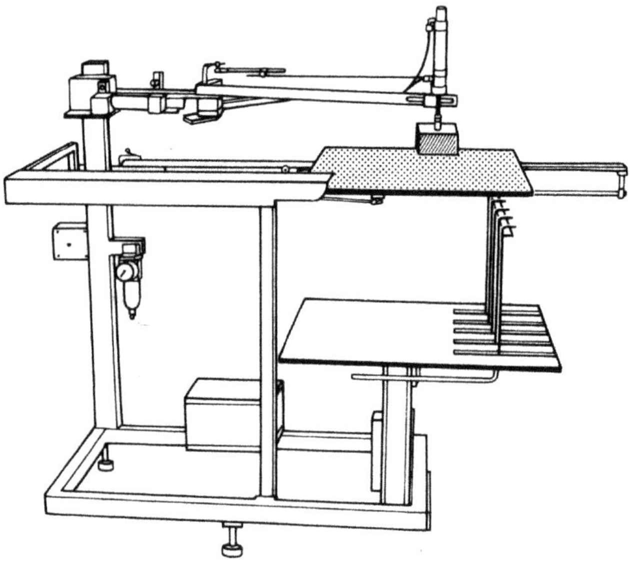

Figure 9.23 Small part stacker.
Figure courtesy: Beisler.

sleeve hemming workstation, cuff, and collar band felling workstation, the flip-type small part stacker has a swivel bar for short parts folded up or folded down. Turntable stacker is another type of small part stacker that stacks small parts concentrically such as cards held in hand. The larger garment part after sewn slided off the back of the machine table over the horizontal bar of a stacker. Pneumatic force then moves the bar away from the machine so that the garment part falls astride it and bar returns to its formal position to await the next part (Carr and Latham, 1999). Large part stackers are of different types, standard, with deflectors, flip-type, etc. Freely stacked large part stackers are common in trouser panel serging workstations. The components are either stacked freely on a double bar mechanism (for better stability) or stacked on a single bar with clamping bar. Rotary stackers are generally used for stacking closed loop type sewn components, for example, elastic loops made for closed waistbands. Elastic tapes are cut and sewn (butt join or overlap join) automatically before glides onto the stacker, which rotates intermittently (Jana, 2003a). Fig. 9.23 shows a small part stacker and Fig. 9.24 shows a large part flip stacker.

9.5 Sewing machines with under bed trimmer

Electronic control has introduced a new form of machine versatility that allows a general purpose machine to become semiautomatic with specific programmable processes such as positioning of needle (up/down), backtracking, counting of stitches, detect bobbin run out, and trimming of thread (Jana, 2015; Glock and Kunz, 2000). Computerized sewing machines or UBT sewing machines have

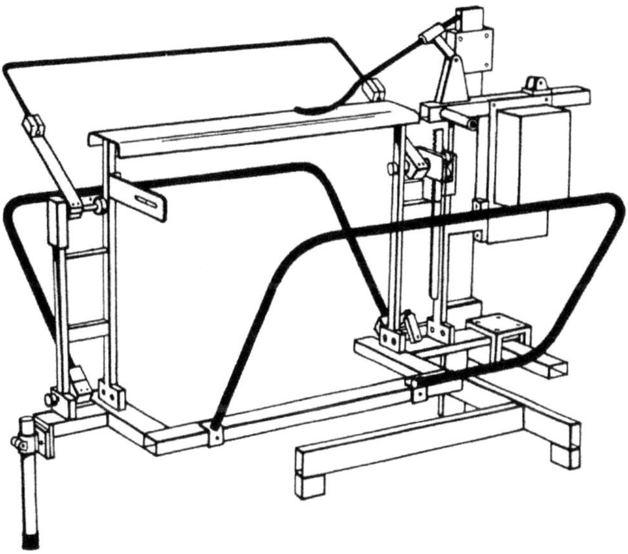

Figure 9.24 Large part flip stacker.
Figure courtesy: Beisler.

memory banks and electronic control panels, which the operator uses to program an operation, which the machine will perform repeatedly (Solinger, 1988), and reprogrammable in terms of stitch patterns, cycle times, and operation of work aids (Glock and Kunz, 2000).

Needle positioner will ensure automatic lifting or lowering of needle whenever the machine is stopped without operators requirement to manually rotate the handwheel. Back-tack feature will ensure automatic preprogrammed number of back-tack stitches at start or end of seam without operator manually and vigilantly operating the back-tack lever. Step programming feature will ensure operator not to slow down to perform a precise stop, as the machine will stop automatically (even if the pedal is kept pressed) after sewing the preprogrammed number of stitches. And the UBT will automatically cut the needle and bobbin threads together leaving a very short tail of thread without any further requirement of thread trimming. The above functions are achieved by combining various hardware and software components such as PLCs, electromagnetics, and pneumatics as described in earlier sections. The following manual actions are being mechanized or automated in computerized sewing machine with UBT.

1. Slowing down the speed of the machine for accurate stop
2. Perform back-tack operation, if required
3. Rotate handwheel to bring the needle up (during any stop there is 50% probability that needle may be down or up)
4. Lift the presser foot
5. Pull the sewn component away from the needle, while continuously rocking the handwheel to release both the needle and bobbin thread
6. Pick up the scissor and cut the thread
7. Move away the cut thread ends manually behind the presser foot.

9.6 Sewing machine with automatic bobbin changer

Automatic bobbin changer works on the principle of first checking the remaining amount of thread in bobbin and once a predetermined amount of remaining thread is reached in bobbin, the bobbin is replaced with a filled one by a robotic catcher. Automatic bobbin changer first made it's appearance in Bobbin Show International (1995), where Juki displayed two position bobbin changer. Only two bobbin casees are being used, while one bobbin is being used in sewing, the spare bobbin is getting winded and ready for replacement. Subsequently, Philippe Mall from Germany and Kinoshita from Japan invented magazine-type automatic bobbin changer, where filled up bobbins are first loaded onto magazines and then magazines are attached to sewing machine. An eight bobbin magazine can lasts up to two and half hours continuous sewing. This is effective for thick thread stitching and for work requiring the lower thread to be replaced frequently. It is optimum for products in which the thread cannot be overlapped, for stitching. In addition to increased efficiency, the auto bobbin changer prevents stitching defects and reduces the mental strain of the worker. Fig. 9.25 shows automatic bobbin checker by Kinoshita, while Fig. 9.26 shows magazine-type automatic bobbin changer by Kinoshita.

The common magazine for lockstitch has eight slots, seven slots are fitted with filled bobbin case and one slot is free for changeover to take place. While the machine is running the empty slot is lined with robotic changeover arm. The remaining thread amount, thread thickness, and stitch length is set at the control box of the auto bobbin checker. The metallic probe of the auto bobbin checker checks the remaining thread

Figure 9.25 Automatic bobbin checker by Kinoshita.

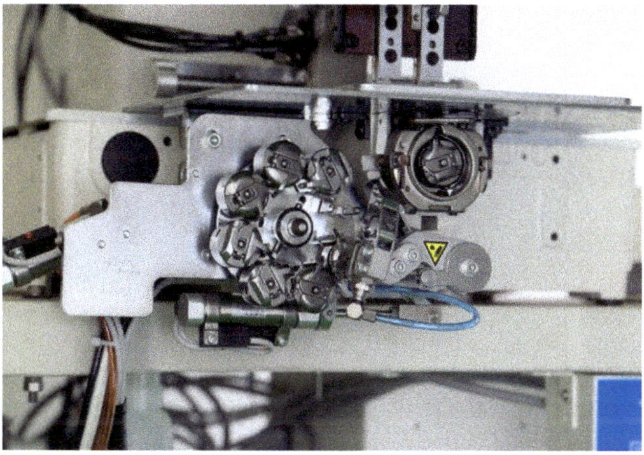

Figure 9.26 Automatic bobbin changer by Kinoshita.

amount at regular intervals during sewing. When the remaining thread reaches the lowest level, the lower thread is cut and machine stops. The robotic arm removes the empty bobbin case from the machine and places it onto the empty slot of magazine. The magazine rotates and aligns the filled bobbin case in line with the robotic arm. The robotic arm removes the filled bobbin case from magazine and fixes onto the machine. The machine is ready to run again.

9.7 Sewing automats for gent's and lady's shirts

The gent's and lady's shirt category includes casual and formal (dress) shirts for men, ladies shirt, and front open tops for women. The following sewing automats were commercially developed by different sewing machine brands during the last quarter century.

1. pocket setter
2. pocket hemmer
3. buttonhole indexer
4. button sew indexer
5. label attaching
6. autojig (collar, cuff, flap runstitch)
7. cuff felling
8. collar cuff topstitch
9. collar and band join
10. sleeve top placket
11. sleeve bottom placket
12. yoke-back join
13. shoulder join
14. cuff join to sleeve

Pocket hemmer for shirt is comparatively a rare automat and only Yuho brand currently offers such automat. The hemmer comes with thread breakage indicator, twofold as well as threefold option and auto heat pressing after threefold hemming. Pocket setter for shirts and ladies top generally works with flat front pattern. The front pattern should be in two-dimensional form and not in three-dimensional (3D) form, i.e., without darts and princess line seam. The front is loaded into the machine, followed by hemmed pocket being loaded into the clamp. The machine automatically creases the pocket, position the creased pocket onto a predesignated position of the front, transfer the pocket and front assembly to the sewing head and attaching is done. Pfaff offers pocket with flap option also in some of its model. Duerkopp, Pfaff, and Yuho offer such automats.

Buttonhole and button sew indexer works with continuous band or conveyor feeding for accurate guiding of front part. The operator simply places the front on the band and rest is done automatically. The band moves the front by a predetermined amount between two buttonhole/sew, the feeding of the button and stacking of the front after sewing is automatic. The preloading device enables overlapped working. Cycle time for button sewing is comparatively faster than button holing, therefore a 2:1 or 3:1 buttonhole:buttonsew machine ratio is common. On average 2000 shirts are possible per 8 h shift with 2:1 buttonhole:buttonsew machine ratio. Fig. 9.27 shows multihead cuff runstitch and trimming automat from Duerkopp Adler.

Brand Label attaching in shirt yoke is again not a very common automat, and Brother displayed such automat during 1999. The label is cut and setting base is transported to sewing area. The label transfer mechanism picks up the label and held by the inner clamp for sewing, a lockstitch sewing head sews the label.

Collar/Cuff/Flap runstitch is a common automat complete with a lockstitch (301) sewing head, with sewn collar/cuff trimming. It is a totally electronic controlled unit managed by a piece of software that allows the sewing of any type of stitch with automatic size changing. To change the model simply replace the low cost and easily made fabric holding plates. The trimming is performed separately from the sewing phase, therefore a different cut can be made, compared to the stitching. While 971-01 (Durkopp Adler) is equipped with a sewing head producing two-needle chain stitch, U-3101-E/01 (YUHO) and UAM 03 (MAICA) produces single-needle lockstitch. Durkopp, MAICA, Weishi, and

Figure 9.27 Cuff runstitching automat by Duerkopp Adler.

Yuho are prominent brands in the market. All these automats comprises of four distinct zones feeding zone, sewing zone, trimming zone, and stacking zone. The product moves sequentially through each zone. This sequential arrangement in a single workstation facilitates high productivity and improved manpower utilization.

Cuff felling is again a simple automat yet not very common now. Duerkopp has demonstrated an automat capable of felling 5500 cuffs per 8 h shift using belt-feed mechanism with a double chainstitch (401) machine. Collar and cuff topstitch are available from Yuho and Duerkopp, a semicircular clamp feeder negotiates the material during sewing.

Automats for collar and band join, sleeve placket make, yoke-back join, and shoulder join and cuff join to sleeve are very specialized and rare. Currently only Yuho offers models for these operations. The machine can be customized for specific requirements. The yoke-back join automats are available with options for center loop, center box pleat, or two side pleats. The cuff join to sleeve automat can accommodate different knife-pleat size and positions.

9.8 Sewing automats for casual bottom wear

The casual bottom wear category consists of casual flat front trousers, khakis, cargos, and jeans. The category characterizes with continuous waistband seam either single-piece folded or two-piece sandwiched, and continuous inseam operation. The following sewing automats were commercially developed by different sewing machine brands during the last quarter century.

1. Welt pocket making
2. Belt loop setter Jeans
3. Decorative sewing in back pocket
4. Hemming back pocket
5. Back pocket setter jeans
6. Topstitching of fly (J-stitch) in trousers
7. Coin pocket attach
8. Waist band (WB) label attaching
9. Automatic bottom hemmer

Welt pocket making is undoubtedly the most iconic automat in this category. AMF Reece, which pioneered pocket welting in the 1950s, introduced the Speedwelt 1000, an electronic version of its old standby (Bobbin Show International, 1995). Productivity is increased because the machine is sewing a pocket, while the operator is loading the panel and pocket. A stacker automatically off loads as well. Welt pocket making or bone pocket making or double bone pocket making as it is called is achieved by a two-needle lockstitch head (301 stitch-type) with center knife cutter and needle feed mechanism.

Belt loop setter is a twin needle pattern tacking (304 stitch-type) head, where operator only loads the jeans; operations such as cutting, folding, and positioning of loops, sewing of loops at jeans waistband and stacking happens automatically. The premade loop roll is already loaded to the machine, identifying and eliminating the belt loop joints happens automatically during die cutting of loops.

Decorative sewing in back pocket, hemming of back pocket, and attaching back pocket in jeans and jeans look-alikes are some of the most viable automats. Decorative sewing is performed by single-needle pattern tacking head (304 stitch-type) with clamp feeding mechanism. The pocket hemming is accomplished by double-needle double chainstitch head (401 stitch-type), the operator only places the pocket on a conveyor, the conveyor transport the pocket, folds downturn, and the sewing and stacking happens automatically. Most of the models come with hot wedge label cutter and automatic label dispenser to the sewing head so that the label is sewn at a predetermined position of the hem automatically. The stacking is commonly done by a magazine or turntable stacker. The production can go up to 14,500 pockets per 8 h.

Pocket attaching automats use sewing jig with a lockstitch sewing head (301 stitch-type) come with variable features. In some models, creased pocket to be loaded, while in some models the creasing is done by the machine. The operator first loads the pocket in a creasing clamp and then loads the trouser leg component in the loading area. Then the operator presses the actuator, the creasing clamp first creases the pocket then comes down and positions the creased pocket onto the trouser leg. A separate jig then grasps the pocket plus the leg assembly, carries to the sewing head and the attaching is accomplished by the stationery sewing head in two nonparallel sewing bursts and bar-tacking at both ends. There are options for using two different color sewing threads in two rows of stitching. Cargo pouch pocket attach option was offered by pfaff (Pfaff Industrial, 2006) model 3588, wherein cargo pouch pocket with or without flap also can be attached automatically. Fig. 9.28 shows piped or welt pocket sewing automat.

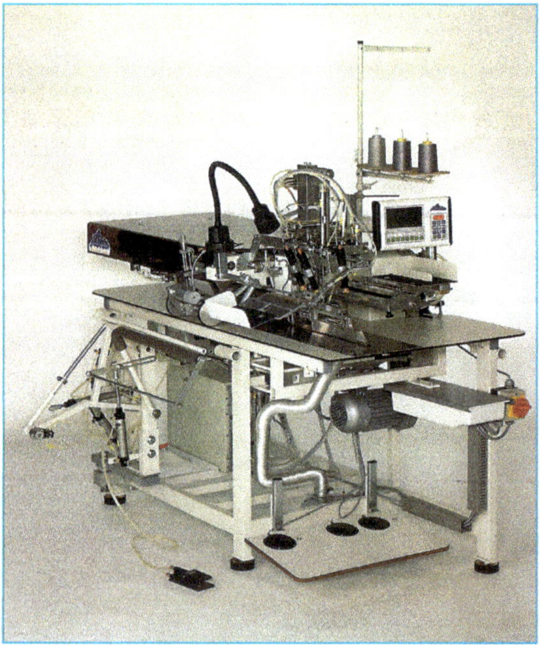

Figure 9.28 Welt pocket automat.
Courtesy: Beisler.

Automats for topstitching of fly in trousers (commonly called "J-Stitch" due to the shape of the stitch profile) are also accomplished by lockstitch pattern tacking head where a jig holds the component and laser beam positions the component accurately. The large part flip-type stacker collects the sewn components. There are options for using two different color sewing threads in two rows of stitching.

Coin pocket attach, brand label attaching in waistband, and automatic bottom hemmer are not very common or widely adopted automats. While Yuho and JAM developed automats for coin pocket attach, Brother developed label attaching to waistband, which was equipped with automatic label loader. Juki announced model ALH-326, for automatic bottom hemming of jeans (Juki Corporation, 1999) with top and bottom roller feed mechanism.

9.9 Sewing automats for formal wear

The formal wear category includes men's and lady's jacket, men's and lady's skirt and pants. From the available records, the following sewing automats were commercially developed by different sewing machine brands during the last quarter century.

1. Welt pocket making
2. Dart and waistband pleat sewing
3. Single- and double-pointed breast darts
4. Serging
5. Side seam and in seam closing
6. Sewing of endless zipper with fly
7. Topstitching of fly (J-stitch) in trousers
8. Felling left fly
9. Premanufactured WB lining and reinforced strip to WB
10. Two step belt loop attach
11. Folding and attaching of pocket facing to pocket bag
12. Attaching pocket bag to WB edge of hind trousers
13. Runstitching of side seam and wing pockets
14. Sewing of sleeve parts and center back of jacket
15. Lapel padding
16. Pocket facing attach in blazer
17. Eyelet buttonhole indexer
18. Autojig runstitch of flap
19. Sleeve setting
20. Collar band and gorge seam sewing in jacket/coat

Out of these, some automats have become extremely popular and currently widely used in the industry, such as welt pocket making, serging, whereas others are discontinued or become made to order. Automation of welt pocket became so extensive, Duerkopp during IMB (1997) displayed as many as nine different models of welt pocket making automats, five models (745-26, 745-28F, 745-28A, 745-28D, 745-24) for jackets and coats and four models (745-22, 745-23, 745-22, 745-26) for trousers. The different variants specialized in straight, slant, with or without flap, matching of plaids, and also automatic insertion of continuous zippers. The mechanism for formal wear is similar to welt pocket making of casual bottom wear with additional features such as check/stripe matching, zipper insertion, and slant shape are incorporated.

With value added features, one brand tried to outsmart others in these fiercely competitive automat categories. Juki in their model APW-298 added a feature of welting patch attach at the back side in slanting welt pocket making for added dimensional stability (Juki International, 2002). Brother in their model BAS-615, BAS-620 introduced a feature of back-tacking done by needle movement, rather commonly used reverse feeding of fabric (Brother International, 2000). This feature negates any possibility of ply shifting, thereby a neater appearance, while sewing difficult to handle fabrics. Duerkopp in their model 745-26, 745-22 has introduced a mechanism to insert endless zipper automatically (Duerkopp Adler, 1997). Fig. 9.29 shows the side seam joining automat from Beisler.

Dart and waistband pleat sewing for trousers and single and double-pointed breast darts for jackets are primarily lockstitch sewing head (301 stitch-type) with stitch condensation at dart point and back-tacking at beginning/end of seam. The loading is done by feeding plate and positioning is done by laser beam.

Serging or single ply overlocking of individual panels of trousers is another common and widely used automat in the industry. The serging ideally should use two thread overlock head (503 stitch-type), edge controller, curve edge sensor for negotiating crotch area and large part stacker. Almost all important brands: Juki, Duerkopp, Brother, Pfaff, Yuho even Pegasus and Yamato developed serging automat. There are serging automats with two sewing heads placed at right angled to each other, where serging of side seam, inseam, bottom serging including crotch and fly curves are integrated.

Figure 9.29 Side seam joining of trouser.
Courtesy: Beisler.

Side seam and in seam closing are again a very important sewing automat, where the sewing head is either double chainstitch (401 stitch-type) or five-thread overlock (504+401 stitch-type). The operator aligns and places both the trouser panel to needle point; photocell senses the presence of fabric and starts the machine. The variable top feed system enables positive control of feeding of top and bottom plies. Electronically controlled contour guide or zippy type of edge guide ensures automatic edge alignment.

Sewing of endless zipper with fly involved two-needle double chainstitch machine (401 stitch-type) with photocell sensing of edge and vertical knife cutter. Topstitching of fly (J-stitch) in trousers is similar to casual bottom wear with single head lockstitch machine with laser beam positioning and jig guided sewing profile. Premanufactured WB lining and reinforced strip joining to WB uses a double chainstitch machine (401 stitch-type) with puller feed. The photocell senses and maintains the edge alignment and vertical knife cuts the lining and strip. A lockstitch head (301 stitch-type) attaches belt loops in two steps. The belt loop attach automat, cuts the loop, and brings it to the machine, which sews the loop to trouser/skirt with two tackings. First tacking is covered by WB, the second tacking can be straight-tack or bar-tack. The sensor identifies and eliminates the belt loop joints and the die cutter cuts the loops. Folding and attaching of pocket facing to pocket bag uses either a lockstitch sewing head (301 stitch-type) or double chainstitch sewing head (401 stitch-type). The facing is automatically folded, automatically positioned on pocket bag and sewn. The lockstitch head is equipped with bobbin thread indicator. Other not so common automats are attaching pocket bag to WB edge of hind trousers, runstitching of side seam, and wing pockets and felling of left fly.

Automats for sewing of sleeve parts, center back of jacket, jacket front edge consists of double chainstitch head (401 stitch-type), step motor driven guide rail feed or clamp feed of parts to ensure zero ply slippage, while superimpose joining two plies. Lapel padding, pocket facing attach in blazer, eyelet buttonhole indexer, autojig runstitch of flap and collar band, and gorge seam sewing in jacket/coat are automats that are often used in jacket manufacturing. Sleeve setting in jacket is an important operation that is not automated. The programmable variable top feed single-needle lockstitch post bed sewing machines are available, where the fullness in sleeve is fully programmed; however, it is not considered an automat as operator is feeding the components, while sewing.

9.10 Sewing automats for knitwear and intimate wear

The product categories under knitwear and intimate wear includes T-shirts, track suits, sweatshirts, cut and sew sweaters, sports bra, structured bra, panties, vests, and men's brief. The following sewing automats were commercially developed by different sewing machine brands during the last quarter century.

1. Elastic ring making
2. Hemming and closing sleeves
3. Bottom hemmer 406/605 or 503 (blindstitch)

4. T-shirt sleeve hemmer
5. T-shirt front placket making
6. Attach rib knit cuff
7. Patch pocket setter
8. Pin tucking sportswear
9. T-shirt side slit sewing
10. T-shirt side seaming
11. Elastic attaching to leg opening of panties
12. Hook tape sewing
13. Eye tape sewing
14. Sewing facings on flat knit with a linked on look

Elastic ring making is common operation and the automat can be used for track suits, panties, and men's brief and other similar garments with elasticated waistband. The machine measures continuous elastic, cuts to size, and join ends with flat or butted seams. Optional heat seal unit may attach label automatically. Automatic detection and elimination of defects such as weaving flaws and stains are possible. This 304 sewing machine head is equipped with automatic bobbin changer.

Hemming and closing sleeves of T-shirt is a double-head sewing automat. The machine has a unique automatic fold-in-half system that allows sleeves to be overlapped during sew cycle, increasing production. It has a self-contained waste removal system. In it the operator places parts to an edge guide, initiates the sewing cycle, and continues loading. The sleeves are automatically trimmed, folded, hemmed by three-thread flatlock head (406 stitch-type). Next sleeve is automatically folded in half, transported to a three-thread overlock head (504 stitch-type), which backlatched, closed the side, and stacked. Stacker indexing is accomplished through programming of the number of sleeves in a bundle. These automats are often equipped with fabric loading unit (Pegasus, 1996).

In automatic long and short Sleeve closing machine, the operator aligns the hemmed or cuffed end and presents it to the presser foot. A photocell senses the beginning edge, drops the presser foot and starts sewing with an automatic backlatch. The electronic active edge guiding has the capability to sew both straight and contoured seams. The system also controls the sleeve until the operator presents the next sleeve. Once the sleeve is closed, the machine stops, thread chain is cut, and the sleeve is stacked automatically. There are also options for automatic turning of sleeve before stacking (Atlanta Attachment model AP264T). Fig. 9.30 shows the rib attaching automat for bottom finishing of sweatshirt.

Bottom hemming automat is either having a flatlock (changeable between 406 and 605 stitch-type) heads or blind stitch overlock head (503 stitch-type). The operator simply loads the tubular or flat T-shirt panels. The active guiding system controls the edge of the cloth across the feeding direction without any friction on the fabric. The hem is automatically formed and overlapped seam length at end of cycle is preprogrammed microprocessor controlled. Klipp-it thread trimmer trims the thread automatically, and stacker unloads the piece. One operator can operate two automats simultaneously. Union special model 94800BHA12A is the most versatile automat and Pegasus also released a low cost version MHG-121, which has no stacker.

Figure 9.30 Rib attaching at bottom hem of sweatshirts (Union Special).

Atlanta Attachment Company, YUHO, Jack, UZU, and JUITA are prominent brands in the market offering integrated front placket setting and center cutting workstations. The integrated workstations have an extended machine table for placing and positioning the body front and the placket panel. The machine is equipped with laser guides to assist in precise placement and alignment of components. Operator is just required to slide the component assembly to the sewing position, and the machine cycle begins automatically. The machine completes the sewing, slit cutting, bar-tacking, and stacking operation automatically. The machine is equipped with automatic thread cutting system. In case of thread break, the machine stops, back-tracks to facilitate the repairs, and eliminates the need to repair offline. The machine has programmable center knife, which operates independent of sewing head. The auto stacker also has unloaded feature. The machine is equipped with bobbin monitor system to facilitate instances of midcycle bobbin run out. The program can be set to select between five different shapes for hidden placket.

Automatic workstation for sewing circular knit bottom bands on sweatshirts incorporates a cylinder arm overlock sewing head with vacuum thread trimmer giving a precise cut. The workstation has electronic controls with a dual electronic active edge guiding system for band and body panel. It is also equipped with pneumatic expansion rollers that facilitates loading of band and body panel to the sewing head. The machine has automatic stacking device that withdraws the sewn panel from the sewing position and stacks them automatically. The operator's task is limited to folding the circular rib knit piece in half to form a band and places it over the expansion guide rollers. The rollers automatically expand for loading the body. The body is loaded over the rollers and band, and a touch of the start sensor expands the rollers to sew position and indexes the parts under the presser foot and sew cycle begins. While one operator tends two machines in tandem, more than 3500 pcs/8 h shift can be produced.

Patch pocket setter in knitwear is comparatively difficult than woven fabric due to stretch in fabrics. In this automat creasing, positioning, and sewing is done automatically. The suction device holds the creased pocket and the main fabric secure during sewing. Yuho (model 3206E) and Brother (BAS-751-KL) are few of the brands offering this automat.

Some of the less popular automats are T-shirt side seaming (U-9705-E/FP), T-shirt side slit sewing (U-3404-E), and pin tucking of sportswear (Pegasus 004). In the lingerie section, Rimoldi offers an interesting automat for elastic attaching to leg opening of panties. In the Rimoldi model GW3-34-2MR-13 there are two 514 sewing head face to face (one of which is mobile) for simultaneously attaching rubber or elastic lace to closed or open leg. In the bra manufacturing hook tape sewing and eye tape sewing are common automats.

9.11 Sewing automats for nonapparel sewn products

The nonapparel sewn product category includes home textiles, car upholstery, shoe upper, luggage, and other allied products such as caps, gloves, sports goods, medical orthotic products, etc. The sewing of home textile segment is highly automated because of the comparatively dimensionally stable fabrics, large size of components, and straighter seam profiles. Curtain, bed sheet, mattress, towel, and mats sewing automats are quite common. Home textiles have specialized automats for material handling, folding stacking also; automats that do sewing are listed here. A separate set of machinery brands such as Texpa, Rimac, Kinna Automatic, Automatex, and MPT caters to home textile segment.

1. Length hemming terry towels
2. Crosscut and hemming of terry towels
3. Length hemming table linen
4. Cross hemming linen table linen
5. Automatic machine for cut and overlock on four sides of table linen
6. Length hemming, crosscutting, label dispensing, cross hemming, inspection, folding, cardboard dispensing, stacking for bed linen
7. Width/length, hemming, crosscutting, label dispensing, closing of three sides, stacking for duvet covers
8. Length hemming, doubling, crosscutting, label dispensing, lateral closing (U-shape or L-shape), stacking of pillow case covers
9. Length hemming, crosscutting, label dispensing, elastic feeding, cross hemming, corner forming, corner sewing of fitted sheets.
10. Automatic pleating machine of curtains with application of tape.
11. Carpet binding/overedging
12. Longitudinal and bilateral binding and overedging on blankets
13. Mattress hemming machine
14. Mattress Handle attaching machine
15. Serge master high speed dual side mattress border serge system
16. Taping machine for mattresses

Automotive seat, furniture, filters, battery insulators, leather goods, leather car seats, convertible soft tops, large area filters, bulletproof vests, parachutes, and stitching applications in the field of aerospace industry, etc., mainly requires two types of automats; single-needle large area pattern sewer and multineedle sewing unit. For example, a single-needle pattern sewer 220mm × 100mm sewing area, automatic label feeding, bobbin detector, thread detector, rotation sensor, and bobbin thread consumption detector are mandatory for seat belt and airbag sewing. Multineedle sewing automats have linear needle drive that ensures highest stitch quality. The needle is not deflected, and the penetration holes are not widened. Large foldable loading table helps to push the sewing material into the system, laser marking device helps exact manual positioning of the processed material. Sewing unit are available in modular structure in steps of 600mm up to approximately 7200mm maximum distance between needles. Double chainstitch (401 stitch-type) height-adjustable top feed system helps to adjust to the material thickness.

9.12 Sewing preparatory machines with automatic control system

Preparatory operations such as marking, creasing, folding, trimming, and turning are part and parcel of sewing operations. These operations prepare the fabric components for quicker and easier sewing operations. Some of the important automats in these category are as follows

- Collar marking
- Collar and cuff turning
- Collar trimming/cutting
- Pocket creasing
- Sleeve placket creasing
- Placket pressing

Collar marking workstation marks at three points along collar band seam line by needle hole. The three points act as reference points for left and right shoulder line and center back point. The operator only loads the ready collar, aligns it, and then presses the actuator. The vertical marking pins come down pierces the collar band edges to create visual mark and goes up. The stacker automatically collects the marked collars and stacks in flat condition. The marking pins/pens can be easily adjusted on a sliding scale for accommodating different collar size. This workstation uses pneumatic components.

After collar topstitch, the raw edge of the ready collar is trimmed by collar trimming/cutting workstation before attaching with collar band. The collar either in open condition or folded condition is trimmed by a vertically moving guillotine cutter. The stacker automatically collects the cut collars and stacks in flat condition. Fig. 9.31 shows sleeve placket creasing machine. Some of the notable brands offering automats in sewing preparatory are MAICA and Ngai Shing.

Figure 9.31 Sleeve placket creasing.

9.13 Future trends

The future trends in sewing kinematics and automation are modularization, digitalization, vision-based sewing, Internet of Things (IoT) and robotics. Modular structure of machine head by Typical Corp (Vetron) and Duerkopp Adler (M-Type) have enabled the use of exchangeable common parts between different machine types resulting easy repair, lesser inventory, and reduced cost of ownership. The main shaft-driven synchronized motion between the needles, take up lever, feed dog, and hook/looper was a worldwide standard kinematics. The Tice technology's double-needle belt loop tacker and multiple head buttonhole sewer has freed the mechanical linkage to the bobbin, opening the doors for unprecedented modularity in sewing (Jana, 2003b). Because Efka showed possibility of real-time machine downtime data in desktop and through SMS a possibility (IMB 2003), networked sewing machine finally poised to become a reality. The IoT functionality are being demonstrated by Brother, Pegasus, Juki, where all the sewing machines are connected to Internet and monitored remotely from any geographical location to analyze production data, off-standard time, break downtime analysis, etc.

In the area of information technology integration, there are already USB-based operator log in Vetron (Typical Corporation, 2015), networked sewing machine (Jana, 2004), Android app-based sewing data exchange (Juki Corporation, 2016), Pfaff SmartSeaming (Pfaff Industrial, 2013) has shown touch screen-based service, and operation and management modules. The mechanical movement of traditional elliptical feed system is now given way to digitally driven horizontal and vertical movement of feed dog resulting various kinds of feed locus that can be selected, while sewing (Juki Corporation, 2016; Brother Industries, 2015). The digital feed is further facilitating decorative stitch, half-stitch at corners, unclogged stitch at crossover seams and many other possibilities (Jana, 2017). With online connectivity becoming affordable, machine to machine communication for computer numerical control–based machine and predictive maintenance will become a norm.

Vision systems integrated sewing has already created ripple in the market. The ability for a vision-based system to look (capture image), process the captured image at split second (much like neurons in our brain), and directing mechanical actions is the reality, which is very similar to a sewing operator handling/guiding fabric during sewing. The Autoseam from Vetron, Brother Vision Sewing, and Softwear Automation have already shown promising prototypes based on this logic (Frye and Jana, 2017).

Time will tell if the recent 3D robotic sewing of T-shirt by the Seattle-based start up Sewbo and Amazon securing patent on "demand-based manufacturing" are really reinventing the wheel or an indication of things to come. There are multiple factors behind the renewed interest in research and development in automation in developed countries after a lull of quarter century. First, during 1990s the developed nations gave up research spending on automation initiative as migration option was lucrative, but after quarter century probably further migration option is looking bleak. Second, the increased processing power of computing and miniaturization of chip size is making things possible earlier thought impossible. Third, the remarkable development in prosthetics is enabling handling dimensionally unstable limp material (such as human hand) easy. Fourth, the political agenda of developed nations toward reshoring and sustainable manufacturing.

Sources of further information

The apparel manufacturing industry has migrated from Europe, United States and Japan to Asia during 1980s and gradually to least developed countries during 2000s. Traditionally, there were three major exhibitions; Bobbin in United States, IMB in Germany, and JIAM in Japan, held once in three years showcasing the R&D of machinery manufacturers. Industry associations such as The Textile Institute, AAMA (American Apparel manufacturers Association), VDMA, International Apparel Federation (IAF), and research institutes such as [TC]2, SATRA, used to compliment the academic institutes such as The Nottingham Trent University-United Kingdom, Manchester Metropolitan University-United Kingdom, Fashion Institute of Technology-United States, Royal Melbourne Institute of Technology-Australia, Institute of Textile and Clothing-HK in nurturing the research and manpower.

However, with migration the investment on R&D by developed countries and machine manufacturers also reduced. Today CISMA exhibition in China and StitchWorld magazine from India are the de facto standards of the manufacturing industry. While the automation initiative in developed countries is restricted to technical textiles, automation initiatives in Asia are more of low-cost improvisation of existing automats. The decade old books such as Apparel Manufacturing Handbook by Jacob Solinger, The Technology of Clothing Manufacture by Harold Carr and Barbara Latham, Apparel Manufacturing Sewn Product Analysis by Ruth E. Glock and Grace I. Kunz and Clothing Technology: from Fiber to Fashion edited by Roland Kilgus are inadequate for providing the latest information of sewing automation.

The books such as Carr and Latham's Technology of Clothing Manufacture by David J. Tyler, Garment Manufacturing Technology edited by Rajkishore Nayak and Rajiv Padhye, and e-catalogs from machine manufacturers are the new sources of latest information on sewing automation, however, due to lack of printed catalog there is dearth of documented information. Some of the important brands that manufacturers sewing automats and e-journals that provide related information are given below.

Brands supplying sewing & preparatory automats

- Atlanta attachment Company, United States (www.atlatt.com)
- Automatex, Sweden (http://www.automatex.com)
- Beisler, Germany (http://www.beisler-sewing.com)
- Brother, Japan (http://www.brother.com/as_oc/ism/)
- Duerkopp Adler, Germany (https://www.duerkopp-adler.com/en/index.html)
- JAM, Italy (http://www.jaminternational.it)
- Juki, Japan (http://www.juki.co.jp/industrial_e/index_e.html)
- Kinna Automatic, Sweden (http://www.kinnaautomatic.com/index.php?page=products)
- KSL, Germany (http://www.ksl-lorsch.de/en/company/)
- MAICA, Italy (http://www.maicaitalia.com/company-profile/)
- MPT, United Kingdom (http://www.mptgroup.com/products)
- Pegasus, Japan (https://www.pegasus.co.jp/en/)
- Pfaff, Germany (https://www.pfaff-industrial.com/de/)
- Rimac, Italy (http://www.rimac.it)
- Rimoldi, Italy (http://www.rimoldiecf.com/it)
- Sipami, Italy (http://www.sipami.com)
- Sunstar, South Korea (http://www.sunstar.co.kr/en/product/sm/)
- Texpa, Germany (http://www.texpa.de/en/)
- Typical, China (http://english.chinatypical.com/Company.asp)
- Vetron, Germany (http://www.vetron-europe.com/en/products/sewing-machines/)
- ViBeMac, Italy (www.vibemac.com/en)
- Yamato (www.yamato-sewing.com)
- YUHO (http://www.yuhomac.com)

Apart from above off-the-shelf brands for sewing automats there are specialized companies that makes customized sewing automats, such as

- Henderson sewing (https://www.hendersonsewing.com),
- DEMA Sewing (http://www.demasewingautomation.com/)
- Global Technical Network Inc. (http://www.gtninc.net/products/products.htm)
- Sewn Products Equipment Company (www.sewnproducts.com)

Media and associations

- www.apparelresources.com
- www.tc2.com
- www.SPESA.com
- www.texprocess.com

References

Anjanappa, M., Datta, K., Song, T., 2002. The Mechatronics Handbook. s.l. CRC Press.
Anon., n.d.-a. Wikipedia. [Online] Available at: http://en.wikipedia.org/wiki/Pneumatics.
Anon., n.d.-b. Noria. [Online] Available at: www.noria.com/dictionary/default.asp.
Anon., n.d.-c. Princeton University. [Online] Available at: http://wordnet.princeton.edu/perl/webwn.
Anon., n.d.-d. Wikipedia 2017 [Online] Available at: https://en.wikipedia.org/wiki/Laser.
Beasock, R.J., Hamid, H.M.N., Clapp, T.G., 1993. Self-Adjusting Fabric Ply Picking Device USA, Patent No. US5324016 A.
Blessing, H., 1977. Facing Ply Separator s.l. Patent No. US4143871 A.
Bobbin Show International, 1995. Bobbin Show International Daily Review. Bobbin Show International, Atlanta, USA.
Brother Industries, 2015. Brother S 7300-A Brochure. s.l. Brother Industries.
Brother International, 2000. Brother IMB 2000 Show Catalogue. Cologne: s.n.
Carr, H., Latham, B., 1999. The Technology of Clothing Manufacture. Blackwell Science, Oxford.
Casten, M., March 13, 1987. Uncle Sam wants a stitch in time in case of military mobilization. Daily News Record.
Casals, A., 1989. Sensor Devices and Systems for Robotics, first ed. Springer-Verlag, Berlin, Heidelberg.
Duerkopp Adler, 1997. Duerkopp IMB 1997 Show Catalogue. Cologne: s.n.
Frye, R., Jana, P., February 2017. Vision sewing: a technology with a future. StitchWorld 24–26.
Glock, R.E., Kunz, G.I., 2000. Apparel Manufacturing Sewn Product Analysis. Prentice Hall, New Jersey.
Groover, M.P., 1987. Automation, Production Systems, and Computer-Aided Manufacturing, second ed. Prentice-Hall, New Jersey.
Hagino, S., Ando, T., 1990. Apparatus for Detecting the End of Cloth-Overlap on a Sewing Machine Germany, Patent No. US4967676 A.
Hughes, F.H., Morton, K.O., LeMere, R., Brown, F.A., 1976. Method and Apparatus for Handling, Positioning and Assembling Fabric Plies USA, Patent No. US4176832 A.
Henderson, F., Jana, P., May 2017. Innovation in these disruptive times. StitchWorld 22–25.
Jana, P., 1999. India: On the Path of Improvisation: Shortcutting the Technology Barrier. The Textile Institute, Chennai.
Jana, P., May 2003a. Automation in sewing: how does it work? And why it didn't? StitchWorld.
Jana, P., 2003b. Trends in Apparel Manufacturing Technology: A Plateau or IT Take-over? Just-Style.com.
Jana, P., May 2004. Networking of sewing machines: beginning of a New Era. StitchWorld 18–20.
Jana, P., May 2014. Automation in sewing room: pocket attaching in shirt. StitchWorld 38–41.
Jana, P., 2015. Sewing equipment and work aids. s.l. In: Nayak, R., Padhye, R. (Eds.), Garment Manufacturing Technology. Woodhead Publishing.
Jana, P., April 2017. Digital feed in sewing machine: a fad or a useful feature…? StitchWorld 28–30.
Jet Sew Technologies, 1998. Bobbin World Press Release. Jet Sew Technologies, Inc., New York.
Juki Corporation, 1999. Juki JIAM 1999 Catalogue. Tokyo: s.n.
Juki Corporation, 2016. DDL-9000C Catalogue. s.l. Juki Corporation.
Juki International, 2002. Juki Corporation JIAM 2002 Show Catalogue. Osaka: s.n.
Kamal, N., 1987. Fabric Picker and Separator USA, Patent No. US4838536 A.

Kilgus, R., 1996. Clothing Technology: From Fibre to Fashion. Verlag Europa-Lehrmittel, Hann-Gruiten.
Liberatore, M.J., 1996. Encyclopedia of Operation Research and Management Science. Kluwer Academic Publishers, Dordrecht, Netherlands.
McPherson, E.M., 1987. Manufacturing Technology for Apparel Automation: Phase II and III Activity. North Carolina State University, North Carolina.
Mittal, R.K., Nagrath, I.J., 2003. Robotics and Control. Tata McGraw-Hill Education, New Delhi.
Noval, O., Dolezal, I., 2002. The Mechatronics Handbook. s.l. CRC Press.
Odrey, N.G., 1992. Maynard's Industrial Engineering Handbook. McGraw Hill, New York.
Pegasus, 1996. Pegasus JIAM 1996 Show Catalogue. Osaka: s.n.
Pfaff Industrial, 2006. Pfaff IMB 2006 Catalogue. Pfaff Industrial, Cologne.
Pfaff Industrial, 2013. Pfaff CISMA 2013 Show Catalogue. Pfaff Industrial, Shanghai.
Rosenberg, J.M., 1983. Dictionary of Business and Management, Wiley, New York.
Solinger, J., 1988. Apparel Manufacturing Handbook, second ed. Bobbin Media Corp., Columbia.
Totterdil, P., 1995. s.l.: s.n.
Tyler, D.J., 2000. Carr and Latham's Technology of Clothing Manufacture, third ed. Wiley. s.l.
Typical Corporation, 2015. Vetron Catalogue 2015. s.l.: s.n.

Further reading

Anon., December 2012. Sewing Automats for Multi Product Multi Process Manufacturing. StitchWorld.
Brother, I.L., 2000. Industrial Sewing Machine Handbook. s.l. Brother Industries Limited.
Coats, 2003. The Technology of Thread and Seams. J&P Coats Limited, Glasgow.
Coats Plc., n.d. [Online] Available at: http://www.coatsindustrial.com/en/information-hub/apparel-expertise/thread-numbering.
Cooklin, G., 2006. Introduction to Clothing Manufacture, second ed. Blackwell Science, Oxford.
Groz Beckert, n.d. Groz Beckert. [Online] Available at: http://www.groz-beckert.com/.
Jana, P., Khan, N.A., 2014. The sewability of lightweight fabrics using X-feed. International Journal of Fashion Design, Technology and Education 7 (2), 133–142.
Juki Corporation, 1988. Basic Knowledge of Sewing. Juki Corporation, Japan.
Pfaff, 1983. Technical Bulletin. s.l. Pfaff Industrial.
StitchWorld, June 2009. Flexible automation from ViBeMac. StitchWorld 40–43.
Stitchworld, February 2012. X-feed from typical: the ultimate feed system. StitchWorld 32–33.

3D body scanning

10

Hein A.M. Daanen[1], *Agnes Psikuta*[2]
[1]Vrije Universiteit Amsterdam, Amsterdam, The Netherlands; [2]Empa - Swiss Federal Laboratories for Materials Science and Technology, St. Gallen, Switzerland

10.1 Introduction

An important aspect of automation in garment manufacturing is related to the emerging 3D whole body scanners to the market. 3D body scanners make a digital copy of the outside of the human body. This digital copy or body scan can be combined with clothing patterns to provide made-to-measure garments. The individually adapted digitized clothing patterns can be further processed, for instance, by automated grading and cutting.

This 3D body scanning technology may prove to be useful in reducing the number of returns in clothing web shops. There is currently no good alternative for physically fitting garments in a retail store. Most web shops ask for the garment size, but the customer often has no cue on which the size fits best, in particular because of vanity sizing (a marketing tool used by several clothing manufacturers to adjust the size indication in the garment to the desire of the customer). Several companies offer to send several clothing sizes to the customer so that they can be fitted at home, but this is an expensive and unsustainable method. Uploading the 3D scan data of the customer over the internet to the manufacturer or retailer may provide the opportunity of virtual fitting to determine the best fitting size or even to make made-to-measure garments. These options are currently investigated in the Dutch national project "fitting garments over the internet."

This chapter is dedicated to 3D body scanning, focused on the clothing industry. The first section describes how manual measurements are used to determine the body dimensions essential for clothing design. The second section is dedicated to 3D scanning devices. The third section focuses on the 3D body scan and processing techniques. The fourth section deals with virtual fitting of garments. Finally, this chapter also discusses standardization initiatives related to 3D body scanning.

10.2 Body dimensions and garment sizing

10.2.1 Manual measurements

The starting point of a fitting garment is the correct knowledge of body dimensions. How to measure the human body dimension unambiguously for clothing is described in ISO 8559-1 (2017). This standard describes in detail what posture the subject should be in when measured, what tools should be used, and how the measurement should be done. This standard is limited to garments only. For technical design (for

Automation in Garment Manufacturing. http://dx.doi.org/10.1016/B978-0-08-101211-6.00010-0

instance, car interiors) other body dimensions are needed than for clothing design. How to measure these dimensions is described in ISO 7250-1 (2008). It is recommended to follow these standards because only unambiguous measurements enable a good comparison between populations and allow for a good link between the body and garment dimensions.

10.2.2 Link between body dimensions and garment dimensions

ISO 8559-2 makes the link between body dimensions and garments (8559-2, 2017). For overcoats, for instance, the links are made as shown in Table 10.1.

ISO 8559-2 provides a comprehensive list of garments and suggestions of primary and secondary body dimensions. If we take the overcoat example, we can plot the secondary dimension against the first dimension for a certain population to learn about the scatter and distribution of the dimensions. Fig. 10.1, for instance, shows the

Table 10.1 Essential body dimensions for overcoats according to ISO 8559-2

Men		Women		Boys		Girls	
PD[a]	SD[b]	PD[a]	SD[b]	PD[a]	SD[b]	PD[a]	SD[b]
Chest girth	Height	Bust girth	Height Waist girth Hip girth	Height	Chest girth	Height	Bust girth

[a] *PD*, primary dimension.
[b] *SD*, secondary dimension.

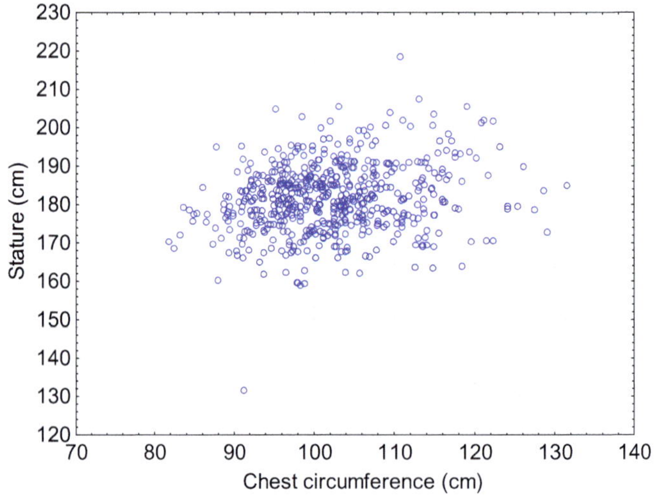

Figure 10.1 Chest circumference (cm) and stature (cm) determined according to ISO 8559-1 for 562 Dutch males.

relation between chest circumference and stature for 562 Dutch males measured in 2000 in the CAESAR (Computerized American and European Surface Anthropometry Resource) (Robinette and Daanen, 2006). This population is representative for males in the Netherlands.

Based on this information a sizing system can be made. If it is decided that the overcoat comes in steps of 10 cm chest circumference a Small can be made for chest circumferences 90–100 cm, Medium for 100–110 cm, and Large for 110–120 cm. The 5% males with chest circumferences <90 cm have to wear a Small; the 3% males with chest circumferences >120 cm cannot find a fitting overcoat.

The tall subjects will have a coat that may just cover the hips, whereas the short subjects may have a coat far under the knee. Therefore, it can be considered to have two lengths. This doubles the amount of garment sizes to 6 but increases the choice by the customers and probably the satisfaction rate.

The challenge is always to have a minimal number of clothing sizes for logistic reasons and a maximal coverage of the population. In the example given above, it is assumed that chest circumference and to a lesser extent the body length determines the garment size. However, it may well be that the customer has a protruding belly and that his waist circumference becomes the determining factor.

When using simple linear measures of the body, simplification of sizing is always occurring. When, however, we use the shape of the body, the fit of the garment may become more sophisticated. Therefore, let us have a look at 3D scanning.

10.3 3D body scanners

One of the first 3D scanners that were used to copy the human body was made in Loughborough (Jones et al., 1989). This was a shadow scanner that could capture the outline of the human body but not the concavities. A few years later, in the early nineties, the company Cyberware discovered the need for 3D human copies in the Hollywood movie industry. Dressers were complaining about difficult access to the actors to test the fit of clothing. The Cyberware scanner enabled a fast copy in a few seconds that could be milled out for clothing fit. Simultaneously, efforts regarding 3D scanning were undertaken in Japan by Hamamatsu, Germany by Vitronic and France by Telmat. A comprehensive review of early systems is available (Daanen and Van De Water, 1998) as well as a more recent update (Daanen and Ter Haar, 2013). Figs. 10.2 and 10.3 show a 3D scanner (Vitronic) and an image of a 3D body scan seen from three different sides (Human Solutions). The horizontal plane is shown as well as some reference points.

10.3.1 Body scan techniques and main applications

The first system on the market was a shadow scanner with obvious limitations in covering concavities such as the umbilicus. Thereafter, most systems became laser-based. A horizontal line was projected on the body that moved from top to toe in about 20 s. A camera under a certain angle captured the image and trigonometry was used to

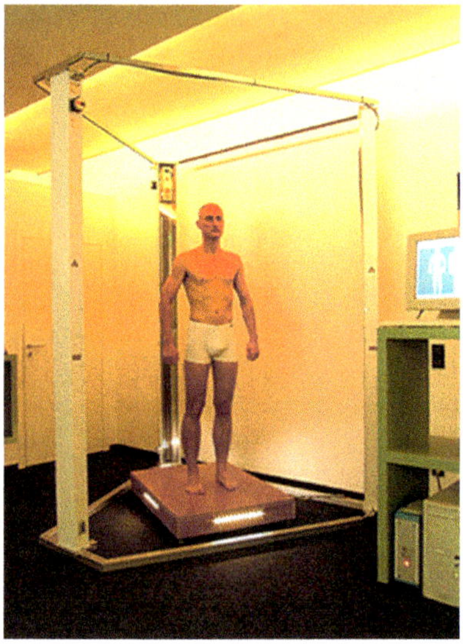

Figure 10.2 3D body scanner.

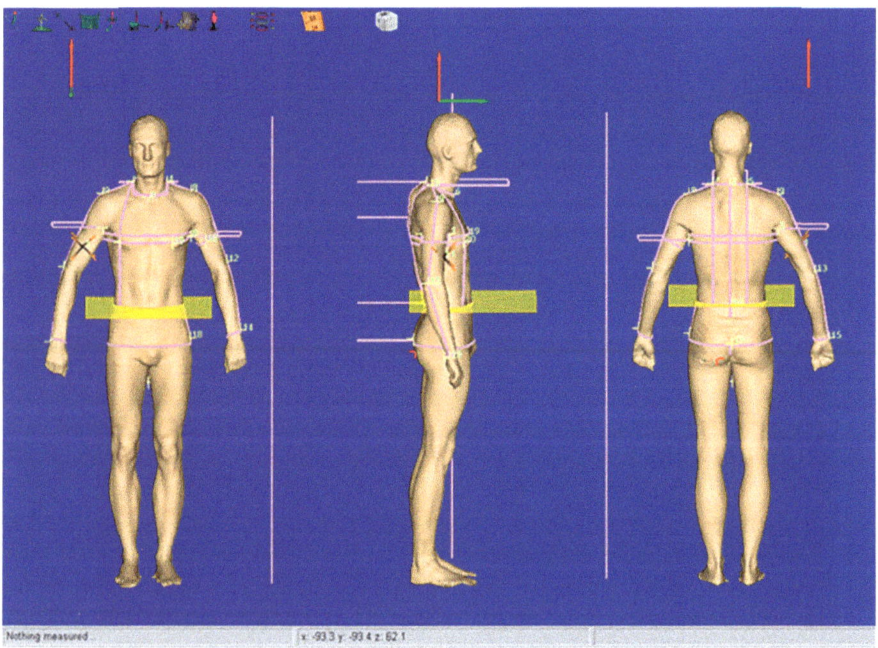

Figure 10.3 Scan seen from three sides with some body dimensions.

calculate the distance of the object to the camera. Typical systems using laser scanning were Vitronic, Cyberware, and Hamamatsu. Laser-based scanners are accurate and reliable, and that is probably why those systems were preferred in expensive 3D scanning surveys such as CAESAR (Robinette and Daanen, 2006) and national surveys in Korea, Japan, and Germany.

Less-expensive systems came to the market that used pattern projection on the body, generally with less precision, but with lower costs (Daanen and Van De Water, 1998). After the development of the Kinect systems for game computers, several companies started using this system in their body scanners, such as TC^2 and Size Stream. Every year a conference is held where the newest scanning systems are shown and their applications are discussed (please visit www.3Dbodyscanning.org for more information).

The main applications for 3D body scanning are discussed in an old but good study (Jones and Rioux, 1997). Quantification of body shape anomalies has many medical applications, for instance, to detect abnormal sizes of the liver, reconstruction surgery, prosthetics, and implant development. The automotive industry uses 3D scanning as an input for digital models to design the car around the vision cones and reach envelopes of the user population. The clothing industry is an important market to optimize garment fit, and more recently the fitness market started using body scans to quantify changes due to physical training (e.g., www.styku.com, www.dexafit.com).

10.3.2 New developments

The newest developments concern the rapid increase of extremely fast stereophotographic scanning systems. Examples can be seen at www.th3rd.nl with some nice examples for the fashion industry. The company 3dMD (www.3dmd.com) focuses on temporal 3D scanning systems. Scan systems up to 60 times per second with high precision are available. These high-speed 3D scanners enable motion capture and may offer an alternative for traditional motion capture systems such as Vicon.

Also, software systems become available that generates 3D models from photographs. When an object is photographed from different angles with some overlap, the software can recalculate this to a 3D image. A low-cost example of such a software system is Agisoft PhotoScan (www.agisoft.com).

Several companies offer apps on a smartphone or tablet that make photos from which body dimensions can be deducted. Institute Biomechanics of Valencia produced an app called Kidsize (www.kidsizesoluition.com) that asks for a frontal and sagittal picture as input, as well as information on age and stature. Stature is used for scaling. The app is developed in an EU project. The profiles are linked to a 3D scan database of 800 children. The tool is thoroughly validated, and a link to garment size is made using 1100 fit tests with children. The fit assessment is done by independent experts and parents (Pierola et al., 2017). Another company that follows this line and employs the same technology is QuantaCorp (www.quantacorp.com). The first draft was evaluated by Vonk and Daanen (2015). A new version is currently under evaluation using the iSense 3D scanning system connected to an iPad (https://www.3dsystems.com/shop/support/isense/videos). 3D about me (www.3dabout.me) uses an app to scan the feet and supply the fitting size of the shoe. They have a database for major brands on the conversion of foot dimensions to shoe dimensions and currently also work for the Dutch military. Metail (www.metail.com) offers virtual fitting on a copy of your body visualized on a smartphone or tablet. They offer the service to photograph the garments and then the customer can see it on her or his body.

10.4 3D body scan

10.4.1 3D scan accuracy

The basic output of a 3D body scanner is a point cloud. A high-resolution scanner produces a more dense point cloud than a low-resolution scanner. High resolution is typically needed if the fingers have to be accurately represented, for instance, for individualized glove design. The quality of a 3D body scan(ner) can be assessed from the resulting scans of calibration objects. A nice overview of different systems is available (Kouchi et al., 2012).

3D whole body scanners available on the market provide scanning accuracy of up to 5 mm according to the manufacturers, e.g., VITUS XXL of up to 1 mm, TC^2 of up to 3 mm, or Size Stream 3D body scanner of up to 5 mm. In multiscanner systems (e.g., 3dMD scanners) or hand-held scanners (e.g., Artec Group scanners) the reported accuracy may be much better down to fraction of a millimeter. Nevertheless, these numbers refer to the geometry size, fitting the small view area of such scanners. The actual accuracy of the final 3D scan of the body consisting of the multiple single frames that have to be brought into a common coordinate system and aligned is generally much better. The quality of such spatial alignment strongly depends on the abundance of the geometric features of the scanned object. Several studies dealt with developing consistent evaluation methods for quality of measurements conducted with 3D scanners (Gonzalez-Jorge et al., 2013; Kouchi et al., 2012; Qiang and Wei, 2009; Qingguo et al., 2014; Robinson et al., 2012).

The proposed use of regular solids with easy-to-measure absolute dimensions (such as spheres, cylinders, cubes or a combination of these corresponding to the total size, and form of the objects anticipated for scanning) for comparison with the dimensions measured through 3D scanning may lead to the underestimation of the accuracy for the frame capturing–based scanners. This problem is avoided by scanners with fixed coordinate system, such as stationary 3D body scanners, where one or more scanning heads or scanners are placed in a common coordinate system defined through calibration (e.g., through simultaneous scanning of board with known geometric pattern) (Stančić et al., 2013) and fixed distance between scanning units (e.g., scaffold supported scanning units). Consequently, the quality of the complete 3D scan in case of stationary scanner depends on the accuracy of the system elements (single cameras) rather than on the mathematical algorithms interacting with the heterogeneity of the object as in case of the frame capturing–based scanners. This fact suggests a need for distinct evaluation strategies of scanners using different capturing principles to enable an unbiased comparison of 3D scanner performance (Psikuta et al., 2015).

Furthermore, the accuracy can be compromised when scanning dark or reflective surfaces or in adverse lighting conditions leading to higher scattering of point cloud. Incomplete surfaces and scanning artifacts are also often the case especially when scanning complex geometries such as a human body. In narrow spaces such as armpits and crouch the body parts shade each other against scanner beam, and consequently, prevent some surface area to be scanned. On one hand this phenomenon can be prevented by setting a posture with legs and arms spread wider than in a standing straight posture, which should minimize body part shading. On the other hand, in many scanners the associated software can deal to certain extent with the remaining surface deficiencies by assuming surface reconstruction based on flat or curved filling. This subsequent postprocessing of 3D scans may finally lead to loss of scan accuracy especially in the areas relevant for chest and biceps girths, and inner leg and crouch length measurements. All these inaccuracies combined with artifacts of breathing and idle movements during scanning process may finally lead to an overall 3D scan inaccuracy of 5–20mm depending on scanner type.

10.4.2 Derivation of traditional body dimensions from 3D scans

For clothing design and evaluation, software is often supplied with the 3D scanner that calculates traditional 1D body dimensions from the 3D scans, such as chest and hip circumference. It has to be realized that these scan-derived dimensions are not equal to manually derived body dimensions (see Section 10.2.1). An example of a comparison between manual- and scan-derived measures for hip circumference is supplied in Fig. 10.4. The anthropometrics were experienced.

Manual determination of hip circumference is difficult because it is hard to establish if the *maximum* circumference is attained. For 3D scanners this is easier because vertical slices can be evaluated step by step. Therefore, scan-derived hip circumferences generally exceed manual determinations. The reproducibility of scan-derived parameters is generally better than manual measures (Robinette and Daanen, 2006) and despite 3D scan inaccuracy described in the previous section. The software systems that generate 1D body dimensions are generally 3D scanner bound (e.g., Size Stream and Human Solutions/Vitronic). The software typically calculates over 100 body dimensions, some including assessed body and segment volumes. Although manual measurements are specified in standards, such as ISO 8859, this is not the case for computer-derived body dimensions. ISO 20685 specifies the accuracy requirements for 1D scan-derived body dimensions regarding 3D body scanning systems (http://www.iso.org/iso/catalogue_detail.htm?csnumber=54909).

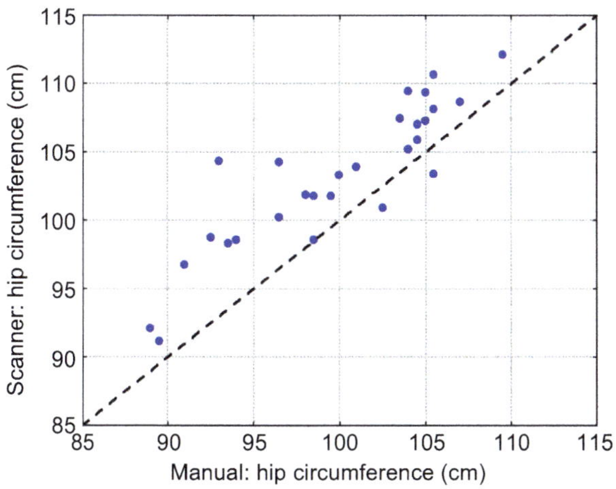

Figure 10.4 Relation between manual- and scanner-derived hip circumference.

10.4.3 Shape analysis

Reducing the 3D scan to 1D-derived body dimensions is not using the full potential of 3D body scanning. Processing the data in 3D will give more and more detailed information on the body.

The point clouds can be converted to a 3D digital copy of a human with color and shading effects, often called an avatar. The points are connected to triangles (also called a mesh), and this mesh is often reduced to the most essential triangles to cover the information of the shape (Fig. 10.5). This means the nose, for instance, has more triangles to cover the shape, than the forehead. These avatars are often used in software for virtual fitting (see Section 10.5.1).

The shape of the human head or body can be quantified using principal component analysis (PCA). When a database of 3D body scans is available, a human model can be made and adjusted to each scan (Allen et al., 2003). Generally, the first principal component is related to stature, the second to weight/body mass (Fig. 10.6), and the third is related to relative arm/leg length. However, also posture components may emerge from the PCA, for instance, the amount of lordosis in the back.

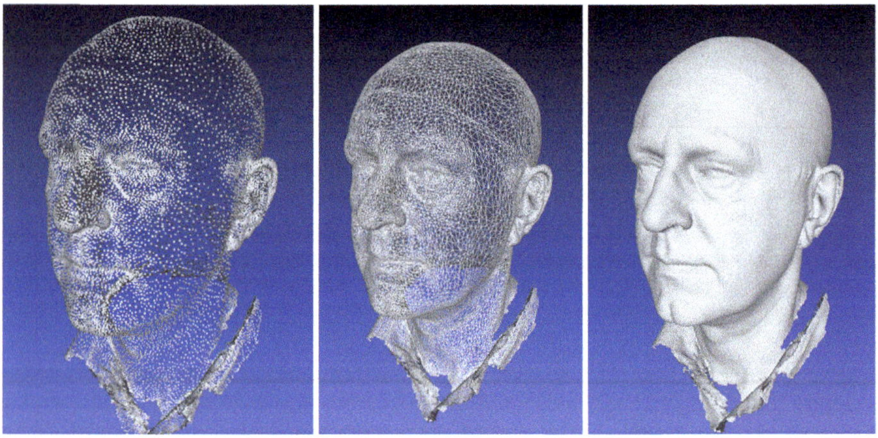

Figure 10.5 Example of a point cloud (left), triangulated mesh (middle), and Gouraud shaded (right) head scan.

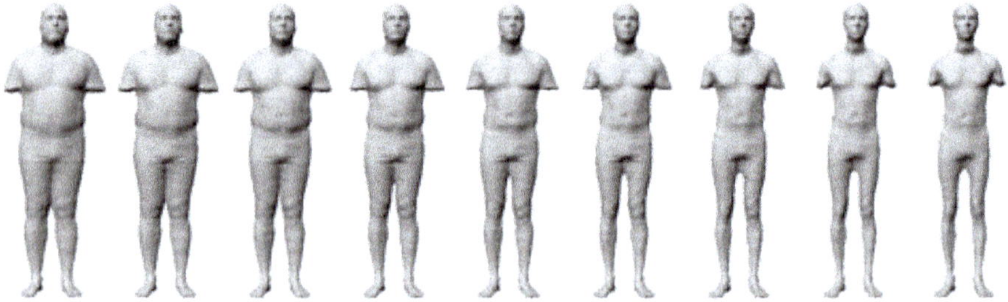

Figure 10.6 Example of model changes for a principal component related to body mass (Azouz et al., 2006).

3D anthropometric databases are getting increasingly common; however, the raw data are not easy to get because of privacy issues. In the Netherlands, the armed forces started to scan each soldier in 2004 so that a database of over 18,000 soldiers is available. Issues on cleaning and repairing databases are published (Haar et al., 2013).

10.5 Virtual fit of garments

10.5.1 Software systems available

In the late 1980s when computer-generated films and video games strived for better performance and higher quality, virtual clothing became a crucial issue. Nowadays, computer-aided design programs are indispensable for the clothing industry, where 3D simulation and animation tools speed up the development process and time to market for apparel products at lower resources and labor needs (Power, 2013). Virtual-try-on clothing models have become very advanced through integration of pattern design software and complex mechanical characteristics of fabric. These models were developed for the fashion industry with aesthetics of fabric pattern and draping behavior as well as assessment of fit as a priority.

The intricate approaches to simulate and animate clothing started in the nineties with several new techniques, which addressed the cloth mechanical properties and the interaction with the environment, such as Lagrange equations of motions and elastic surface energy (Fontana et al., 2004; Volino et al., 2005), mesh-to-mesh collision detection and response including several layers (Magnenat-Thalmann and Volino, 2005; Volino et al., 2005; Volino and Magnenat-Thalmann, 2005), and the mass-spring modeling to speed up the simulation of complex clothing (Meng et al., 2010) to name some of them. All these techniques require numerical methods and high computation capacity. The greatest advantage of this kind of software is twofold: (1) the realistic body shape either as dimensioned avatar or direct 3D scan can be used for garment fitting, and (2) in some software packages, the pattern design changes in 2D construction model are updated almost simultaneously in the 3D simulated virtual stitched garment (Power, 2013). Moreover, to simplify the construction process for complex body shapes, an advanced method to construct 3D garments on a virtual human body was developed at Technical University of Dresden where the pattern construction flattens automatically in 2D (Hlaing et al., 2011).

The developed techniques to simulate garments and integrating these in computer-aided pattern design process unfortunately do not guarantee accuracy of the simulation. The first validations of virtual cloth draping ever made were only qualitative by comparing the photographs of the real garments and pictures of simulated ones (Wu et al., 2011). Ernst et al. (2012) approved the simulation quality of complex garments by comparing several simulation software providers, such as Lectra, Gerber, and OptiTex to judge the realistic look and the fit of garments. The quantitative evaluation of the 3D try-on software was attempted for the first time by Mert (2016), who compared the results from the Fashionizer (MIRALab, University of Geneva, Switzerland) and actual garments scanned on the shop window manikin as reference

Table 10.2 Selected suppliers of body scanning software for virtual design

Main supplier	Web address	Country
Optitex	www.optitex.com	Israel
Gerber	www.gerbertechnology.com	USA
Lectra	www.lectra.com	France
Gemini	www.geminicad.com	Romania
DCSuite	http://www.physan.net/eng/DCsuite/product_qual.asp	Korea
Clo3D	www.clo3d.com	Korea
Tukatech	www.tukatech.com	USA
Assyst	http://www.human-solutions.com/group/front_content.php?idcat=214&lang=2	Germany
Fashionizer	www.miralab.ch	Switzerland

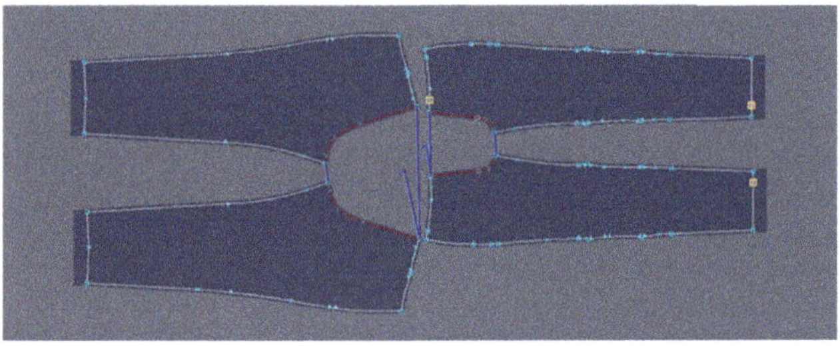

Figure 10.7 Digital pattern of a pair of trousers.

using the distance between the body surface and the garment as a quantitative measure of 3D simulation quality. It is recommended that other 3D try-on software should undergo such a validation procedure including variety of garment designs and fabric properties.

In a recent study, the neoprene properties of the triathlon suit worn by the 2016 Olympic medalists were modified in the DCSuite system and investigated for fit effects (Vedder and Daanen, 2015). The impact is considerable and enables the optimization of pressure on the skin, but again validation of the models is necessary.

The main suppliers of software that allows for virtual design of garments on a 3D body are given in Table 10.2. Most systems started out as a software package for the design of garments and later included 3D scans for virtual fitting. Fig. 10.7 shows an example of a digital pattern of trousers.

10.5.2 Virtual fit

A typical virtual fit simulation process includes usually several steps. After importing or creating the 3D representation of the body for which the clothing fitting should be done (Fig. 10.8(a)), the pattern design is prepared in 2D space or in another dedicated software and imported to the virtual 3D try-on software (Fig. 10.8(b)). The

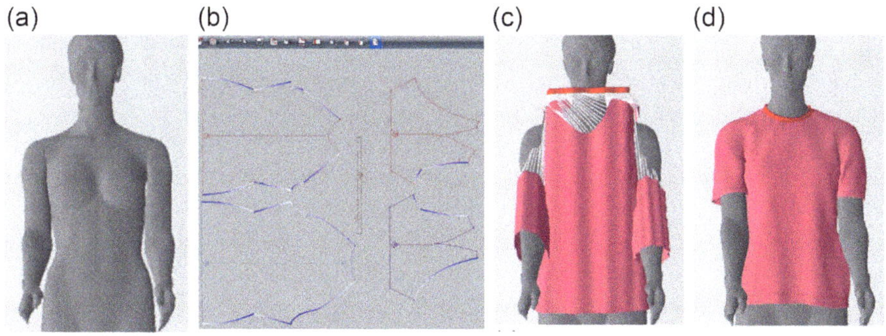

Figure 10.8 Typical simulation process including subsequent steps, such as importing a scan or avatar (a) and clothing pattern (b), define the stitches, virtual stitching (c), and the simulation result (d).
Example from Vidya software by Assyst, Germany.

Figure 10.9 Example of a virtual fit (www.tukatech.com). Left side: complete view; Middle: visualization of tightness of fit; Right side: transparent view.

corresponding stitching edges are then assigned and the pattern pieces are placed in 3D space in their roughly approximate position (Fig. 10.8(c)). Finally, the garment pieces are stitched, collisions between the body and the garment computed in an iterative cycle, and final simulation result of the draped garment displayed (Fig. 10.8(d)).

Many providers offer additional features allowing quick assessment of the garment fit on the model. They include distance between the body and the garment (e.g., Lectra, Assyst), volume of the garment (e.g., Optitex), pressure exerted on the body if the garment is tight, and tension within the fabric if it is pulled between body apexes or along its girth (e.g., Gerber, Optitex, CLO3D). An example of a virtual fit is supplied in Fig. 10.9.

Until now body dimensions needed for clothing pattern development are measured in a standing straight body posture regardless of the type of garment being designed. However, some functional garments are used and required to provide wearing comfort and supportive or protective function in another specific body posture for most of the time, for example, biking or crouching during firefighting. Scientist from MIRALab (University of Geneva), Geneva School of Art and Design and Artanim (Geneva, Switzerland) used Fashionizer and some firmware animation tools to record and simulate the realistic movement of the body and simultaneously visualize the tension distribution and its change during movement on 2D pattern. This method has a potential to adjust 2D clothing pattern to minimize the tension and improve the wearing comfort and protection in a given body posture or movement. Nonetheless, only few software systems offer animated avatars among their tools with very limited choice of movement (e.g., walking in CLO3D).

10.6 International standardization activities

10.6.1 International Standardization Organization

In the International Standardization Organization (ISO) one group is active in the field of anthropometry (ISO/TC 159/SC 3/WG 1 Anthropometry) including 3D anthropometry and one group on clothing sizing systems (TC 133: Clothing sizing systems—size designation, size measurement methods, and digital fittings).

The anthropometry group is extending the ISO 7250 document on basic human body measurements for technological design with a new Part 3 that contains a worldwide database of human body dimensions. This is important for the design of clothing and equipment. Other important documents from this group are ISO/DIS 20685-1 Ergonomics—3D scanning methodologies for internationally compatible anthropometric databases—Part 1: Evaluation protocol for body dimensions extracted from 3D body scans and ISO/WD 20685-2 3D scanning methodologies for internationally compatible anthropometric databases—Part 2: Evaluation protocol of surface shape and repeatability of relative landmark positions. The documents are available from ISO in Geneva.

The clothing sizing group TC133 has the following documents under direct responsibility:

- ISO 5971:2017. Size designation of clothes—Tights
- ISO 8559-1:2017 Size designation of clothes—Part 1: Anthropometric definitions for body measurement
- ISO 8559-2:2017 Size designation of clothes—Part 2: Primary and secondary dimension indicators
- ISO 18163:2016 Clothing—Digital fittings—Vocabulary and terminology used for the virtual garment
- ISO 18825-1:2016 Clothing—Digital fittings—Part 1: Vocabulary and terminology used for the virtual human body

- ISO 18825-2:2016 Clothing—Digital fittings—Part 2: Vocabulary and terminology used for attributes of the virtual human body
- ISO 18831:2016 Clothing—Digital fittings—Attributes of virtual garments

More information is available at: https://www.iso.org/committee/52374.html.

10.6.2 The North Atlantic Treaty Organization

Proper fit of military clothing and equipment is important for the health, safety, and operational performance of the North Atlantic Treaty Organization (NATO) soldier. Secular trends in body dimensions, new operational demands, and integration of new devices in the clothing system aggravate the issue of clothing fit. There is a considerable difference between the investigated NATO countries in approach and methodology of clothing fit, which lead to difficulties in international cooperation. It is the expectation that the NATO forces will benefit from the new techniques that lead to improved garment fit. Also, interoperability may be enhanced if knowledge is shared regarding methodology to optimize fit within NATO countries. Therefore, NATO approved a Research and Technology Group (RTG) in the medical and human factors (HFM) panel on the topic of 3D body scanning for clothing fit and logistics.

The objectives of this HFM RTG are to investigate the need and requirements regarding:

- The identification of research gaps and opportunities in the area of 3D body scanning for clothing fit and logistics;
- The facilitation of integrated research initiatives in 3D whole body scanning and processing;
- The facilitation of integrated procedures for clothing deployment to ensure similar fit over NATO countries;
- A review of novel approaches for 3D scanning and processing.

More information is available at https://www.cso.nato.int/activity_meta.asp?act=7482. In February 2019 a publically available report will summarize the findings.

10.6.3 Institute of Electrical and Electronics Engineers

Institute of Electrical and Electronics Engineers (IEEE) has started a 3D body processing technology working group to draft new standards on 3D body scanning and data processing. The proposed standard addresses the fundamental attributes that contribute to 3D body processing quality of experiences, as well as identifying and analyzing existing metrics and other useful information relating to these attributes. It defines a standardized suite of objective and subjective methods, tools, and frameworks for assessing 3D body processing quality of experience attributes, and it specifies methods, tools, and frameworks to facilitate standards-based interoperability, communication, security, and comparison among 3D body processing technologies such as 3D/depth sensors, scanners, digitization, simulation and modeling, analytics, and animation/visualization for solution providers as well as for consumer-facing companies such as in retail, health/wellness, sports/athletics, and medical industries. More information is available at: https://standards.ieee.org/develop/wg/3DBP.html.

10.7 Conclusion

3D whole body scanning enables the generation of an accurate digital copy of the outside of a human being for increasingly lower costs. The 3D scans can be imported into dedicated software that allows for the determination of virtual fit of garments. The face validity is good, but there is a need for more detailed validation of the models that are employed, in particular the simulation of material properties. The increased use of 3D scanners leads to large databases of different user populations. These databases can be employed to optimize sizing systems of garments, so that a maximum part of the user population can find a fitting garment that comes in a minimal number of garment sizes.

Most 3D scanners are used in clothing settings to derive 1D body dimensions, such as hip circumference, so that manual measures become obsolete. However, it is good to realize that manual measures differ systematically from scan-derived measures and that the real potential of 3D scanner lies in the processing of human shapes. PCA offers a nice technique to evaluate the differences between humans in a certain population. Also the principal components can be used in the design process.

Several organizations (ISO, NATO, IEEE) have started expert groups to establish standards on 3D body scanning and scan processing. These groups generally consist of a mixture of clothing scientists, IT experts, and anthropometrics. A realistic scenario for future clothing shopping may be that a customer uploads his/her 3D scan to the web shop of the retailer and gets a 3D view of the selected garment on his body. Tools may be available to change the design or materials according to customers' desire. When satisfied, the customer may proceed to the payment section and receive a made-to-measure or selected garment at home.

References

7250-1, I, 2008. ISO 7250–7251: Basic Human Body Measurements for Technological Design – Part 1: Body Measurement Definitions and Landmarks. ISO, Geneva.

8559-1, I, 2017. ISO 8559–1: Size Designation of Clothes – Part 1: Anthropometric Definitions for Body Measurement. ISO, Geneva.

8559-2, I, 2017. ISO 8559–2: Size Designation of Clothes – Part 2: Primary and Secondary Dimension Indicators. ISO, Geneva.

Allen, B., Curless, B., Popović, Z., 2003. The space of human body shapes: reconstruction and parameterization from range scans. In: Paper Presented at the ACM SIGGRAPH 2003 Papers, SIGGRAPH '03.

Azouz, Z.B., Rioux, M., Shu, C., Lepage, R., 2006. Characterizing human shape variation using 3D anthropometric data. Visual Computer 22 (5), 302–314. http://dx.doi.org/10.1007/s00371-006-0006-6.

Daanen, H.A.M., Van De Water, G.J., 1998. Whole body scanners. Displays 19 (3), 111–120.

Daanen, H.A.M., Ter Haar, F.B., 2013. 3D whole body scanners revisited. Displays 34 (4), 270–275. http://dx.doi.org/10.1016/j.displa.2013.08.011.

Ernst, M., Detering-Koll, U., Güntzel, D., 2012. Investigation on Body Shaping Garments Using 3D-body Scanning Technology and 3D-simulation Tools. Lugano.

Fontana, M., Rizzi, C., Cugini, U., 2004. 3D virtual apparel design for industrial applications. Computer-Aided Design 37, 609–622.

Gonzalez-Jorge, H., Riveiro, B., Vazquez-Fernandez, E., Martínez-Sánchez, J., Arias, P., 2013. Metrological evaluation of Microsoft Kinect and Asus Xtion sensors. Measurement. 46 (6), 1800–1806. http://dx.doi.org/10.1016/j.measurement.2013.01.011.

Hlaing, E., Krzywinski, S., Roedel, H., 2011. Development of 3D Virtual Models and 3D Construction Methods for Garments. Lugano.

Jones, P.R.M., Rioux, M., 1997. Three-dimensional surface anthropometry: applications to the human body. Optics and Lasers in Engineering 28 (2), 89–117.

Jones, P.R.M., West, G.M., Harris, D.H., Read, J.B., 1989. The loughborough anthropometric shadow scanner (LASS). Endeavor 13 (4), 162–168.

Kouchi, M., Mochimaru, M., Bradtmiller, B., Daanen, H., Li, P., Nacher, B., Nam, Y., 2012. A protocol for evaluating the accuracy of 3D body scanners. Work 41 (Suppl. 1), 4010–4017. http://dx.doi.org/10.3233/WOR-2012-0064-4010.

Magnenat-Thalmann, N., Volino, P., 2005. From early draping to haute couture models: 20 years of research. Visual Computer 21, 506–519.

Meng, Y., Mok, P., Jin, X., 2010. Interactive virtual try-on clothing design systems. Computer-Aided Design 42, 310–321.

Mert, E., 2016. Effect of Air Gap Thickness and Contact Area on Heat Transfer through Garments in Real Life Situation (Ph.D. thesis). University of High Alsace, Mulhouse, France.

Pierola, A., Epifanio, I., Alemany, S., 2017. Child t-shirt size data set from 3D body scanner anthropometric measurements and a questionnaire. Data in Brief 11, 311–315. http://dx.doi.org/10.1016/j.dib.2017.02.025.

Power, J., 2013. Fabric objective measurements for commercial 3D virtual garment simulation. Clothing Science and Technology 25 (6), 423–439.

Psikuta, A., Frackiewicz-Kaczmarek, J., Mert, E., Bueno, M.A., Rossi, R.M., 2015. Validation of a novel 3D scanning method for determination of the air gap in clothing. Measurement 67, 61–70. http://dx.doi.org/10.1016/j.measurement.2015.02.024.

Qiang, Z., Wei, W., 2009. Calibration of laser scanning system based on a 2D ball plate. Measurement. 42 (6), 963–968. http://dx.doi.org/10.1016/j.measurement.2009.02.004.

Qingguo, T., Yujie, Y., Xiangyu, Z., Baozhen, G., 2014. An experimental evaluation method for the performance of a laser line scanning system with multiple sensors. Optics and Lasers in Engineering 52, 241–249. http://dx.doi.org/10.1016/j.optlaseng.2013.06.002.

Robinette, K.M., Daanen, H.A.M., 2006. Precision of the CAESAR scan-extracted measurements. Applied Ergonomics 37 (3), 259–265. http://dx.doi.org/10.1016/j.apergo.2005.07.009.

Robinson, A., Mccarthy, M., Brown, S., Evenden, A., Zou, L., 2012. Improving the quality of measurements through the implementation of customised reference artefacts. In: Paper Presented at the 3rd International Conference on 3D Body Scanning Technologies, Lugano, Switzerland.

Stančić, I., Musić, J., Zanchi, V., 2013. Improved structured light 3D scanner with application to anthropometric parameter estimation. Measurement. 46 (1), 716–726. http://dx.doi.org/10.1016/j.measurement.2012.09.010.

Ter Haar, F.B., Reulink, H.G.B., Daanen, H.A.M., 2013. 3D scanning of Dutch military – secular trends in PCA for 18,000 soldiers. In: 4th International Conference on 3D Body Scanning Technologies, Long Beach CA, USA, 19–20 November 2013.

Vedder, K.J., Daanen, H.A.M., 2015. Virtual fitting of a triathlon swim suit. In: Proceedings of Digital Fashion Conference 2015. Digital Fashion Society, Seoul.

Volino, P., Magnenat-Thalmann, N., 2005. Accurate garment prototyping and simulation. Computer-Aided Design and Applications 2 (5), 645–654.

Volino, P., Cordier, F., Magnenat-Thalmann, N., 2005. From early virtual garment simulation to interactive fashion design. Computer-Aided Design 37, 593–608.

Vonk, T.E., Daanen, H.A.M., 2015. Validity and repeatability of the sizestream 3D scanner and Poikos modeling system. In: 6th International Conference on 3D Body Scanning Technologies, Lugano, Switzerland, 27–28 October 2015. http://dx.doi.org/10.15221/15.293.

Wu, Y., Mok, P., Kwok, Y., Fan, J., Xin, J., 2011. An Investigation on the Validity of 3D Clothing Simulation for Garment Fit Evaluation. Venice.

Computer-aided design—garment designing and patternmaking

11

Yamini Jhanji
Technological Institute of Textile and Sciences, Bhiwani, India

11.1 Role of computers in textile and apparel industry

The advancement of technology over the last few decades has led to massive application of computers. This ranges from simple computer applications at homes, schools, and small scale setups to sophisticated state-of-the-art computers with technological upgradation in large scale setups, retail outlets, and industries, where it is used for manufacturing management applications such as inventory control, scheduling, ordering, sales analysis, and other information purposes. Linking production segments with order, sales, and inventory information through electronic data interchange systems are advantageous and important in a competitive industrial climate.

Textiles and computers are believed to be genetically related with each serving as bread and butter for the other. The relationship between the two can be backed to the times when computers are believed to be from punch card system used in jacquard weaving machine. Textile industry is no exception from other competitive fast growing industries where computerization has taken over majority of the cumbersome, slow-paced, time-consuming, labor-intensive, and less-productive manual operations (Nayak and Padhye, 2015a,b). Textile units that excel in today's competitive business environment are well versed and open to adapt the new technologies offered in the form of different software packages. The role of IT evolution after computer revolution has been rightly understood by leading textile cocorporates in domestic and international markets.

Apparel industry does not remain untouched and unaffected by the revolution offered by the digital world and computer technology. Computers are generously being used in apparel, accessory, and home textiles industries for varied operations. The apparel industry has witnessed tremendous improvement and upgradation in terms of equipment utilization, automation, precision, productivity, and computerized operations. Accordingly, majority of the fashion manufacturing processes, especially those used for sampling and design development have been computerized. Many software are handy for fashion designers to perform various tasks namely fashion research, fashion design and illustration, pattern design, patternmaking, textile design, garment construction, production management, marketing, and sales (Wang et al., 2005; Istook and Hwang, 2001; Nayak et al., 2015; Yan and Fiorito, 2002).

Automation in Garment Manufacturing. http://dx.doi.org/10.1016/B978-0-08-101211-6.00011-2

11.2 Introduction to computer-aided design

Computer-aided design (CAD) is the utilization of computer technology in real or virtual designing. The design of geometric models for object shapes is often called computer-aided geometric design. The output of CAD often must convey symbolic information such as materials, processes, dimensions, and tolerances according to application-specific conventions just like in manual drafting of technical and engineering drawings. CAD system is dedicated to design, analysis, and manufacturing in various industrial contexts. Analysis component is referred as computer-aided engineering (CAE), CAM stands for computer-aided manufacturing and the entire process of CAD, CAE, and CAM as computer-integrated manufacturing (CIM). Robotics and Internet communication along with CAD and CAM constitutes CIM technology (Zoran, 1995; Parthibaan and Mahalingaam; McAllister, 1983). However, the acronym CAD/CAM is often used to refer to all the three functions: CAD, CAM, and CAE. These technologies are supplemented by some methods of artificial intelligence, solid modeling, and feature-based design (Wang et al., 2005; Nayak et al., 2016).

CIM is referred to an open system as this system facilitates the direct transfer of production data to commercial electronic data processing system, which can further be processed (Collier and Collier, 1990a,b). CIM in the apparel industry refers to the integration of manufacturing processes at different stages with technological intervention of computers. The integration thereby helps to achieve multiple benefits such as easing the cumbersome processes at the optimum quality standards, simultaneously exploring new fabric qualities, innovative ideas, and fulfilling the consumers' requirements. The basic principle of CIM is the development of directions and techniques aimed at integration of technical, production, and marketing processes and procedures, the underlying basis of integration being the use of computer and communication technology.

CAD is a powerful and effective tool in the hands of engineers and designers catering to research, innovation, new product design and development (McCartney et al., 2000). CAD software can not only be utilized to increase the quality and productivity of design but also used for the design improvisation. The system also serves to improve communication through documentation and to create database for manufacturing. CAD has powerful and user-friendly tools, which aid creativity and visualization (Wang et al., 2003). CAD find its versatility in both two-dimensional (2D) and three-dimensional (3D) working environment by designing curves and figures in 2D spaces, surfaces, and solids in 3D objects.

CAD is an important state-of-the-art industrial tool extensively used in many design applications, such as mechanics, automotive, shipbuilding, aerospace, industrial and architectural design, textile, apparel, and fashion world (http://www.ehow.com/about_5344814_importance-cad-fashion-designing.html). The worldwide acceptability and adoption of CAD software package is attributed to benefits such as lower product development costs and a greatly shortened design cycle. CAD provides flexibility to designers to lay out and develop work on screen, obtain its physical replica by printing and saving it for future editing, and retrieval (Nayak et al., 2015). A complete CAD/CAM system used in apparel industry would comprise of the following:

- Graphics package for art design work
- Input hardware, digitizer, scanner
- Interactive hardware workstation
- Pattern design software (PDS) system
- Grading software
- Lay planning software
- Output plotter and printer
- Single, medium, and high ply computer-driven bulk cutters
- Production and management control software
- Computerized sewing and pressing equipment
- Overhead transport system

11.2.1 Benefits of computer-aided design in designing and garment manufacturing

CAD software facilitates the creation and visualization of designer's digital image, thereby saving time by limiting the need for tailoring and physical sample preparation.

The software allows the designers to visualize designed fabric in virtual environment using virtual models (Stylios et al., 1996), where the designers can experiment with different fabric choices, colors, textures, to foresee the drape and fit before actual garment construction (Fontana et al., 2005a,b).

Computer-generated sketches are more reliable and easy to interpret, minimizing any mistake encountered during garment construction. The use of computer software results in significant material and time saving, eliminating the need for preparation of sample prototypes and generation of virtual storyboards, mood boards to be showcased to buyers and retailers.

CAD software provides innumerous advantages in designing and cutting room operations streamlining the entire process with overall enhanced productivity, accuracy, and efficiency. Greater flexibility in pattern designing, grading and marker planning, reduction in waste percentage, increase in quality of cutting, and reduction in sample making time are some of the benefits of application of CAD/CAM systems in garment manufacturing (Luo and Yuen, 2005; Zoran, 1995).

A case study is reported to explain the practical benefits obtained in each manufacturing step in garment industry by switching from manual method to CAD/CAM technology. Various work processes such as time-consumption in pattern designing, grading and pattern alteration, waste percentage in cutting, and lead time were compared for two units—one with CAD/CAM systems and the other with manual operations. Time taken to complete different activities in these two units showed that the unit with CAD/CAM systems could perform patternmaking, grading, and marker planning in less time compared with the other where all operations were manually executed (Table 11.1). It was further observed that the waste percentage in cutting could be reduced with 10% material saving with the CAD/CAM systems as

Table 11.1 Time required to complete various patternmaking activities using computer-aided design (CAD)/computer-aided manufacturing (CAM) systems

Activity	Time (minutes) required to complete activity	
	In unit with CAD/CAM systems	In unit with manual operations
New pattern creation	15.22	100
Grading (four size)	5.67	120
Marker planning	5.92	100

CAM, computer-aided manufacturing.

compared to the manual operations. The unit with CAD/CAM systems showed production of 15,856 pieces/day, whereas the unit with manual system had much lower production of 4013 pieces/day. Furthermore, the lead time for the unit with manual method was higher (55 days) as against the unit with CAD/CAM systems (39 days). Table 11.1 shows the time required to complete various patternmaking activities using CAD.

11.2.2 Historical background of computer-aided design

In 1970, markers were first made on computers by Dallas team leading to foundation of Camsco. The manual marker making process was cumbersome and time-consuming with operators traveling backward and forward along long laying up tables placing card pattern pieces on paper and drawing round them. It was quite a difficult job because of different pieces and sizes involved gap between pieces and the target of higher marker efficiency was hard to achieve. The ability to computerize the pattern placement within a given area was not a difficult task. The real difficulty was that these pieces were odd shapes containing punch holes, notches, stripe lines, grain lines, etc., which had to be catered by the computer system. This problem was solved by electronically tracing the irregular shapes and related information and feeding this data into the computer—a function referred to as digitizing.

Another software system was developed by America-based team under the name of Hughes AM1 system, which was later sold to Gerber who are considered as the pioneer in this field, till date. A French company, Lectra came into existence in 1980 catering to the needs of garment industry as far as software intervention was concerned. Thereafter another company Microdynamics came up with PC-based system. Another development came with the inception of Assyst team who targeted at multitasking operations by means of PDS. The potential and versatility of CAD was initially explored by the media industry even before fashion designers began integrating the technology in computer drawings and designing. CAD, also known as computer-aided drafting, first hit the fashion scene in 1987 and has been gaining popularity

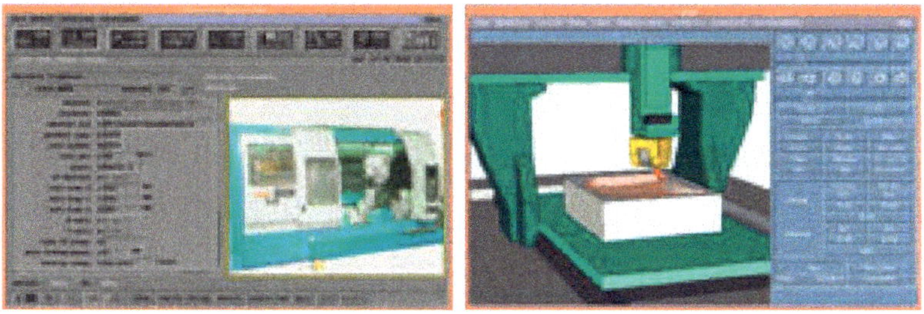

Figure 11.1 Computer-aided design/computer-aided manufacturing application in mechanical field—library of machine tools and graphical simulation.

ever since its inception in sketching, drawing, fashion illustration, textile, and apparel industry, designing of tools and machinery (Wang et al., 2005; Collier and Collier, 1990a,b; Luo and Yuen, 2005).

11.2.3 Application areas of computer-aided design

Application areas of CAD software is almost endless and span from routine work to the most complex, intricate ones: to create blueprints for a building, in aerospace, in industrial practices, and to create layouts for a fashion piece. CAD is often used in the drafting by architectures and contracting or technical designing by engineers. CAD/CAM techniques have become established in fields such as electrical, electronic, mechanical, fashion, textile, and apparel engineering (Collier and Collier, 1990a,b; Zoran, 1995; Yan and Fiorito, 2002).

The prime consideration in certain applications such as design of electrical circuits and printed circuit boards and the design and layout of manufacturing is to facilitate design creation and alteration by CAD function. There are still some applications that demand contribution of both CAD and CAM functions equally. The most notable examples are in the aerospace industry in which many products are designed on, analyzed by, and manufactured under the control of a linked CAD/CAM technology. The applications further cover all types of manufacturing operations, such as milling, turning, wire EDM, punching, etc.

Construction and architecture sector is another avenue for CAD systems. Applications range from simple building design, to large scale projects and interior designing (Figs. 11.1 and 11.2). CAD is also widely used in computer animation production for special effects in movies, advertising, and technical manuals. Research area also utilizes this versatile software in computational geometry, computer graphics (both hardware and software), and discrete differential geometry.

In addition to the above applications, the other area where promising and appreciable application of CAD technology has been found is the apparel or fashion industry. The use of CAD in the fashion industry has become more crucial and wide spread owing to the need to increase efficiency and make design process easier. CAD software are integrated at each step of traditional clothing production process and then interfaced with different kinds of machines to produce an excellent work of fashion

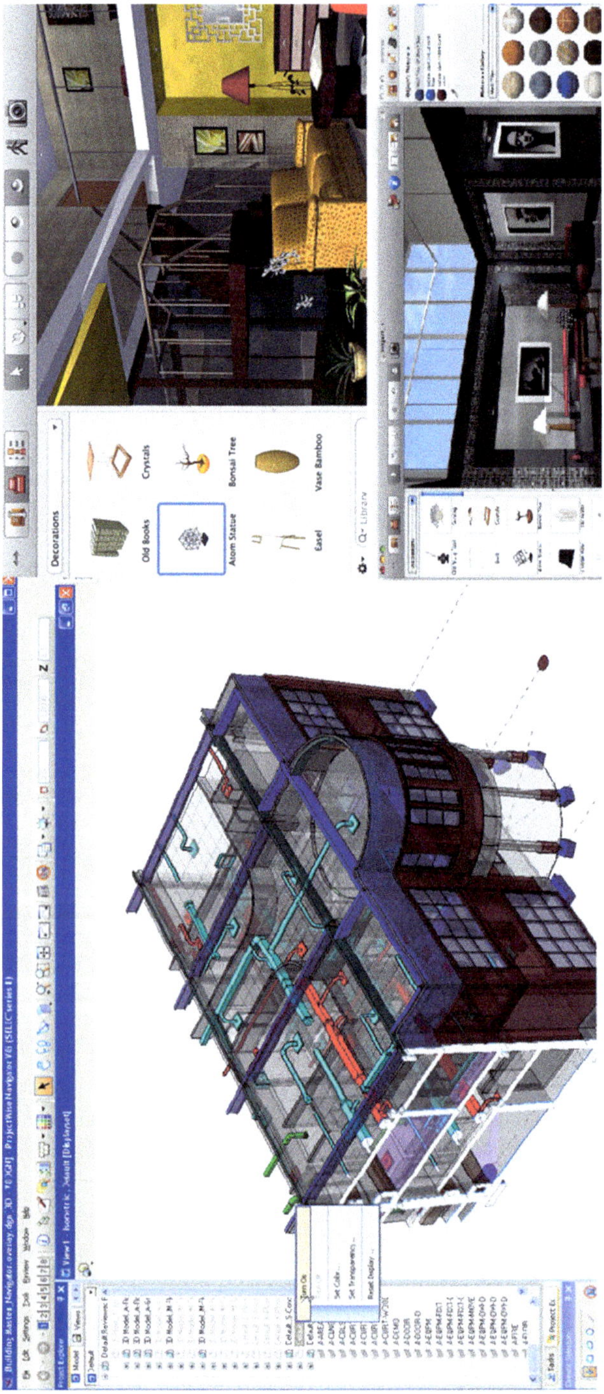

Figure 11.2 Various construction projects using computer-aided design technology.

art. The CAD application ranges from fashion design to garment manufacturing (http://www.tc2.com/fashion-design-cad-software.html).

CAD technology is also finding applications in woven and knitted fabric textile design, accessory, and home textile industry. All the operations during garment manufacturing can be made cost-effective, accurate and considerable reduction in lead time can be achieved with the intervention of CAD technology (Nayak and Padhye, 2015a,b; Gupta, 2005). The initial high cost of installation is offset by the numerous benefits the system offers in holistic manner at each step, streamlining the entire process. CAD system is not only used by designers for creating their sketches, and design illustrations, but also used for the promotion of the designed product by designing billboards, invitation cards for the fashion show, dress tags, carry bags, advertisements for print and electronic media, pamphlets, brochures, catalogs, and posters (http://www.communitycollegetransferstudents.com/computers-and-fashion/).

There are various CAD software containing many interactive tools, that can easily support designing such promotional items. The largest usage of CAD systems is for pattern design and development (Pattern Design System), (Pattern Generation System), pattern grading, and lay planning, where most of the savings are achieved. It becomes imperative today for all companies subcontracting a large supplier to have capability of handling digital patterns. Lay planning can save on material and it can justify the investment on CAD systems. Fabric spreading, pattern cutting, and sewing operations also witness wide spread use of different CAD software packages, which provide user-friendly and time-saving environment to operators. Computer controlled machines on the same lines as CNC machines used in metal manufacturing are operative in the apparel industry. These machines perform fabric laying and cutting automatically and can be directly controlled from the CAD systems. The latest innovation to revolutionize the digital world is the 3D visualization, which is the creation of 3D virtual models or avatars, 3D mapping, virtual shopping, and tailored and custom-made garment designing. The application of CAD in different garment manufacturing operations is shown in Figs. 11.3(a)–(f) and 11.4.

11.3 Different software used in designing and garment construction

The wide spread use of computers in the apparel and accessory industry has revolutionized the industry with a gamut of options as far as apparel designing, patternmaking/grading, fashion illustration, and accessories designing are concerned. Gerber, Lectra polygon, Apparel CAD, CADTERNS, CAD Fashion, Fashion CAD, Design concept 3D, Assyst Bullmer, Investronica, and APS-ethos embroidery software, AutoCAD, CorelDraw, Illustrator and Photoshop, Koledo and Optitex, U4ia, and color matters are some of the renowned software catering to all the major operations in apparel industry. AccuMark by Gerber Technology Design and merchandising system is a versatile tool

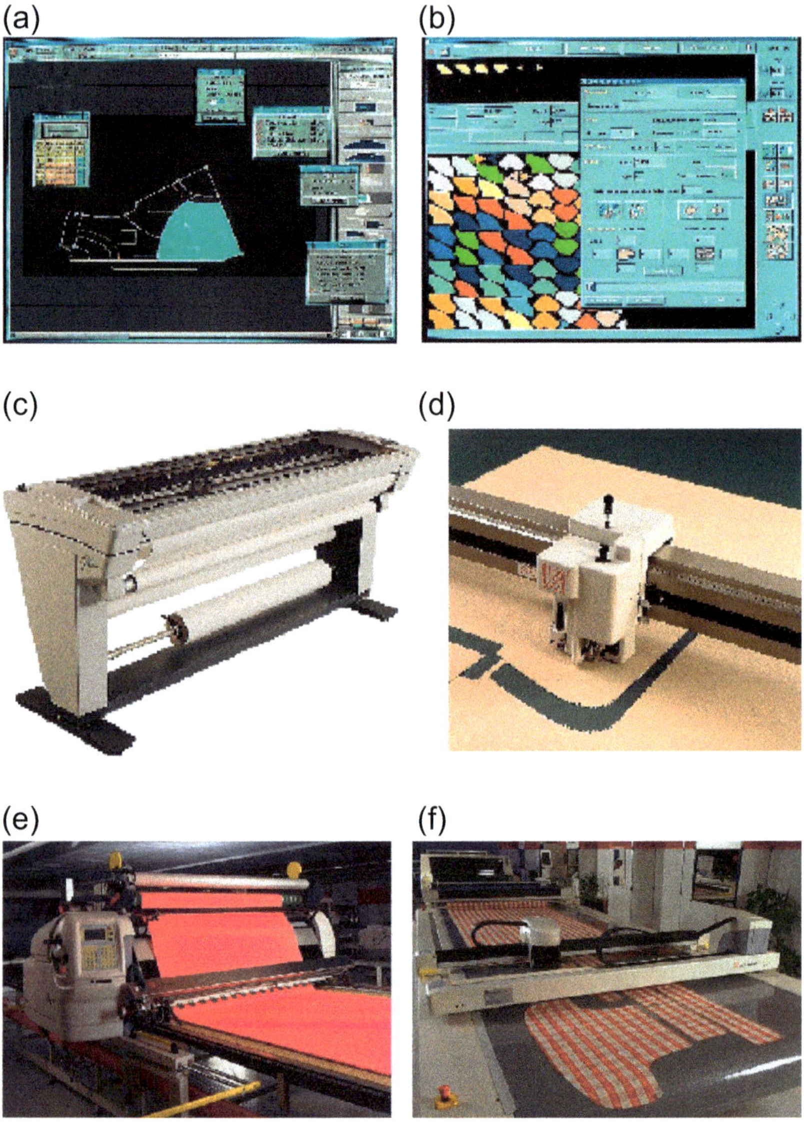

Figure 11.3 Application of computer-aided design in cutting room operations. (a) Pattern design, (b) lay planning, (c) cutter plotter for stencil design, (d) pattern cutting, (e) fabric laying in cutting machines, and (f) the cutting process.

for every aspect of the garment manufacturing process (http://www.gerbertechnology.com/fashion-apparel/development-pre-production/accuplan/). Apparel Innovator (Plural Technology) caters to the needs of apparel manufacturing and retailing organizations targeting integration of product development and other business processes. Aps-Ethos (CAD/CAM Technology) is flexible embroidery and cutting software, which can be tailored as per the individual needs of companies worldwide. BlueCherry

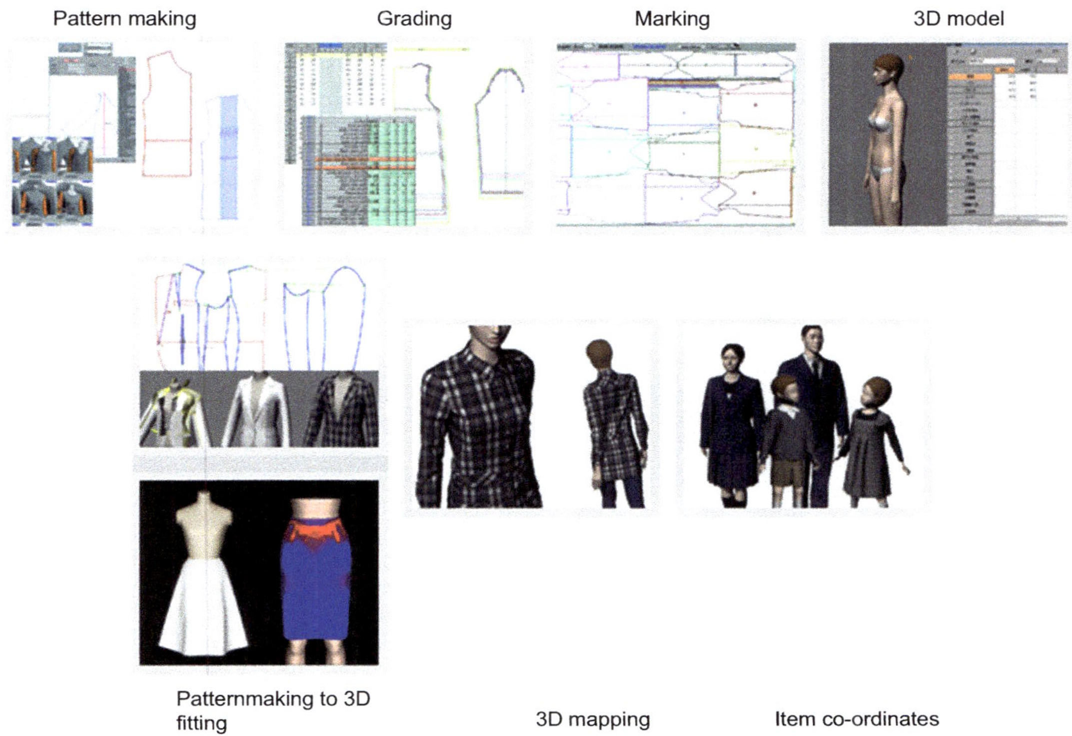

Figure 11.4 Application of computer-aided design in patternmaking, grading, and creating three-dimensional (3D) avatars for visual fit.

(computer-generated solutions) offers several salient features such as Enterprise Resource Planning (ERP), product life cycle management, Omnichannel Planning, Warehouse Management, Shop Floor Control, and eCommerce. A variety of software providers are thus providing complete solution for apparel and accessory industries right for the design initiation stage to the promotion and marketing phase (Nayak and Padhye, 2015a,b).

Three types of software packages are frequently used in garment industry:

1. *CAD-based software for the fashion industry*—Gerber, Lectra, and Optitex are some of the companies making expensive yet very specialized professional software based on CAD technology (http://fashion-incubator.com/3d_pattern_cad_software).
   C-DESIGN Fashion (C-DESIGN) allows the creation of ready-to-wear merchandise and enables quick and efficient release of technical files. Vetigraph is a complete CAD/CAM solution provider for apparel and fashion industry with facilities for grading patterns, plotting of created markers, and automatic cutting. Zedonk Software (F2iT) is an affordable apparel and production management software for apparel and accessory designers and manufacturers. Inventor (Autodesk Design) software is equipped with tools for producing, validating, and documenting complete digital prototypes. Apso (Tricycle Product development) is the software catering to home textiles. PolyNest (Polygon) software functions include digitizing, grading and pattern design (Luo and Yuen, 2005).
2. *Scaled down CAD software for home sewers or custom clothing makers*—the software is based on CAD technology but are made more affordable for domestic users by limiting the available options in the software. Wild Ginger Pattern Master, Fashion CAD, and Telestia Creator are some examples of scaled down CAD software catering well to the needs of small garment units, boutiques, and custom apparel makers. Measurements need to be input to the system to obtain the pattern block. Style libraries are often available in these programs for any stylization in the basic block. CAD-based programs are "Windows" based, therefore, for running on a "Mac" system, Windows installation on the system is a must.
3. *Vector graphics/drawing software*—any type of required pattern can be drawn with this software, which is quite user friendly and enables the user to work with a highly accurate ruler and pen. However, the software is not created specifically for pattern drafting. Vector graphics find their utility when design illustrations, sketches, and designed fabrics, are needed in both print and on the web. Vector graphics maintains clarity without being pixilated when subjected to scaling, reducing, rotation, or stretching. A variety of tools within a vector program speeds up the process over paper drafting. Adobe Illustrator, CorelDraw, Inkscape, Illustrator Draw, and a free iPad version of Illustrator are some of the versatile vector drawing software. Adobe Illustrator, CorelDraw or Freehand, and AutoCAD are Graphics (drawing) packages, whereas Adobe Photoshop is image editing program.

Automated Computer-Aided Design (AutoCAD)—is the first CAD system that provides a complete solution for drafting and designing. This package creates technical drawing where measurement precision is of prime concern. The package enables learning many other CAD packages because most concepts and commands introduced by AutoCAD are utilized by other systems.

CorelDraw— another vector graphic package has been developed and marketed by Canada-based software company, Corel Corporation. CorelDraw is the first software to combines vector graphics software and a photo editing program. Contrast and color balance can be easily adjusted by users with full range of editing tools available in the software. CorelDraw also facilitates handling of multiple master layers from within the main program. The software is one of the professional graphic design software with graphic design tools, intuitive layout designs, and search capabilities to find graphics. Application area of CorelDraw includes creating simple technical drawing, figures, designing textile fabric, embroidery, mood board, garments, and garment features using grids and guides.

Adobe Illustrator—is an excellent CAD application as far as detailed technical drawing and fashion illustrations is concerned. The software can be successfully used for text and vectors graphics editing made up of lines and curves defined by mathematical objects. Adobe Illustrator serves as a perfect format for creating detailed technical drawing and fashion and hence providing design documents to be retrieved for future referrals. Creation and storage of different silhouettes, basic garment shapes, accessories, and trims is achievable with the use of symbols and brush stroke libraries provided in the software.

Adobe Photoshop— developed and published by Adobe Systems is a pixel-based editing program. The software with its salient features of commercial image manipulation and image editing is considered to be current market leader worldwide. Photoshop is extensively used for editing and correcting digital photos and preparing images for mood/story boards, magazines, and posters. The list of functions performed by this software is long with the ability to create print designs for fabrics, home textiles, apparel designing, and digital design portfolio. Several selections, drawing, editing, and color change tools in the Photoshop enable the user to edit and modify the images as per the requirement.

Adobe offers different selection tools: marquee tool, lasso tool, and magic wand tool. The marquee tools helps in selection of rectangles, ellipses, rows, and columns. The lasso and polygonal lasso tools help in drawing both straight-edged and freehand segments of a selection border. The magnetic lasso tool is useful for quickly selecting objects with complex edges set against high-contrast backgrounds. The magic wand tool helps in selecting a consistently colored area (for example, a red flower) without having to trace its outline.

Color modification in Photoshop can be achieved by two ways: by permanent alteration of the pixels in the active layer and by the use of an adjustment layer. Adjustment layers enable experimentation with color and tonal adjustments without permanently modifying the pixels in the image. Dodge tool, burn tool, and sponge tool are the image shading tools in Photoshop. Dodge tool is used to lighten the image based on a traditional photographer's technique for regulating exposure on specific areas of a print. Burn tool darkens specific areas of the image by increasing the exposure to darken areas on a print (burning). Sponge tool is used to change the color saturation of an area and can be used to increase or decrease contrast by moving gray levels away from or toward the middle gray.

Background filling tools such as paint bucket and gradient are the coloring tools in Photoshop, which are used to color area underside the image. Paint bucket tool fills adjacent pixels that are similar in color value to the pixels clicked. The gradient tool creates a gradual blend between multiple colors. Various kinds of drawing

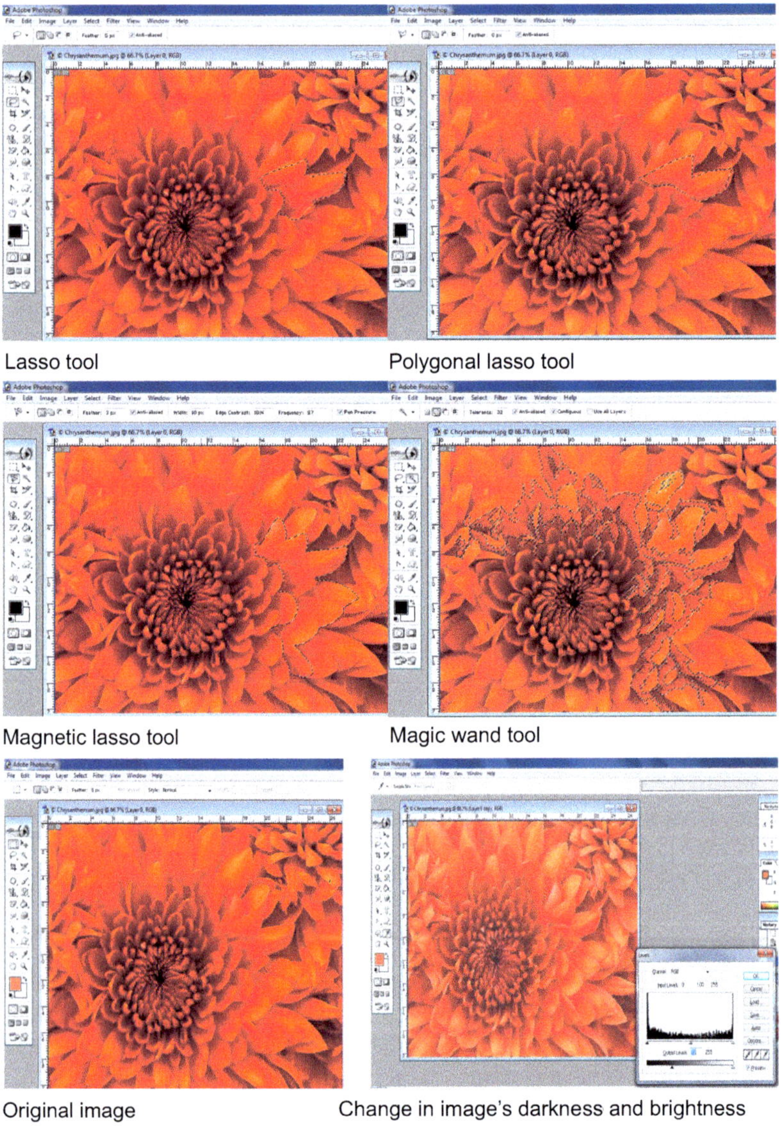

Figure 11.5 Different tools used in Adobe Photoshop.

tools in Photoshop are available, which help to create multiple numbers of images such as rectangle tool, rounded rectangle tool, ellipse tool, and line tool. The pen tool, freeform pen tool, polygon tool, custom shape tool, add anchor point tool, delete anchor point tool, and convert point tool are also available in Photoshop. Multiple shapes in a layer can be drawn and interaction of overlapping shapes can be specified. Shapes can be edited once they are drawn. Fig. 11.5 shows the different tools used in Adobe Photoshop for selection, drawing, color editing, etc.

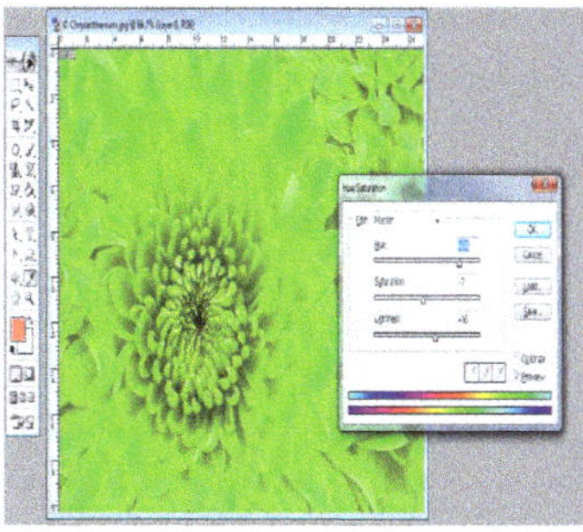

Change in image's hue, saturation and lightness

Color change of image

Paint bucket tool

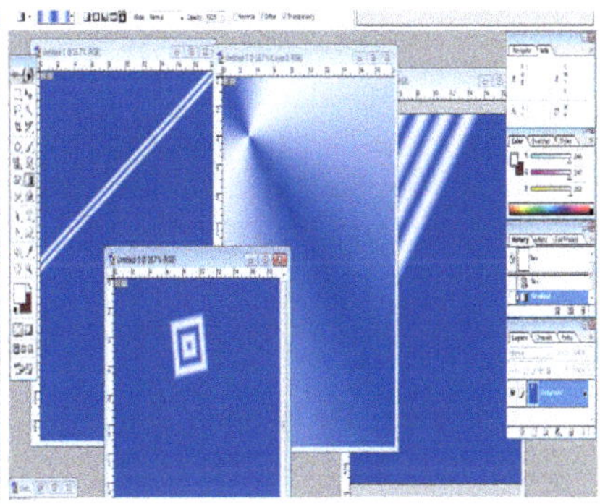

Gradient tool

Dodge, burn and sponge tool

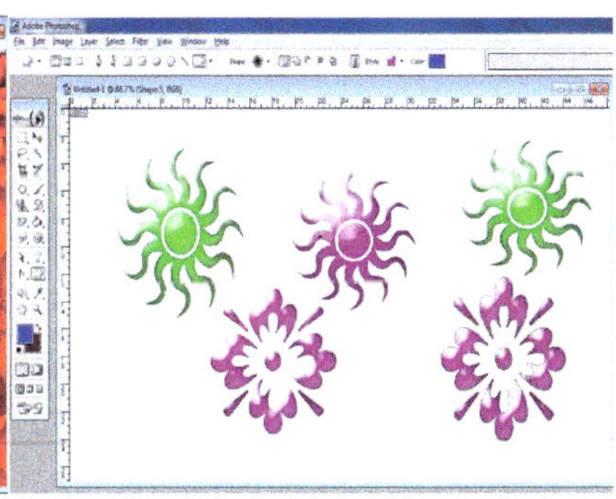

Custom shape/drawing tool

Figure 11.5 Continued.

11.4 Computer-aided design for fabric design

Woven and knit fabric designing for varied applications such as apparels, accessories, and home textiles can be successfully accomplished with CAD systems, which not only permit the user to visualize the pattern in various fabrics but also alter the appearance. Various fiber and yarn attributes; and weave and knit characteristics within a selected range, can be visualized in computer without producing physical trial samples. High resolution monitors can be used to view the design. Separate flat screen displays can thus be used to produce high quality photographs of the fabrics so obtained. Replicas of the displays can also be produced on plotters. Many of the available programs are inexpensive and run on microcomputers.

Several textile designing software serve as a crucial guide for designers improvising their capability and creativity. These software packages enable the designer to experiment with a number of textures, colors, and patterns to reach at the targeted designs along with the availability of sketch backgrounds in concept boards, tools for repeating patterns, texture mapping, and product rendering. Technical textile designing software come with inbuilt yarn (graphical and technical), fabric (woven and knitted), and motif designing options. Designing software nowadays are integrated with dobby and jacquard looms.

Wonderweaves, Reach technologies, Summagraphics, and Prostyle are a few companies providing textile designing software. Designing of woven fabric starts with correct selection of yarn construction parameters on the system. Designer can choose the color options for fibers, number of strands in the yarn and twisting them together without much hardship using the fabric design software. Next task is to weave the fabric from the chosen yarn parameters, after selecting the weave or to designing a stripe or a check pattern. The process has become rapid and accurate owing to the realistic image on the screen. Vast color options are also available that facilitates designers to view many different color schemes and effects in minutes. Mass production can thus be started once satisfactory design is obtained. The process is of course less time-consuming and exhaustive owing to the elimination of making full sample lengths to realize the design effect.

Design of knitwear garment patterns can also be done by variety of designing programs available these days. These programs ensure quick mass production as well as one-off supply where garments are customized as per individual consumer's choice by their direct linkage to automatic knitting machines. Software programs can also be coupled with graphic packages enabling the production of silk screens for printed textiles. Designer can thus simply peel off the layers of color printing sequence and plot them individually to be readily used as screens.

TUKA studio is the complete software suite, which caters to the needs of textile technologists as far as fabric, print, and textile design are concerned. It is an intuitively designed software system that allows the user to plan complex weaves, color schemes, and prints. The software has been designed for industrial garment making; however, simplified interface is equally beneficial for novice designer. TUKA studio comprises of seven different modules—color separations, repeats, weaves, jacquards, colorways, knits, and storyboards as discussed below.

- Color separations module is used to reduce colors and clean design quickly and efficiently.
- Repeats module creates repeat motifs with an array of tools for designing, creating repeats, color management, scale, and resolution.
- Weaves module creates plaids, dobby patterns, weave effects and enables the user to define and design the type of weave required using multiple colors.
- Jacquards module uses drawing tools to create designs from scratch and can even modify existing designs already processed through the Color Separations module.

- Colorways module creates hundreds of print colorways using standard codes or personal seasonal palettes. Custom colors can also be created with a single selection.
- Knits module defines knit dimensions, use multiple colors, and design according to the thickness of the knit fabric required.
- Storyboards module enables presentation development intended for multimedia digital presentations, professional designs, storyboard printouts, or export to the internet.

Pointcarre is a software used to create print artworks (Fig. 11.6). Customized palettes or the integrated Pantone textile library color picker enables the color of pattern to be changed quickly and easily. A variety of repeats such as straight, half drop, mirror, and with or without overlap can be created using the software (http://www.pointcarre.com/print.html).

Pointcarre knit allows any designer to draw on screen, on virtual graph paper, with square or rectangular grid, at the right scale, and in repeat. Design size can be modified to a specific number of rows and columns as per the knitting machine gauge. The knit structure can also be simulated and visualized on the screen (Fig. 11.7). A jacquard and structured stitch library is available in the software, which allows simulation of knitted fabrics in a realistic way. Stitches can be added and color variations can be

Figure 11.6 Repeat print art work created using Pointcarre software.

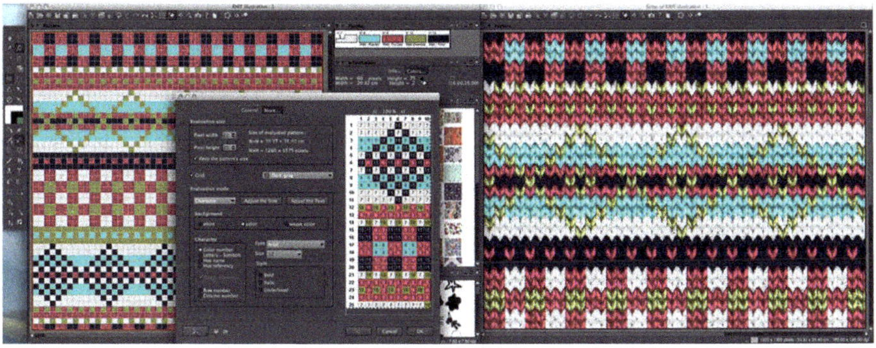

Figure 11.7 Knit fabric designing and stitch selection.

experienced by dropping the colors directly in the palette of designs (http://www.pointcarre.com/print.html. Weave Design software is also available for woven fabric designers to easily create checks and stripes or jacquard fabrics (Fig. 11.8).

Effect of newly designed cloth can be visualized by dressing a computer-generated human image with the software designed fabric. Color scanner can be used to scan a photograph of model dressed in type and style of garment best suited for designed fabric. Fashion designer can dress various types of garments using a gamut of materials and colors, which serves as a design guide for the next merchandise line in the upcoming seasons. A more realistic and effective process is solid 3D imagery where a model form or mannequin is digitized into the memory of 3D program. Garment can therefore be directly designed and draped on the mannequin on the computer monitor. The model on the screen is first seen as wire frame; however, it can later be filled to give a more realistic form and finally completed with shading. Newly designed fabrics can be used to create garments on this 3D form, which can be shown to buyers without having the need to produce physical sample lengths for designed fabrics (Fig. 11.9).

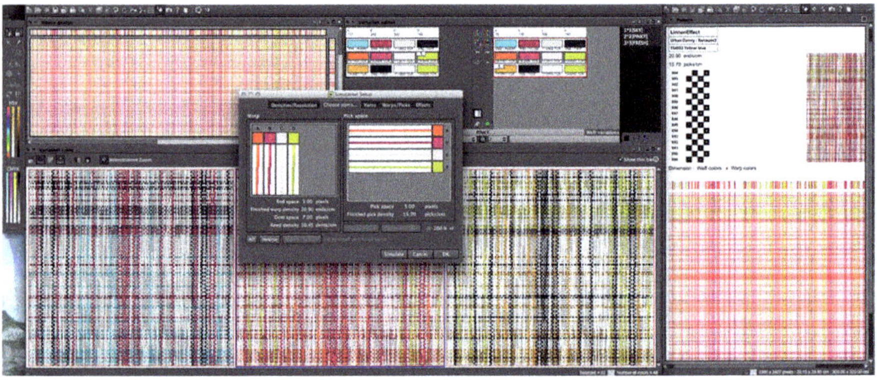

Figure 11.8 Woven fabric designing.

Figure 11.9 Visualization of designed fabric on computer-generated model.

11.5 Computer-aided design for apparel design

An area in which CAD has a significant impact is apparel design. 2D-CAD systems allow interactive manipulation of garment pieces faster and more easily than in the normal drafting process for patternmaking. Pattern pieces for a variety of designs, stored in the database, can be retrieved by the designer to create the desired styles. The grading of patterns has also been computerized for some time as this step is numerically based and therefore amenable to computer programming. Once the patterns are graded and nested, the pieces can be retrieved and quickly arranged in a marker layout by the computer, thus decreasing the time spent in these preproduction steps. Software programs now allow automatic placing of pattern pieces in a configuration to 3D-CAD systems are now being developed, which allow a user to design and view garments, make changes, and add visual effects. It has been very difficult to develop accurate 3D representations from the 2D databases inherent to 2D-CAD systems.

The 3D systems lack the ability to predict and represent garments as they would actually appear in a draped form, although they can present a 3D representation created to simulate draped garments. Use of this capability could dramatically shorten the response time between buyers and manufacturers because a photograph of the computer image can be available quickly without having to wait for physical samples. Accurate draping predictions as described earlier can be important in this process to minimize fabric use, as opposed to trial and error methods. CAD/CAM systems based on 2D databases are being used for 2D patternmaking, allowing flat pattern pieces to be designed, graded, placed for cutting, and the numeric code generated to drive cutting devices. The gap in this continuous step is the ability to link 3D original design with 2D patternmaking. The approaches to drape prediction and visualization described earlier could bridge that gap and allow a designer to create an artist's sketch of a garment, specify the fabric characteristics, visualize the draped appearance, alter the design if necessary, unfold the pattern, produce pattern markers, and electronically drive cutting devices.

Graphic art (GA) software programs are playing a crucial role in manufacturing process particularly for clothing, shoe, and embroidery designs. These programs are versatile in providing a wide range of drawing and painting options by making the use of electronic pen on graphic pad.

Computer Design Inc. (CDI), an American-based company, is a famous name as far as 3D software are concerned. The company is pioneer in softwares, which are capable of converting 3D garment to 2D patterns and realistic 3D solid graphics. The software providers are striving hard toward improvement in memory capacity and datas accessing speed.

Designing of woven or knitted fabric is followed by garment or accessory designing. GA software serve as a working aid for the pattern technologists to create a garment with accurate silhouette and design details as dictated by software in the sketch form and ensuring that their design creation is in close proximity with the software's recorded and devised sketch.

Embroidery software are also available to enhance and give new wings to designer's creativity. The designers can create and feed the designs so created into computer-driven embroidery sewing machines. GA software is complete package providing solution to garment, accessory, and embroidery designers as designers can readily make working sketches to cater the needs of production managers and retail buyers. The software programs are generally referred to as pattern design softwares owing to their capability of construction and designing of garment patterns once the design has received approval for sampling and showcasing to the buyers (Fontana et al., 2005b).

11.6 Computer-aided design for designing process

The exhaustive and nerve wrecking efforts of the designer to ultimately reach their dream creation has eased much in today's fashion world where CAD has become a major part of the design process. There are varied and wide applications of CAD/CAM in fashion designing. The idea or concept generation where designers' utilize their imagination to establish a design idea marks, which is the beginning of fashion designing process. This initial step is very crucial and time-consuming for the designer as a smart idea may lead to smart product. Inspiration for any innovative idea can be picked up by designers from any source such as print media, websites, nature, flora and fauna, street fashion, and so on. CAD serves as a helping hand for the designers in concept generation by facilitating drawing and illustrating each of their thought in software, which is much easier rather than conventional method of using paper. Thus, CAD plays a crucial role in enhancing the potential of the designer and bringing to light what was thought improbable.

The concept generation is followed by sketching, where the designer is expected to combine the knowledge of upcoming trends and creativity to achieve a unique design keeping the target consumer in mind. The perfect sketch as proposed by the designer can be easily achieved with the use of multiple tools available on the software menu to draw and erase, or do and redo, according to the designer's desire.

CAD fashion design system such as eTelestia, allows the designer to design sophisticated, appealing collections, and present the collection in the most professional way. The system provides user-friendly tools for designing a wide range of designs. Models of different proportions can be used as a background guide for drawing original designs (http://www.tc2.com/fashion-design-cad-software.html). Designers can easily save, store, and create their collection libraries in the software. A brand new collection can also be created by incorporating modifications to existing designs. The flexibility in selection of own color for particular styles (Fig. 11.10) or use of the available color libraries for each season is one of the many other advantages the system offers. Fashion design software is also provided with the Virtual Fitting (moulage) option (Fig. 11.10), which enables creation of model in any size measurements and can thus be used as a base to draw own designs. Technical drawings can easily and accurately be drawn and clear guidelines for the pattern development process within the system can fully be explored. The system provides clear communication basis to the fashion designer and the pattern technologist. A variety of fabric prints can be applied and fitted to the created designs. The software can be upgraded with fashion design libraries to get new innovative ideas

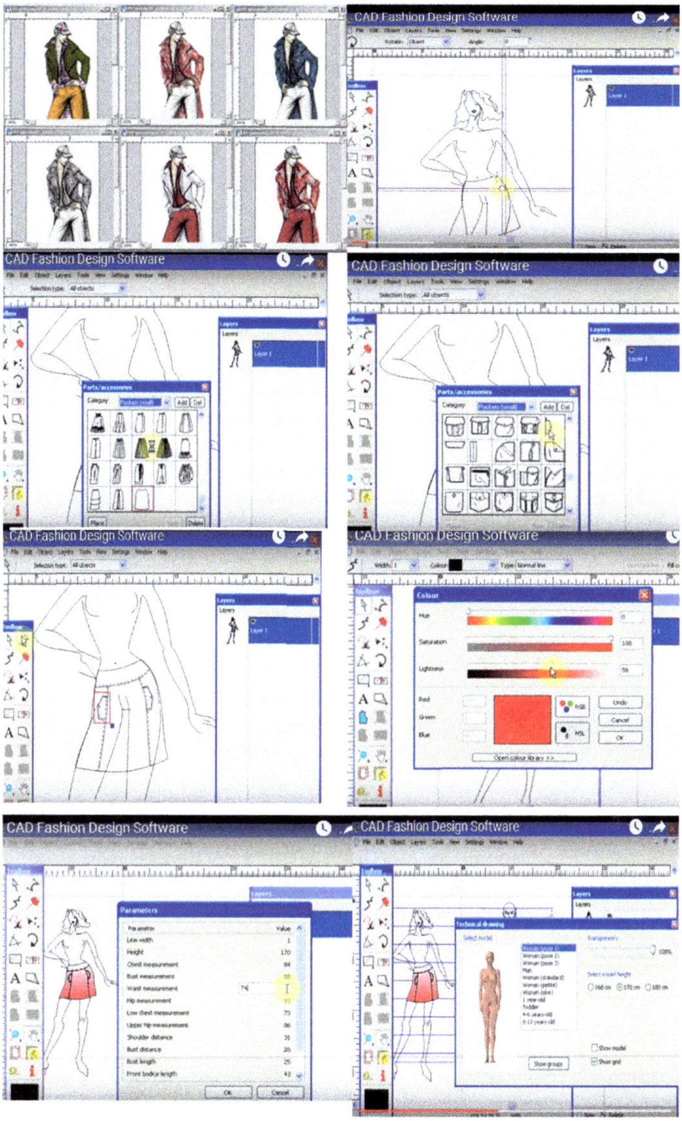

Figure 11.10 Style of skirt, pocket style and positioning, color change, and visualization on model using design computer-aided design software.

for the collections. The CAD-based software is equipped with advanced tools to create and modify designs, makes the designer's task easier by arranging and organizing designs in layers and allowing them to create own style library, and working on multiple designs simultaneously.

After the designer is sure about the idea behind the collection he/she needs to draw sketches of the dresses and accessories. The designer starts drawing from rough sketches that he/she later needs to trim to decide about the final sketches. Sketching is one of the most important steps in designing because the designer has to work through his/her creativity and knowledge of the trends to create something unique yet to be acceptable. Every designer must have to be a very good illustrator. The production stage in the fashion field is important because it is the point where the final product is showcased and everything that has to do with a particular design needs to be established. As the production starts, the designer or garment producer tries to visualize in mind the final product, which look and keep a track of any flaws in the sketches that can be modified. The foresightedness and good vision of designers is a must so that the final product may not prove a ruin. CAD software allows the designer to design the 3D image of the product on computer screen, visualize it and see the pros and cons of the design idea. The designer can design the entire attire with prints, colors, trims, accessories, and embellishments on computer to look for any flaws and rectify them.

Promotion is the next step following production. Fashion designer's final product has to be made approachable to final consumers for which promotion is of course very crucial. CAD software again comes in picture during promotion through designing of promotional materials and items. The promotional materials designed can be displayed in billboards, magazines, brochures, posters, catalogs, fashion shows, and invitation cards to draw attention of both fashion leaders as well as fashion followers alike. Fashion designing with the use of CAD has numerous advantages. CAD aids the designers to create an even and detailed design collection when designer wishes to create a collection that is connected in terms of pattern and flow. Designers can look their creations for any flaws and also in different colors, patterns draped on different models on the computer screen before being displayed to the general public. The software also ensures time saving process, with less utilization of money and resources owing to visualization of fabrics and designs before putting into final production stage.

11.7 Computer-aided design in patternmaking

The manual method of patternmaking is time-consuming and requires patternmaker's skill and expertise. The use of computers in patternmaking not only results in time-saving but also ensures high precision and accuracy. PDS system is considered to be versatile software, which enables multiple functions to be performed. Some of the basic functions performed by PDS system are listed below:

- Pattern outline can be changed to draw a set of patterns from the existing patterns.
- Pattern positioning on the screen can be adjusted by freely and readily moving the patterns.
- Patterns can easily be copied, paired, or made into multiple clones.
- Different patterns can be manipulated by joining or fusing them together.
- Style interpretation and pattern adaptation can be done by addition or omission of styling details, darts, and seams.

PDS is 2D that makes use of flat pattern cutting technique rather than that of modeling for pattern construction. Flat manipulation technique has proved its effectiveness over the years in speeding up the process of pattern construction and standardizing company's fit and sizing requirements. PDS system requires every kind of pattern style and shape as their memory input to create a library of tested and verified material in memory. This in turn can serve as data for any design manipulation to be carried out in future. PDS system shows the ability to memorize the previous pattern construction process and can thus automatically perform pattern designing (Luo and Yuen, 2005). PDS systems are menu driven, which can be on the monitor itself or on a digitizer board, and the items on menu can be selected with stylus.

Pattern generation software (PGS) such as Conex from Assyst and PGS from Investronica are also available to pattern technologists for quick and accurate pattern generation (Kim and Kang, 2003). The system comprises of size table, an alteration table, and style listing. Relevant information related to garment to be constructed need to be fed to the system such as lapel width for different styles. The full set of patterns in all sizes can be created by the system in fraction of seconds. The pattern pieces so created can be transferred to CAD system from PC. 3D Modaris software by Lectra uses traditional pattern design tools to make pattern generation faster: creation from scratch or from existing patterns, grading, reproducing traditional or advanced methods, and checking, using state-of-the-art techniques (http://www.lectra.com).

11.7.1 Pattern digitizing/scanning

Scanning when compared to digitizing is a faster and many a times accurate method to register the outline of pattern into the computer memory. The software providers make sure that their version can provide rapid generation of digital copies as the entire garment manufacturing process needs to be time saving so as to target shorter lead times. The software is equipped with efficient digitizer to create a digital copy of outline patterns in seconds with even the most complicated pieces. Moreover, digitizer should be able to accurately digitize all types of contours and internal parts or tight curves, grade them, and assign piece attributes or any information needed for marker making. The patterns after being entered into the system need manipulation for the next stage. During pattern manipulation, the block pattern can be styled or existing styled pattern can be altered.

11.7.2 Pattern drafting

Scaled down versions of actual patterns that are either drafted on computer or scanned can be made using pattern drafting software such as Adobe Illustrator, which is quite user friendly with simplified drafting tools. Drafting by means of computers permits the ability to copy, paste, and repeat the digital patterns as per future orders eliminating the necessity of tracing pattern for adjustments and manipulations (Fig. 11.11). Lines and curves can be measured down to millimeters, thus enabling pattern movement and matching notches much more quicker and accurate (Kang and Kim, 2000; http://clothhabit.com/pattern-drafting-software/).

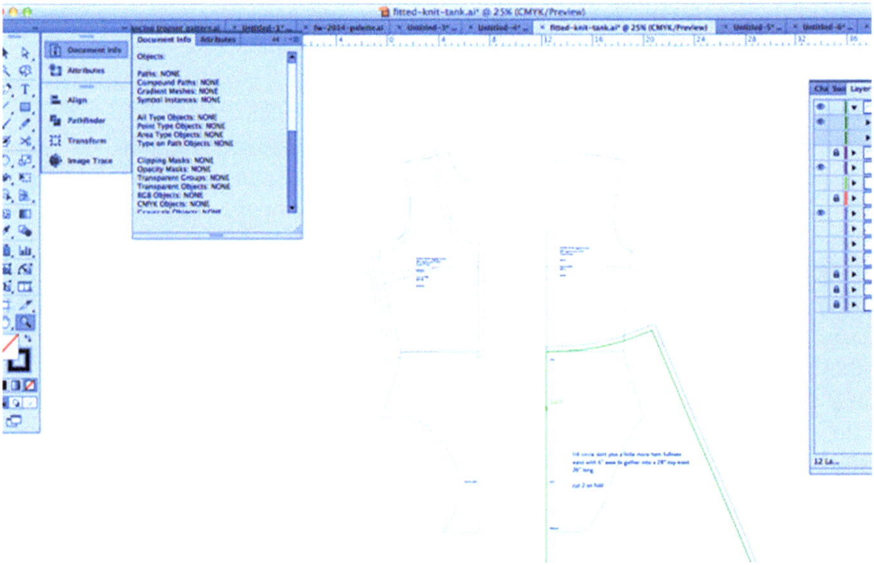

Figure 11.11 Computer-aided design technology for patternmaking.

11.7.3 Pattern grading

Pattern Grading, when done manually, is associated with complicated calculations and laborious work to scale a pattern into different sizes. CAD-based software provide quick adjustment of pattern sizing so as to assist fitting the entire range of merchandise, including complex size variations. Different pieces and values can be graded simultaneously. The software offers features such as simple notch and split parts grading, angle grading for complex shapes, alteration grading for short-term sizes, and variable grading for larger sizes. One such pattern grading software is Telestia Creator Pattern Grading CAD software based on the Telestia pattern grading methodology that grades patterns accurately into 3D shapes with perfect fit (Fig. 11.12) (http://www.tc2.com/pattern-grading-cad-software.html).

The grading software has simplified the grading process by the elimination of all the hard work. It ensures easy and quick pattern grading without affecting the shape, fit, and balance of the original design. The software provides option of variety of increments by allowing the selection of grading points on the block, while it grades each one automatically. Standardized grading rules can be used but as per need own grading rules can also be inserted. Style interpretation is also possible when grading difficult design details on a garment. Telestia Creator Pattern Grading software provides a fully featured CAD program, build in grading rules, grade tables for style interpretation, and advanced tools to create and modify patterns. Furthermore, the system enables easy drafting by making use of pattern grading template, has simplified tools and functions to draw, modify and adjust the patterns. The software can work on multiple designs simultaneously; own pattern library can be crated. Last but

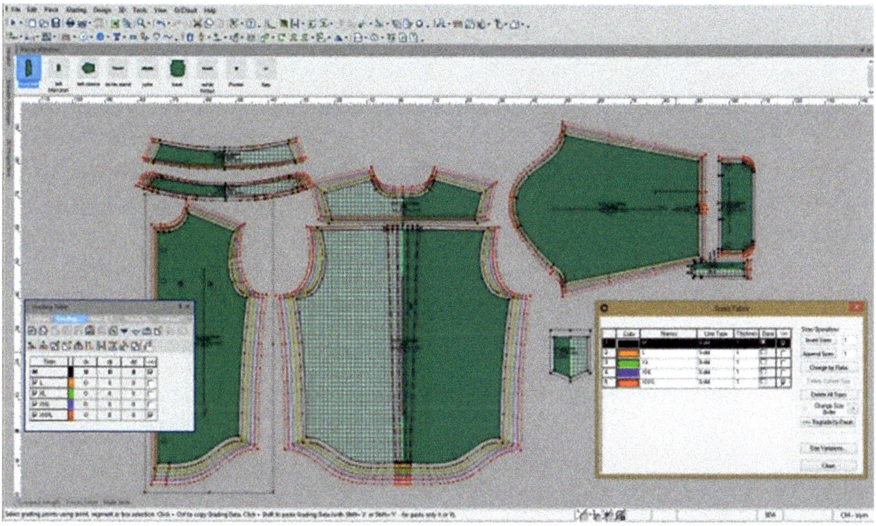

Figure 11.12 Computer-aided design technology for pattern grading.

not the least, the full size patterns can be printed using any compatible printer and ready patterns can be scanned as well for adaptation.

Currently available CAD software are capable for multiple functions such as patternmaking, grading, interactive, and automatic marker planning with minimal operator skill. Richpeace is one such CAD-based software that permits to draw a pattern of required size of front bodice block with the help of drafting and grading system by using tools such as: intelligent tool, rectangular tool, and point tool as shown in Fig. 11.13. Pattern grading, to achieve different sizes of patterns from standard or base size can also be done accurately on the software. Edit size table allows filling of different sizes to be graded (Fig. 11.14). Sizes can be filled in required column and the base size can be selected by clicking on base size option. Different colors for different sizes can be chosen for pattern identification.

Requirement of different parts can also be filled (X, Y coordinates) in the grade table. Pattern symmetry tool enables to see the pattern grading (Fig. 11.15). Different patterns required for completing the garment such as collar, cuffs, pocket, back bodice can be drawn and seam allowance can be added by the use of seam allowance option tool from tool bar. The file can be saved on desktop that can be used for marker making. Garment marker system (GMS) module in the software that enables the creation of marker. Richpeace marker making system offers automatic marker making and time nesting features. Marker cut planning can be calculated based on material, color, quantity of order, maximum patterns that can be accommodated in marker, and maximum plies that can be spread in a lay. Markers with different size ratios can easily be worked out by replacing sizes in the existing marker with any required size. Marker option and define marker in menu bar enables the marker making process in software. Width and lengths of marker are defined. Pattern file is selected in GMS module and marker is completed by filling quantity options for different patterns. Marker area appears on

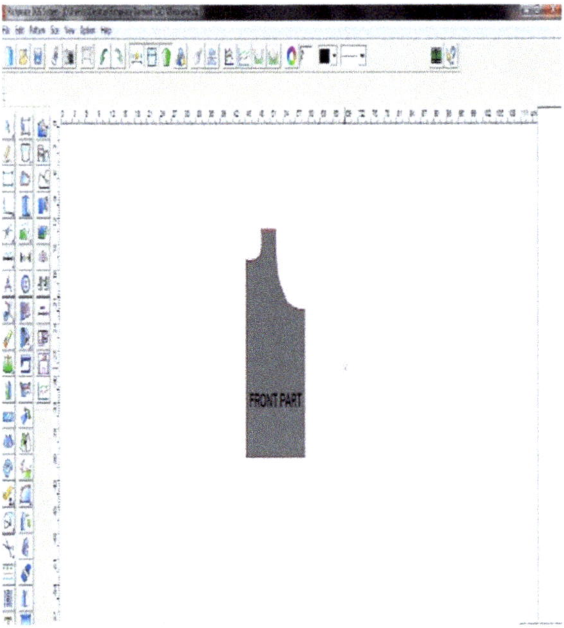

Figure 11.13 Patternmaking in Richpeace software.

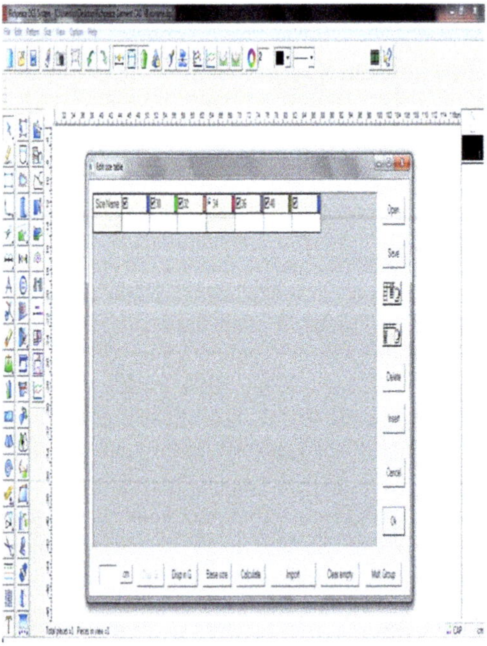

Figure 11.14 Edit size table for grading.

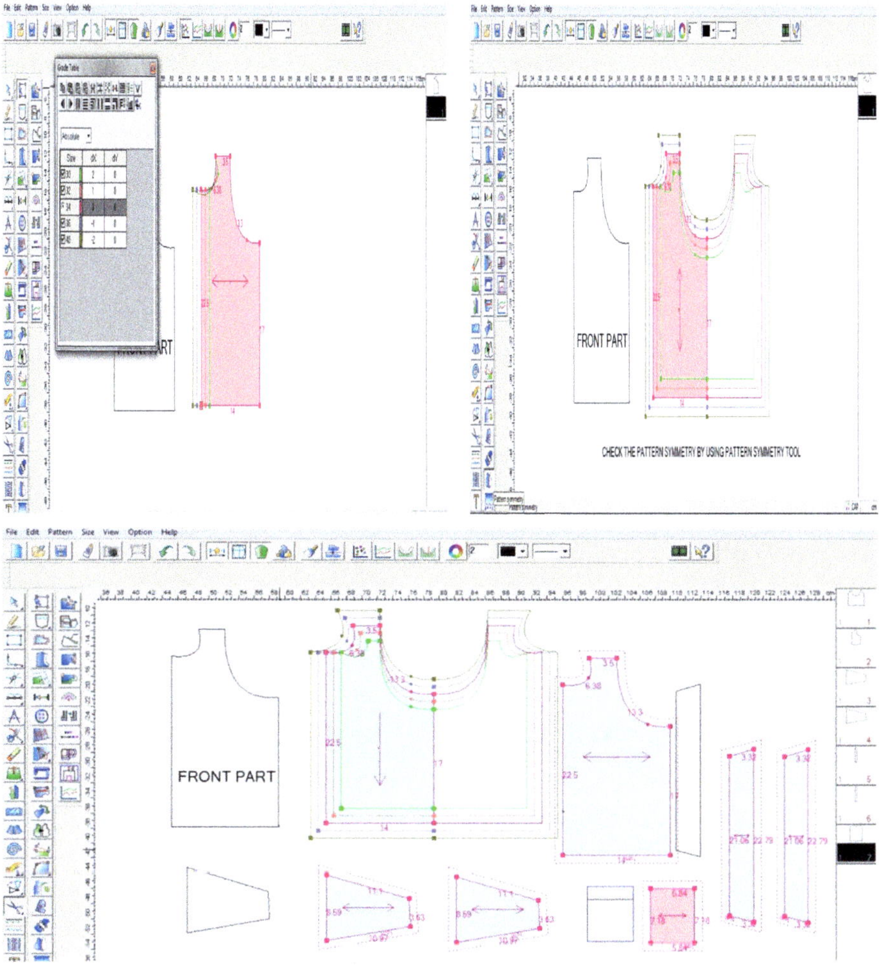

Figure 11.15 Pattern grading on Richpeace.

the screen. Miniature patterns are displayed on screen above marker area (Fig. 11.16). The patterns are selected, dragged, and dropped in marker area thereby creating an interactive marker. The system calculates marker efficiency automatically. Auto nesting option enables maximum efficiency to be achieved automatically. However in this case, it needs to be ensured that all patterns are laid in marker area.

11.7.4 Core of patternmaking suite

The section discusses the multidimensional aspects of PDS system with respect to pattern manipulation, finishing, and grading to develop a complete library of patterns varying in sizes, shapes, and styling details. The last part of section discusses some of the disadvantages encountered with the use of this system.

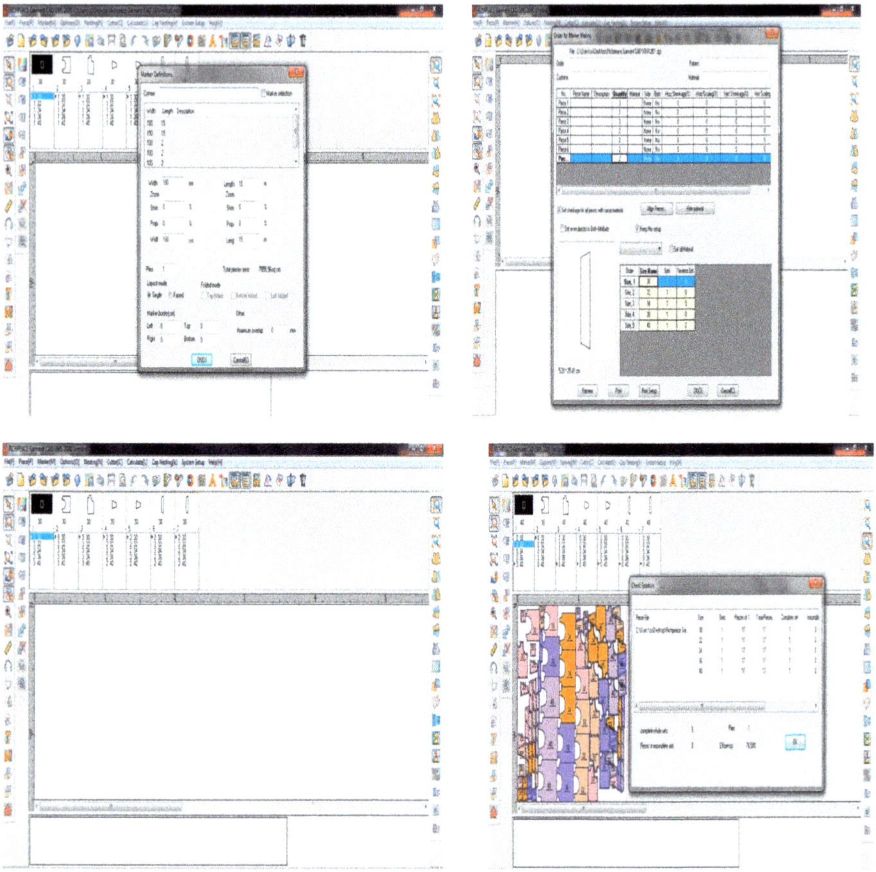

Figure 11.16 Interactive and automatic marker making on Richpeace.

The advantages offered by PDS over manual method of patternmaking are innumerous, which has encouraged widespread use of the system in highly competitive and fast-growing garment industries. The system ensures elimination of several manual steps in the design process with a superior design experience that can be fully tailored and customizable as per user needs. Apart from minimizing the manual efforts and lead time, the system can also be advantageous in the following aspects:

- Visualized measurements of different styles and internal features such as notches, buttons, and drill holes can be checked and altered.
- Draft pieces from scratch, digitize existing hard copy patterns, or even edit and finalize existing digital patterns.
- Easy mobility of pattern as it is corrected for shape, adding or adjusting details.
- Darts, seam allowance, special corners, advanced measurement techniques, pleats, curves, and facings can be tailored in accordance with customers needs across a wide range of merchandise types.

- User-friendly working environment with movable and easy to understood toolbars.
- No hardware lock in, including importing and exporting to and from all major CAD formats.

In a nutshell, it can be concluded that the PDS system is performing multiple functions and their menus can broadly be divided into (1) pattern manipulation, (2) pattern finishing (adjusting seam allowance, location of seams, and notches), and (3) pattern grading.

There is no doubt that PDS has many advantages to offer as far as pattern design and construction process is concerned. However, certain limitations can also be encountered with these systems such as: the PDS system, which is not very user-friendly with complex menus and submenus and requires complete proficiency in the hands of operator to carry out the process flawlessly. The system is purely a flat medium and does not make use of 3D modeling (Luo and Yuen, 2005).

11.8 3D fashion design and development software

3D pattern system unlike the 2D pattern system relies on modeling for pattern design and development. The system requires 3D digitizer enabling the object placement and taking space into computer memory. The digitizer in this system uses three measurements (X, Y and Z coordinates) producing a wire frame image of the model, which later on can be filled into a solid form. This 3D representation by rotation can be viewed from different angles. The model form here is simulating the actual size and can be effectively utilized for modeling the garment obtained from computer-generated cloth on this model. Virtual garment styling and stitching software are revolutionizing the fashion arena with their benefits offered to designers, manufacturers, retailers, and buyers alike.

Drape modeling, 3D visualization of designed garment in draped form, is one of the key technologies in CAD garment design systems (Hardaker and Fozzard, 1998; Stylios et al., 1996). It is inevitable for the designers to assess the design, fabric suitability, and the accuracy of garment patterns in the computer environment. With the ever increasing demand to visualize the drape and fit before physical sample creation, most of the CAD companies are attempting to develop 2D to 3D design system. This system enables 2D pattern pieces to be virtually sewn together by computer and the garment fit and drape can be analyzed. The software allows design iterations on styles so as to perfect the fit/look eliminating the need to create multiple physical samples and go through multiple rounds of fitting. Measurements for fitting model can be fed into the software and redrafting of the block to form a new block for a style, can be successfully accomplished in no time. In the absence of standard fitting model, virtual model to designer's measurements can be chosen (Fontana et al., 2005b). Physical samples can also be taken for comparison with the computer version.

The software also enables to work with variety of fabrics in terms of their physical properties. A virtual sample can be created for a realistic representation by visualizing the garment with fabric designs/textures and drape characteristics. Modifications can be made in two or three dimensions (Stylios et al., 1996). Any alterations applied to selected patterns on the virtual avatars are automatically translated into 2D patterns and vice versa. Virtual draping has also opened the avenue to check the fit on the dynamic dress forms in different body postures. Bends and drape of different areas of various garments such as T-shirts, trousers, aprons, skirts, and jackets can thus be analyzed, where dynamic models simulate an individual involved in physical activity. 3D virtual software such as Modaris 3D Fit fashion, Accumark APDS-3D by Gerber, PAD System, Maya cloth, Syflex LLC system, e-fit by Tukatech are some examples that are used for 3D visualization.

Form fitting apparels can also be made using 3D modeling feature of Optitex for checking the drape of garments (http://www.tukatech.com; http://optitex.com/solutions/odev/pattern/). Virtual models or avatars can be created by 3D scanning technologies applied to different parts of the human body. The system enables the measurement of any surface area of the human body. Some areas where digitization of human body has been implemented are: anthropometry, cosmetics, fashion, and beauty. 3D body scanning process vary in the data collection mode (Istook and Hwang, 2001) as point clouds, surface models, textured models, and unprocessed data as per the requirement of the targeted application. LASS, Modaris, Tuka Teck are a few companies providing these software. TUKA3D is advanced and user-friendly 3D apparel design and development software. The software has provision of customized virtual fit model and building life-like virtual clothing samples. Tukatech's 3D apparel design software eliminates the need for trial and error in physical sample creation, ensuring that any design fits right the first time (Stylios et al., 1996; http://www.tukatech.com).

3D virtual sample making is made possible by the software where 2D patterns can be converted into 3D apparel samples (Fig. 11.17), that accurately simulate the weight, stretch, and color of fabric. The software's virtual fit sessions with animation (Figure) allow the user to bypass physical sample making and thus dramatically reducing the time and cost associated with product development. Fit model can be scanned using the Styku body scanner (Fig. 11.17). Hundreds of measurements can be extracted from a body scan that can then be used to create a 3D avatar that is an exact replica of the fit model. Full-motion simulation animates the virtual fit model to visualize garment's drape in motion. Models can be animated to show different motions such as running, dancing, cycling, posing, and walking on the runway. The effect of motion on tension of different fabric portions can also be visualized by highlighted colors (Fig. 11.17). Users with an extensive library of fabric presets can visualize garments with cloth properties that replicate actual fabrics. Realistic fabric simulations can be built with pleats, darts, tucks, easing, textures, and transparencies.

Virtual storyboard and virtual photo shoot can also be created for online sales, and marketing. The system offers greatest benefit of showcasing complete collection without manufacturing a single garment, thus the made on demand business model is actually becoming reality.

Another 3D software package for simulation is Vidya. This software has salient feature such as: size and fit monitoring and optimization in 3D, variant monitoring, simulation of body measurements by fashion manikins, both in static and dynamic state. Fashion manikins with real size and fit can serve as perfect replica for human body (Nayak and

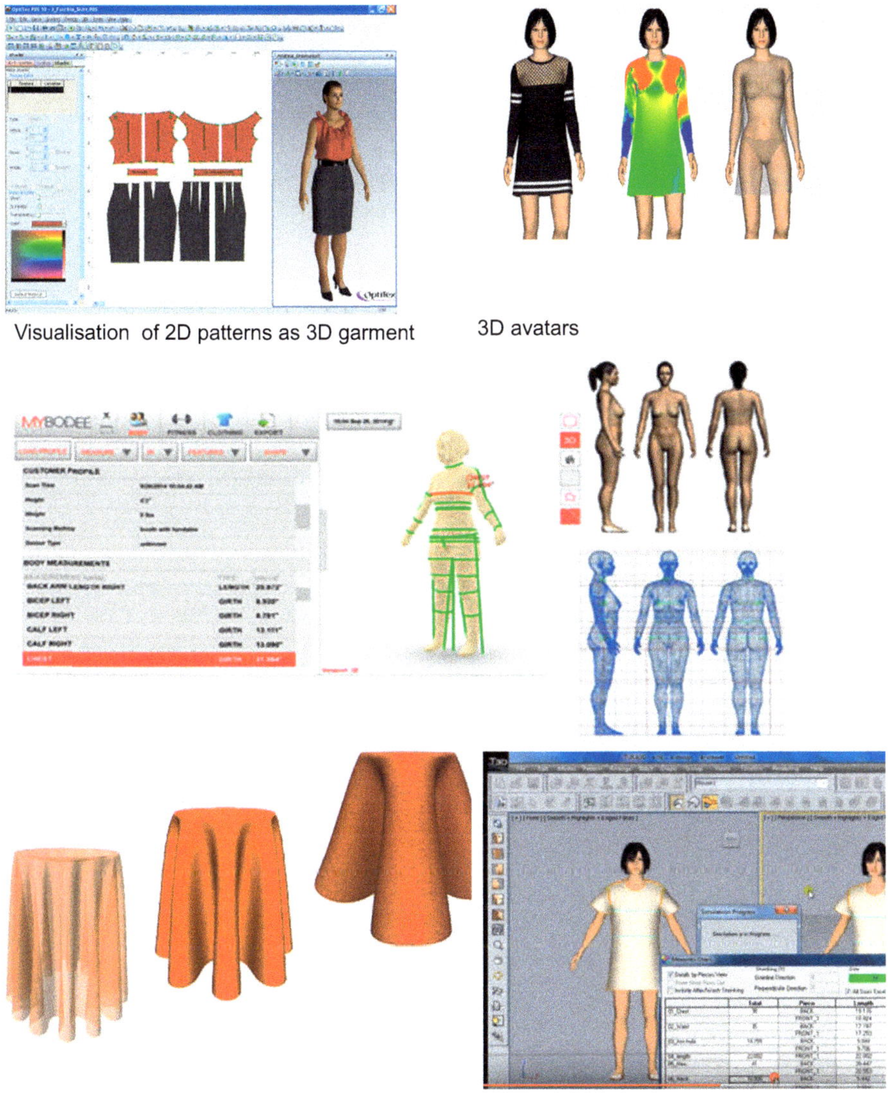

Figure 11.17 Creation of 3D avatars and visualization of garments's drape and fit in dynamic and static postures.

Padhye, 2017; http://www.human-solutions.com/vidya/front_content). Basic patterns can be customized to match the customer's body measurements or proceed with the new measurements and ready-made sizes (Fig. 11.18). The software enables variant monitoring for as many patterns as needed. The software allows for quick color (overlay colors), materials (texture/elasticity/drape), and prints changes (Fig. 11.19).

SDS-ONE APEX3 is another innovation in 3D-CAD software, which features 3D simulation for product evaluation from any angle using the fabric manikin (Fig. 11.20).

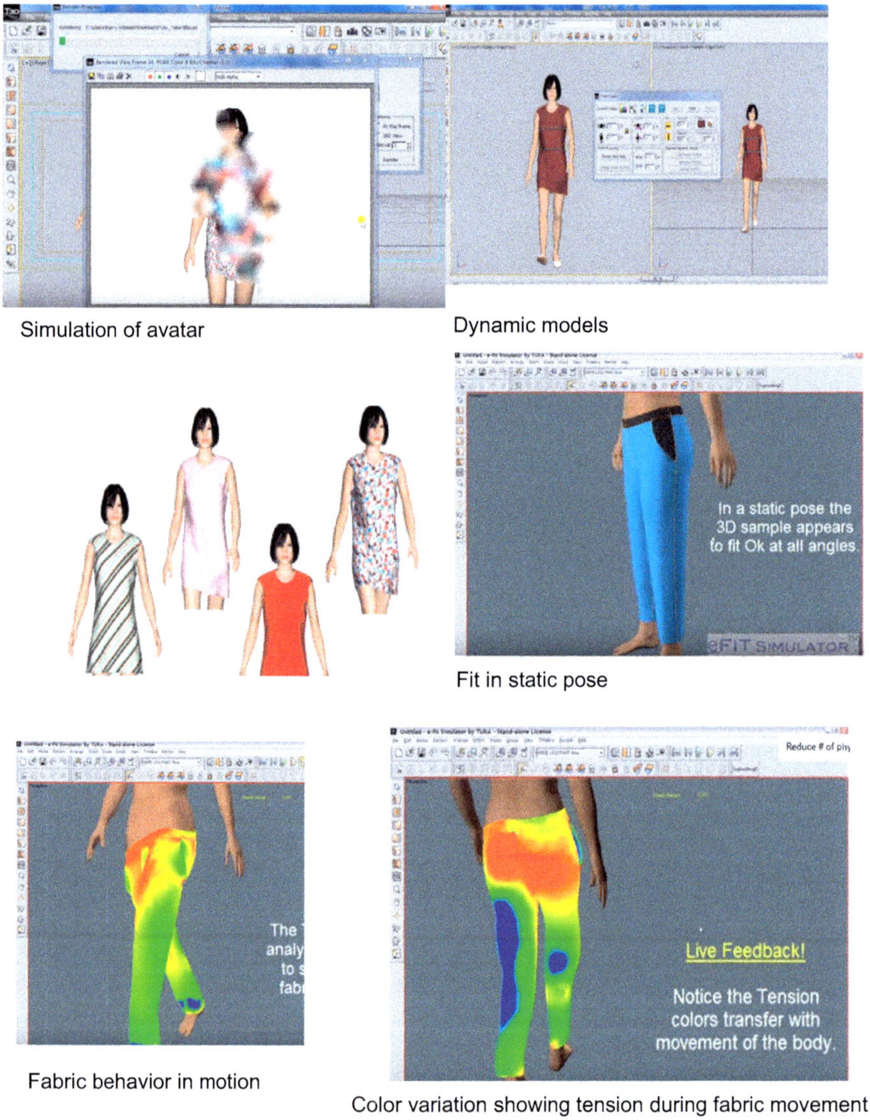

Figure 11.17 Continued.

3D item database allows 3D images to be created easily. 3D model of the software is an original model created by inputting body measurements, skin color, hair color, hair style, and with adjustment of body pose. Once the sewing line is set, patterns created using the measurement table or patternmaking, grading and marking software can be converted to virtual garment to visualize drape and fit on the computer monitor in 3D Fitting Simulation (Hardaker and Fozzard, 1998). Fabric tension and pressure points

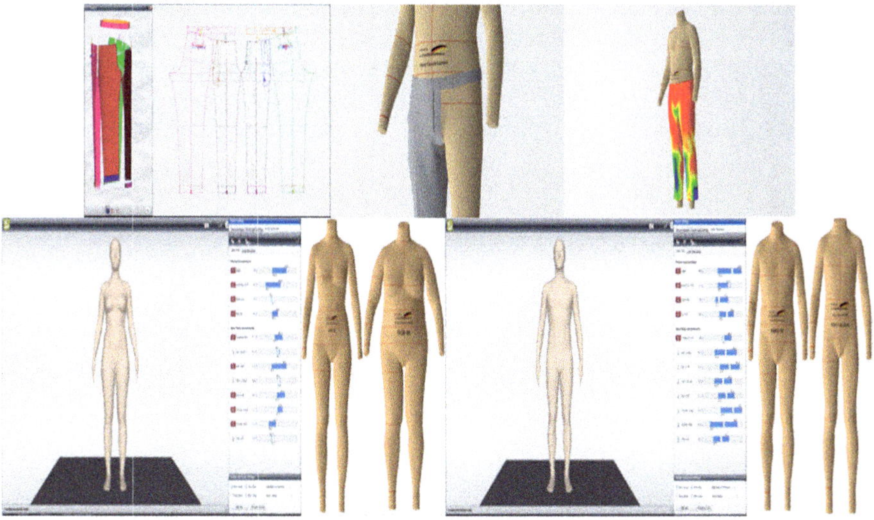

Figure 11.18 Size and fit monitoring by Vidya.

Figure 11.19 Variant monitoring by Vidya.

Figure 11.20 Fashion manikins.

can be displayed in color for reference in product development. 3D Mapping is also possible with scanned or simulated fabric mapped onto the 3D garment.

3D mapping software such as Pointcarre can be effectively used to make realistic presentations of seasonal collections in dynamic ways by dragging and dropping fabric designs on photographs of accessories (Fig. 11.21). Mapping finds application

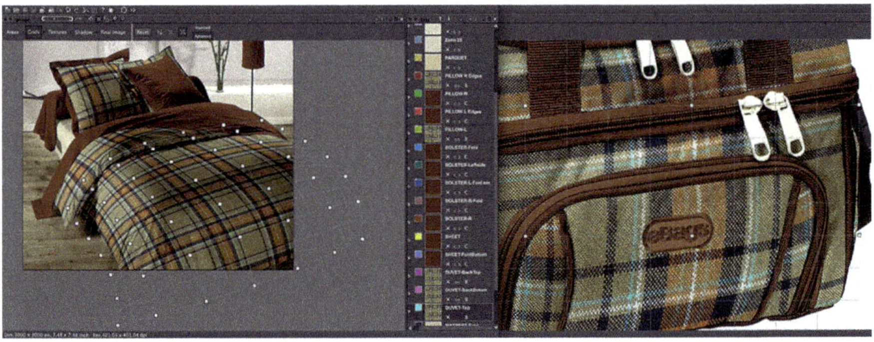

Figure 11.21 Home textiles and accessory designing using three-dimensional mapping.

in apparel, upholstery, and home fashion textiles. The process starts with selection of fabric design, dragging and dropping it on the picture of any apparel, accessory to be designed, and adjusting the fabric on the model with a simple grid (http://www.point-carre.com/print.html).

Another commercially popular 3D software based on 3D visualization technology is AccuMark 3D. The software is an enhanced design module to visualize samples in 3D. AccuMark 3D is an integrated solution that enables sample departments to create more styles in less time and accelerate development time. True to life digital rendering of designs is possible with the software, thereby greatly reducing the costly sample remakes and results in increased production efficiency.

11.8.1 Online mass customization

Garment styles, appearance, fit, and suitability can be assessed by internet systems for showcasing to buyers and consumers. CAD system along with sophisticated cutting technology enables incorporation of drape models on popular shopping websites for virtual shopping. The prospect consumer creates an account, log into the websites, select the garment to be ordered, and is made to fill an online questionnaire pertaining to body measurements and appearance such as height, weight, shoulder width, etc. (Niki, 2005). Archetype is one of the software that facilitates virtual shopping where the order is automatically passed through software and based on the online questionnaire filled by the customer. In this appropriate standardized pattern pieces for the garment can be selected and manipulated. The information is then fed to the CAD package incorporating the required changes and production of automatic marker is accomplished. Fabric cutting, assembling, and dispatching are done direct to the customer's home. Many technologies are being used to realize the retailer's concept of mass customization such as body scanning, bar codes, and laser cutters (Nayak et al., 2008; Istook and Hwang, 2001). AccuMark Made-to-Measure (Gerber) is made-to-order solution for tailored clothing and custom-fitted garment manufacturers. The software accelerates delivery of custom garments by using measurements derived from either high tech, 3D body scanners or traditional tape measures to create patterns.

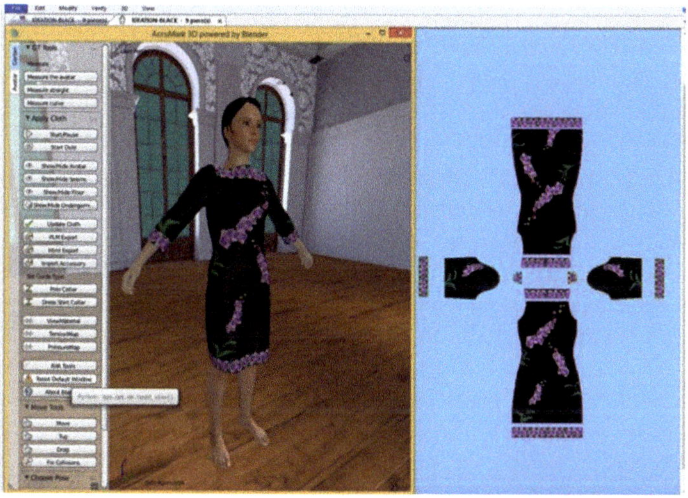

Figure 11.22 Patternmaking by flattening of the three-dimensional garment.

3D flattening software are also finding applications in apparel and accessory manufacturing industries. The software helps in creation, editing, stylization of patterns obtained by flattening a 3D garment (Fig. 11.22). The software tends to be working backward, i.e., flattening of 3D garment into 2D patterns and has provision of storing the patterns to create pattern library within the software.

11.9 Computer-aided design in cutting room operations

CAM deals with smooth and efficient operations at each stage of garment manufacturing process particularly in all cutting room operations i.e., marker planning, spreading, and cutting. Computers are finding application in several departments such as plotting, spreading, cutting, and surface ornamentation such as printing and embroidery (Nayak and Padhye, 2015a,b).

Traditional paper-pen and other tracing and marking techniques have been efficiently replaced by the latest plotter systems. The plotters are connectable to PCs and can plot any type of files. ALYS plotter (Lectra Company) is equipped with communication features that enables remote maintenance and reduces downtime. Marker libraries to obtain efficient marker, an interface to ERP systems for transfer of order information and planning results, marker making stations, and other CAM equipment can be quickly accessed. Marker making stations can directly create new marker, which can be retrieved for future use. Lectra, Zuki, and Gerber are pioneers in computer-integrated cutting devices. AccuNest, SDSR-ONE APEX, SMARTmark are some of the software that are designed to generate automatic and interactive markers, coordinating with cutting tools and equipment. Instant creation of markers is possible with SDSR-ONE APEX's marking software that features ultrahigh speed automatic marking. Fully automated markers and markers in sequence can be created by the software. Efficient cutting can thus be achieved with markers transferred to pattern computer-aided manufacturing (P-CAM) fabric cutting machine.

SMARTmark an add-on module to TUKAcad maximizes the placement of markers and ensures highest achievable marker efficiency. The software enables accurate estimate of fabric yield in accordance with style based on actual ratio to be cut thereby saving 3%–15% fabric cost per marker. The software has some additional features such as nesting by sizes, defining single size, or ratio markers. Style details such as notches and seam allowance can also be modified. Fabrics that have stripes or plaid lines can be input directly on the marker screen. This saves time in placing pieces where stripes and lining need to be aligned in pattern pieces.

AccuNest is an automated nesting software that optimizes the material yield. The software makes use of powerful algorithms for analyzing multiple nesting solutions and delivers the one with the highest material utilization. Marker making process can be dramatically accelerated with added advantage of reduced costs by the use of this software. AccuNest is much faster and more efficient as compared with manual marker making. It automatically generates costing and production markers for fast, accurate material calculations. AccuNest takes care of all material constraints applied to pieces such as directional rotation, tilt allowances, and spread constraints to ensure quality and minimize costly mistakes in the cutting room, maximizing quality control, and productivity. Peak production demands can be met without any additional labor cost as the software automatically generate nests 24 h a day without human intervention.

AccuNest is available with two levels of computing power: AccuNest Professional Edition employs single-core computing technology, which analyzes multiple algorithms and optimal marker strategy is delivered without human intervention.

AccuNest Multi-Core (MC) utilizes the power of multiple computer processors to automatically generate markers that deliver better material utilization (Fig. 11.23). Cut part quality can be improved with this system owing to control on both the direction of cut and the start point to prevent material from moving during cutting. Cutting time can be reduced by monitoring the exact order of cutting and resorting to more efficient cut path with the aid of CutWorks and ToolPath modules. CutWorks is a modular software that offers a complete design, nesting, and cutting solution for flexible materials.

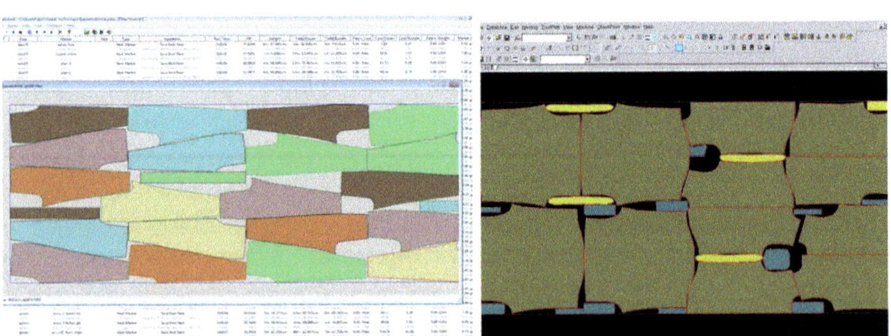

Figure 11.23 Automatic marker generation by AccuNest.

Material utilization and nesting speeds can be improved by CutWorks, which includes manual and automatic nesting modules.

AccuMark MTM is a versatile software with multiple functions. The software simplifies the business processes by importing order information from various sources, including mainframe computers or web pages and automatic generation of cut data. AccuMark MTM has been integrated with Gerber's AccuNest automated marker making system for automatic generation of markers, thus maximizing productivity and material utilization. The software intended for marker making can also generate interactive marker plan where operator is required to drag and drop the miniature patterns in marker area displayed on monitor and marker efficiency gets updated with positioning of each pattern in the designated area (Fig. 11.24).

Spreading aims at formation of lay to meet the bulk production in a garment industry as per size, style, and color requirements. Traditional spreading methods are manual leading to increased labor fatigue and time-consumption. Patternmaking softwares are integrated into the latest spreading machines that enables fabric layers to be directly fed. ProgressBrio55 (Lectra) is one such software that communicates directly with the CAD workstations and cutting machines making the spreading process much more accurate, quick and results in significant material saving as compared to manual spreading. Fast loading of fabric rolls and perfect fabric alignment without any tension or slackness can be achieved with the cradle feed system.

AccuPlan is a powerful spread and cut planning tool that enhances the production efficiency by utilizing existing libraries and databases. The software has resulted in automating the planning process, thus the work orders can be downloaded from the ERP system, importing cut work orders and streamlining the entire process right from calculating marker efficiency, spreading parameters to submit cut files, and cut tickets to the

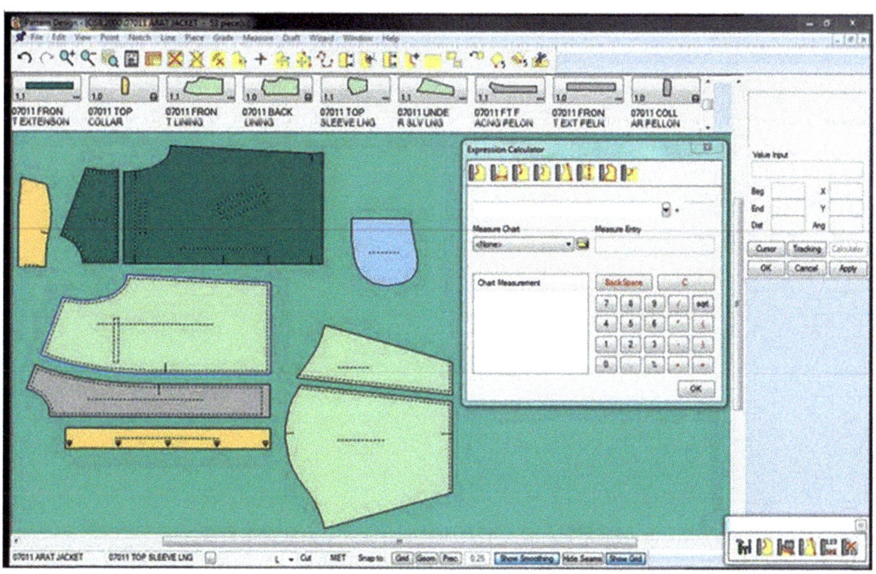

Figure 11.24 Interactive marker planning.

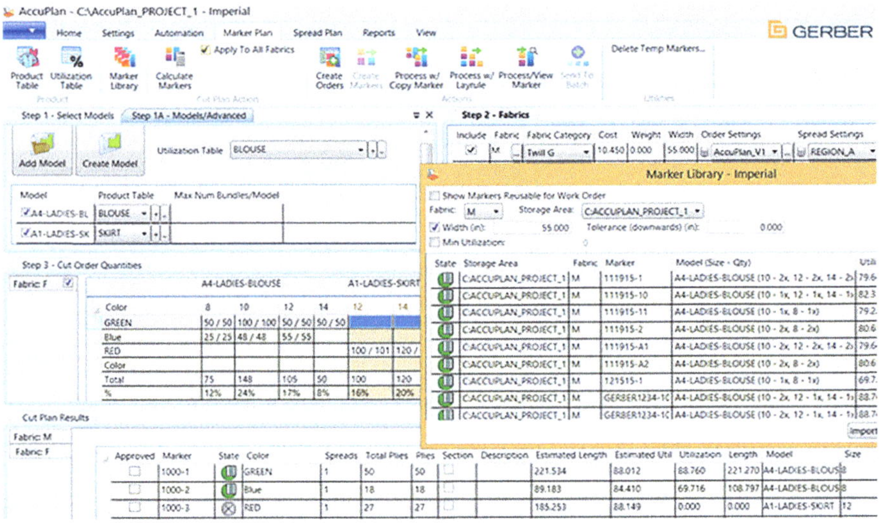

Figure 11.25 Spreading and cut planning by AccuPlan software.

cutting Room. The automation has resulted in significant reduction in operational costs and improvement in productivity. Material requirement for each work order can be calculated by the software, which allows processing multiple orders simultaneously and generating reports based on cost, cut plan results, and fabric consumption (Fig. 11.25). The software is adaptable in accordance with changing market trends and customer demands.

Another cutting software is Telestia Creator Pattern Cutting CAD software, which is based on the Telestia pattern cutting methodology. The software provides accurate patterns with perfect fit, which can be printed in life size on any standard printer. Telestia Creator Pattern Cutting software is equipped with advanced tools to create and modify patterns, drafting, drawing, modifying, adjusting the patterns, and adding variable width seam allowance.

The separation of garment components into separate cut pieces is very crucial step as these patterns are the blueprint for the final garment and improper garment fit can be the outcome of errors and inaccuracies during pattern cutting. The conventional cutting tools are restricted in their use depending on the number of fabrics plies in the lay, shape, and size of patterns to be cut and fabric characteristics. Cutting machines integrated with the computerized instructions and protocols for the multiply fabric cutting have become part and parcel in technology driven, profit oriented apparel, and accessory manufacturing units (Niki, 2005). SDS-ONE APEX is the software frequently used in cutting room owing to system's capability to offer smooth CAD-to-CAM workflow and the machine data serves as output to various cutting machine formats including Shima Seiki's own P-CAM fabric cutting systems. The system provides a user-friendly operating environment consisting of a pressure-sensitive cordless pen, digitizer tablet, and trackball unit with duplicate keys for frequently-used functions, allowing simultaneous use of both hands for maximum productivity (http://www.shimaseiki.com/product/cadcam/sdsone_apex/). The system is versatile in selection of cutting widths and thickness, ranging from high-volume multiply machines to single-ply machines for sampling and made-to-measure clothing applications.

The sewing process can be performed more efficiently with the incorporation of computer-integrated sewing machines in the sewing room. The state-of-the-art computer-integrated machines are equipped with programmable needle positioning, presser sensors, needle bar sensors, threading systems, and programmable sensors for stitch length regulation. Brother, Juki and Tazima are some of the machine manufacturers providing a wide range of computer-integrated sewing machine models.

Ornamentation on apparel and accessory surface to create intricate multicolored motifs quickly can be performed by computerized embroidery machines. Designs as per motif requirement and fabric types can be easily and quickly fed into the software. Feeding of design is generally quick and easy.

11.10 Conclusion

The advancement in technology has paved the way of computer in variety of domestic, industrial, manufacturing, designing, and marketing applications. The intervention of computers in fast growing competitive industrial environment is advantageous as far as productivity, cost-effectiveness, and time saving approaches are concerned. Textile and apparel industries are not aloof of this technological upgradation with CAD/CAM system implemented for different operations an added boon for the industry. CAD finds its practical utility right from design initiation stage through lay planning, spreading, patternmaking, cutting, and finally sewing. 2D pattern design software based on flat patternmaking techniques have been successfully explored for patternmaking, grading, and creating pattern libraries within software for future retrieval. However, a further progress made in the field is 3D software that works in virtual environment permitting 3D visualization of drape and fit on virtual model or avatar thereby eliminating exhaustive process of physical sample generation. 3D scanning, mapping, customized avatars in accordance with specific anthropometric and facial features, customized garments are some of the innovative and exciting avenues available with CAD software packages. Enhanced productivity, competitiveness, and shorten delivery schedules can be guaranteed by linkage between design and manufacturing operations along with other preproduction steps of patternmaking, grading, and marking. The proficiency and skill of operators to adapt to an electronically controlled system of production is of concern to many as is the pros and cons of the system by affecting the employment opportunities with most of the operations being computerized.

References

Collier, B.J., Collier, J.R., 1990a. CAD/CAM in the textile and apparel industry. Clothing Textile Journal 8 (3), 7.

Collier, B.J., Collier, J.R., 1990b. CAD/CAM in the textile and apparel industry. Clothing and Textiles Research Journal 8, 7–13.

Fontana, M., Carubelli, A., Rizzi, C., 2005a. Cloth assembler: a CAD module for feature-based garment pattern assembly. Computer-Aided Design 2 (6), 795–804.

Fontana, M., Rizzi, C., Cugini, U., 2005b. 3D virtual apparel design for industrial applications. Computer-Aided Design 37 (6), 609–622.

Gupta, S., February 2005. Design innovation and new technologies. Clothesline 58–62.

Hardaker, C.H.M., Fozzard, G.H.W., 1998. Towards the virtual garment: three dimensional computer environments for garment design. International Journal of Clothing Science and Technology 10 (2), 114–127.

Istook, C.L., Hwang, S.J., 2001. 3D body scanning systems with application to the apparel industry. Journal of Fashion Marketing and Management 5 (2), 120–132.

Kang, T.J., Kim, S.M., 2000. Development of three-dimensional apparel CAD system: part 1: flat garment pattern drafting system. International Journal of Clothing Science and Technology 12 (1), 26–38.

Kim, S.M., Kang, T.J., 2003. Garment pattern generation from body scan data. Computer-Aided Design 35 (7), 611–618.

Luo, Z.G., Yuen, M.M.F., 2005. Reactive 2D/3D garment pattern design modification. Computer-Aided Design 2–172.

McAllister, April 1983. Isaacs III: are you ready for robots? Textile World 35–45.

McCartney, J., Hinds, B., Seow, B., Gong, D., 2000. Dedicated 3D CAD for garment modelling. Journal of Materials Processing Technology 107, 31–36.

Nayak, R., Padhye, R., 2015a. Garment Manufacturing Technology. Elsevier.

Nayak, R., Padhye, R., 2015b. Introduction: The Apparel Industry. Garment Manufacturing Technology, pp. 1–17.

Nayak, R., Gon, D.P., Khandual, A., 2008. Application of LASER in apparel industry. Man-Made Textiles in India 51, 341–346.

Nayak, R., Padhye, R., Wang, L., Chatterjee, K., Gupta, S., 2015. The role of mass customization in the apparel industry. International Journal of Fashion Design, Technology and Education 8 (2).

Nayak, R., Kanesalingam, S., Wang, L., Padhye, R., 2016. Artificial intelligence: technology and application in apparel manufacturing. In: TBIS-APCC 2016. Binary Information Press, Textile Bioengineering and Informatics Society, pp. 648–655.

Nayak, R., Padhye, R., 2017. Manikins for Textile Evaluation. Elsevier.

Niki, T., August 2005. From CAD to cutting. Clothesline 86–90.

Parthibaan, M., Mahalingaam, G., Robotics in Textile Industry – A Global Scenario.

Stylios, G., Wan, T.R., Powell, N.J., 1996. Modelling the dynamic drape of garment on synthetic humans in a virtual fashion show. International Journal of Clothing Science and Technology 8 (3), 95.

Wang, C.C., Wang, Y., Yuen, M.M., 2003. Feature based 3D garment design through 2D sketches. Computer-Aided Design 35, 659–672.

Wang, C.C.L., Wang, Y., Yuen, M.M.F., 2005. Design automation for customized apparel products. Computer-Aided Design 1–30.

Yan, H., Fiorito, S.S., 2002. Communication: CAD/CAM adoption in US textile and apparel industries. International Journal of Clothing Science 14 (2), 132–140.

Zoran, S., 1995. Computer-aided processes in garment production – features of CAD/CAM hardware. International Journal of Clothing Science and Technology 7 (2/3), 81–88.

Advancements in production planning and control

12

Sweta Patnaik, Asis Patnaik
Cape Peninsula University of Technology, Cape Town, South Africa

12.1 Introduction

Throughout the process of globalization, the evolution of the textile economy opened new and unforeseen pathways. Fast fashion and global value chains altered the structures and philosophies of production and consumption (Gereffi and Memedovic, 2003; Azmeh and Nadvi, 2014). The increase in labor costs, new development policies, changes in consumer habits, and political tensions were some of the factors that mitigated the advantages of low-cost competition based on the exploitation of low wage work (Distler et al., 2014; The Economist, 2012). The internet as well as automation, information and communication technologies disseminated information in real-time all around the world.

At the moment, specialized consultants and analysts estimate that the end of competitive advantage of low-cost manufacturing is near (Bruno and Pimentel, 2016). Outdated technology, inefficient production, unskilled worker, physical infrastructure, and precarious communication only add additional costs; regional political instability introduce threats and increase risks; and long distances increase energy costs (Sirkin et al., 2011; The Economist, 2012; Eloot et al., 2013; Euler Hermes Economic Research Department, 2013; Foresight, 2013; Distler et al., 2014). Modular automation and the robotization of apparel attract investments from governments and even large global buyers who used to invest in the emigration of production to countries with low production costs (Book et al., 2010; Owano, 2012).

Automation in the production industry has undergone a number of developmental phases and has played an increasingly important role in the apparel manufacturing operation. It has not made many inroads into the apparel industry until recently as it was cost prohibitive, where, garments could be produced cheaper by a human in a low-cost country than it could by utilizing high-tech machinery (Nayak and Padhye, 2015a). But now with decreasing automation costs and the functions being more flexible, accommodating various customer needs, more companies will be able to adopt this technology. The use of automation robotics and advanced tools, equipment and computer software for various operations will be the key for success of garment industries in the future.

Increasing popularity of fast fashion in retail has made greater speed of merchandise movement to market, which is a top priority in global sourcing, and an increasing efficiency namely in labor will be the key to low cost. First, world countries are accepting this to reduce labor costs and compete with other countries (Nayak and Padhye, 2015b). Automated machines will takeover the routine jobs done by workers, which is not to say

Automation in Garment Manufacturing. http://dx.doi.org/10.1016/B978-0-08-101211-6.00012-4

that the workers will be completely eliminated, they will instead be used for other functions such as managing machine operations or in programming a process. This will help in contributing toward low labor cost, an efficient process, and high quality products.

Manufacturing of high quality clothing benefits from advanced designs, computer-based information systems (IS), advanced planning and scheduling (APS), material requirements planning (MRP), and automated manufacturing technology. Products are sold with high value addition leading to consumer satisfaction, widespread global marketing with the ever changing market trends. As mentioned by Michelini and Razzoli (2013), the buyers get an idea about the quality, reliability, and value/price ratio; manufacturers are confronted by demanding challenges, with pressure on prices, new fashion, customization, wide product mix, quick response with dependable delivery dates, to name a few. To achieve these, certain strategies need to be considered such as:

- Work-cycles and production schedules should match with the available resources
- Maintenance of the manufacturing processes to achieve optimum quality with zero-defect production targets
- Delivering the product mix with just-in-time schedules as mentioned by the clients

Traditionally the spinning and weaving were introduced at home and then became a cottage industry. Postindustrial revolution further developed them into organized sectors with machineries and automation. Since then, numerous advances and developments have occurred over the past decades in spinning, weaving, knitting, and garmenting processes that reduces cost and improves the quality. Advancements in garment manufacturing have occurred in the form of robotic handling devices, cutting table with automatic loading system, smart transportation systems, Virtual TryOn, 3D garment design (Thakkar, 2009), MRP, and APS. Robotic handling devices (Michelini and Razzoli, 2013; Anon., 2015) are being used in the garment industry for automated handling of cut fabric parts where they are collected and delivered to the automated manufacturing system. The other named systems and processes help in increasing the efficiency and accuracy of production, which will be the major area of discussion of this chapter.

This chapter discusses on the advancements in production planning and control (PPC) in the context of automation in garment manufacturing. The reasons for adopting automation, the strategies followed along with the current and future trends in the textile and clothing industries have been explained. Traditional garment manufacturing was labor-intensive and was cost-effective due to less demand on the quality and slow production cycle. Due to increased demands for quality clothing, automation is the approach that offers interesting possibilities and potentials for high-tech and better quality garments with lower cost and quick response to the consumer market in terms of current designs and orders. Hence, the scope of using advanced tools and software relating to PPC has also discussed in detail in this chapter.

12.2 Automation in production systems

Automation can be defined as the use of technology with the application of electronic, mechanical, and computerized systems to operate and control production. Automation in production can be of two different categories: (1) automation of manufacturing systems in the factory and (2) computerization of the manufacturing systems. Garment manufacturing industries have adopted the automation, robotics, and application of advanced tools and software for PPC, MRP, and effective control of flow of materials in production and supply chain. Hence, the recent advancements can be classified as automation of manufacturing systems and advancements in production planning. The following section describes the role of manual labor in production systems and the importance of automation.

12.2.1 Manual labor in production systems

On asking the question, if there is any place for manual labor in the modern production, the answer is yes. Even in the highest-automated production systems, human labor is needed in one or the other form. They will be required to manage and maintain the plant or even any advanced production systems. In addition, there are many reasons or situations in which manual labor is usually preferred over automation in garment manufacturing (Sullivan, 2009):

- Tasks are too technologically difficult to automate—certain tasks are difficult to be automated such as problems with physical access to the work location, adjustments when required in the task, and when it demands hand-eye coordination. Example is automobiles seen in final assembly lines.
- Short product life cycle—if a product needs to be designed and brought into the market within a short period and the product needs to be in the market is for a relatively short period, then instead of designing a new automatic manufacturing system, manual labor is preferred for a faster production over an automated system. Tooling for such production can be done in less time and less cost than automated production.
- Customized product—on designing and producing one of a kind product with unique features, manual labor may have an advantage over automated because of its versatility.
- Coping with ups and downs in demand—changes in the input demand of a product changes the output production, and this is possible only during manual labor mode of production. Automated manufacturing system might have a fixed cost involved, where, if output is reduced the unit cost of the product increases.
- Reducing the risk of product failure—introduction of a new product to the market is always uncertain, where some products will have short and some have long product life cycles. Therefore, using manual labor reduces the probable losses associated with a big investment in the form of automation.

12.2.2 Reasons for automation and advanced tools

Companies mostly undertake projects in manufacturing automation for various reasons as described below:

1. To increase labor productivity—automation of any manufacturing operation, irrespective of the size, increases production rate and labor productivity. This means that the amount of output per hour is increased with automation.
2. To reduce labor cost—higher investment in automation in a global competitive market is justifiable on the basis of the reduced labor cost and increased productivity by automation.
3. To mitigate the effects of labor shortages—there is either a general shortage of labor or costly labor charges in most advanced nations, which led to automated operations.

4. To reduce or eliminate routine manual tasks—it can be argued that a similar kind of work done by a worker over and over makes it monotonous, boring, and possibly irksome. Automating such tasks serves a purpose of improving the general work condition.
5. To improve worker safety—by adopting automation into some risky operation can make the worker safe. Furthermore, it gives the worker to divert his attention in to a supervisory role, and the work is made safer.
6. To improve product quality—automation not only results in higher production rates but also leads to greater uniformity and conformity to quality specifications, thereby leading to an almost zero-defect rate.
7. To reduce the manufacturing lead time—automation reduces the elapsed time or lead time between the customer order and product delivery, thereby providing the manufacturer with a competitive edge.
8. To accomplish processes that cannot be done manually—certain operations are incomplete without the use of machines. These processes need precision, which cannot be achieved by manual operations.
9. To avoid the high cost of not automating—the benefits of automation are not seen directly but often in unexpected and intangible ways, such as higher sales, improved quality, and better brand image. Companies who do not go for automation, see themselves at a competitive edge with regard to their customers, employees, and competitors.

12.2.3 Strategies for automation and production systems

Following the USA* Principle is a good first step in any automation project (Sun, 1994; Sun and Riis, 1994; Hax and Majluf, 1996; Johnson and Scholes, 1997; Porter, 1998; Segal-Horn, 1998; Bruno and Pimentel, 2016). *USA stands for: **U**nderstand the existing process, **S**implify the process, **A**utomate the process. The USA principle is a common sense approach to the automation projects.

There are 10 strategies as described below, which are practical and relevant in the context of garment manufacturing. Many researchers and practitioners argue that integration and flexibility are the important concepts governing the development of automation systems and review of automation strategies (Scheines, 1988; Hannam, 1996; Owen, 2000). The success of an organization relies significantly on the formulation and execution of its strategies (Sun, 1994; Sun and Riis, 1994; Hax and Majluf, 1996; Johnson and Scholes, 1997; Porter, 1998; Segal-Horn, 1998). These are referred to as strategies because some of them are applicable whether the process is a candidate for automation or just for simplification.

1. Specialization of operations: This involves the use of specific equipment designed to perform one operation with the greatest possible efficiency.
2. Combined operations: Production occurs in a sequence of operations. Complex products require a lot of processing steps, therefore, ongoing for combined operations reduces the number of machines through which the product was routed. This strategy might save a lot of set up time, and material handling.
3. Simultaneous operations: Logical extension of combined operation is performing simultaneous operations that are combined at one point. Therefore, as operations are performed simultaneously, the total processing time is reduced.

4. Integration of operations: This strategy involves linking several workstations together into an integrated mechanism by the use of automatic handling devices to transfer the products between the stations. This reduces the movement of the products through various machines.
5. Increased flexibility: This strategy is adopted to obtain maximum utilization of equipment for medium volume situations by using the same equipment for manufacturing a variety of products. It also involves the usage of flexible automation.
6. Improved material handling and storage: This strategy is adopted to reduce work-in-progress and shorter manufacturing lead times.
7. Online inspection: Inspection is done on completion of a product to assess the quality of work. Incorporating online inspection permits corrections to the process as the product is being made.
8. Process control and optimization: This involves a range of control schemes that are intended to operate the individual process and related equipment efficiently. The individual process times can be reduced and the product quality is improved.
9. Plant operations control: This strategy is concerned with the control at the plant level. It therefore attempts to manage and coordinate the plant more efficiently.
10. Computer-integrated manufacturing (CIM): This leads to the integration of the factory operations with engineering design and business functions of the firm. CIM involves extensive use of:

    - Computer applications
    - Computer databases
    - Computer networking throughout the enterprise

The 10 strategies described above act as a checklist of the possibilities for improving the production system through automation and should not be considered mutually exclusive.

12.2.4 Advantages of automation

The benefits that are obtained from automation include:

- Fast planning and scheduling
- Reduced production time—increased efficiency in shop floor operation
- Improved data collection
- Control of production at multilocations
- Reduced waste as the operations can be paperless
- Less human error and more repeatability
- Increase in accuracy
- Less employee costs—less safety issues and reduced payroll
- Higher production—increase in profitability

12.2.5 Disadvantages of automation

The disadvantages of automation include:

- Less versatility—limitation on flexibility and variety
- Larger initial investment—higher installation costs
- Increase in unemployment
- Unpredictable costs—maintenance costs, training employees and expensive spare parts

12.3 Automation in manufacturing systems

Automated manufacturing systems are applied on a physical product replacing the labor-intensive work. The mode of manufacturing involves the use of automated machines and robotics where the level of human intervention is reduced. The automated operations may involve: processing, assembling, inspection, or material handling, where more than one automated system can be used for the operations of the same product. Examples of automated manufacturing systems include:

- Automated machine tools
- Transfer lines performing a series of machinery operations
- Automated assembly systems
- Manufacturing systems using industrial robots to perform operations
- Automatic material handling and storage systems
- Automatic inspection for quality control

Maleki (1991) reported that automation can be classified into three basic types based on the function of production volume and product in a manufacturing enterprise. The three types of automated manufacturing systems are: (1) fixed automation, (2) programmable automation, and (3) flexible automation. Rolstadas and Anderson (2000) added that automation can be employed in the preassembly, assembly, and postassembly stages of apparel manufacturing. Many researchers and practitioners argue that integration and flexibility are the important concepts governing the development of automation systems and review of automation strategies (Scheines, 1988; Hannam, 1996; Owen, 2000). According to the National Research Council (1991), an integrated manufacturing system can embrace various aspects of an organization from product concept to product launch and from customer order to after-sales support. Such a system involves the application of a wide range of tools, techniques, and technologies (Hannam, 1996; Russell, 1998, 2000). These applications can be broadly divided into soft and hard automation as depicted in Figs. 12.1 and 12.2, respectively.

12.3.1 Fixed automation

As the name indicates, it is a system in which the sequence of processing operations is fixed in terms of equipment configuration. The sequence for each of the operation is simple, involving a linear or rotational motion or perhaps a combination of two, for example, the feeding of a rotating spindle. Features of fixed automation are (Sullivan, 2009) as follows:

- Higher initial investment cost
- Higher production rates
- Reduced flexibility toward accommodating product variety

The product is economically viable only if it is produced in very large quantities and at high production rates. The higher initial cost can be widely spread of a large number of units being produced, thus making the unit cost low. Examples

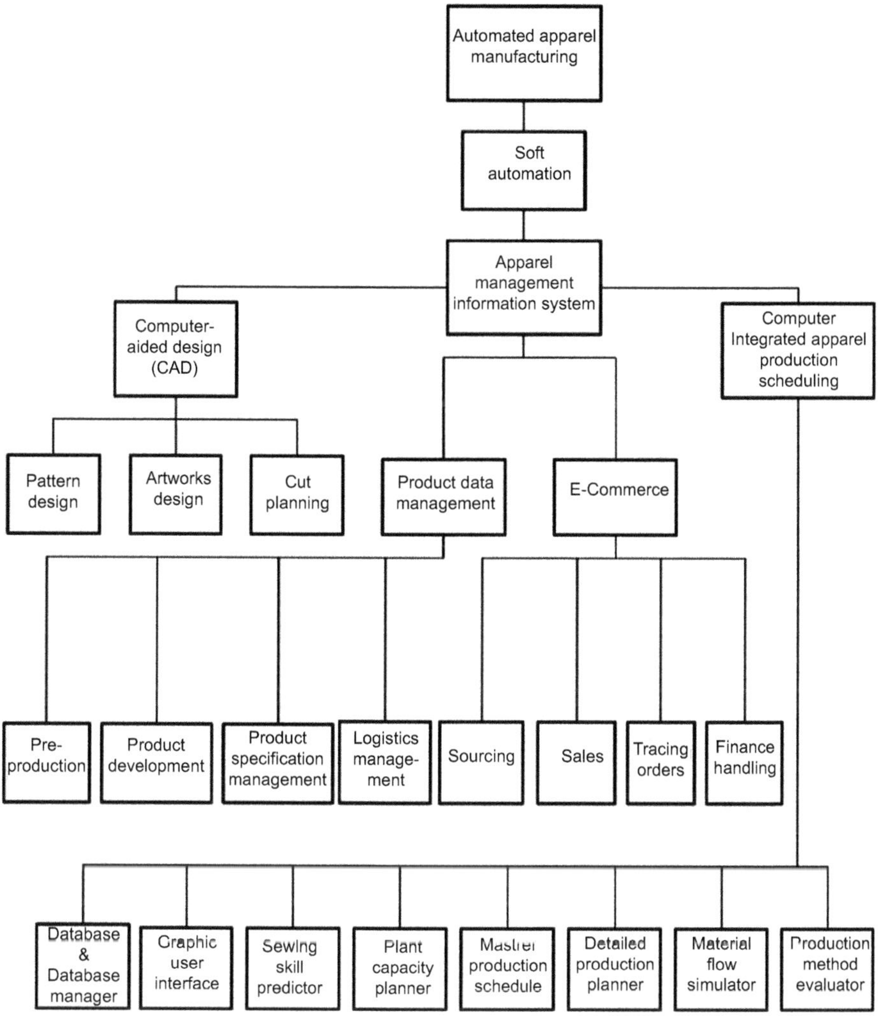

Figure 12.1 Classifications of soft automation in apparel manufacturing.
Reproduced from Chin, K.-S., Pun, K.-F., Lau, H., Leung, Y.S., 2004. Adoption of automation systems and strategy choices for Hong Kong apparel practitioners. The International Journal of Advanced Manufacturing Technology 24 (3/4), 229–240.

of fixed automation can be automated assembly lines in a garment manufacturing industry.

12.3.2 Programmable automation

In this type of automation, the production equipment is designed with the capability of changing sequence of operations to accommodate different product configurations (Sullivan, 2009). The whole sequence of operations is controlled by a program, which

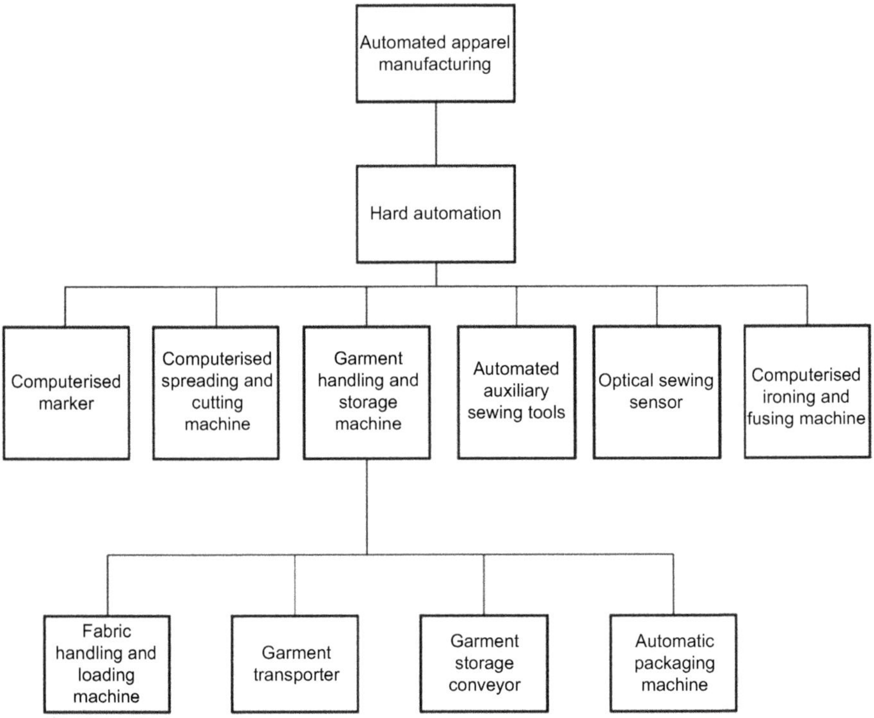

Figure 12.2 Classifications of hard automation in apparel manufacturing.
Reproduced from Chin, K.-S., Pun, K.-F., Lau, H., Leung, Y.S., 2004. Adoption of automation systems and strategy choices for Hong Kong apparel practitioners. The International Journal of Advanced Manufacturing Technology 24 (3/4), 229–240.

is a set of instructions coded that can be read and interpreted by the system. Similarly, to produce new products, new programs can be created and entered into the equipment. Some features of programmable automation are as follows:

- Higher investment cost for the equipment
- Lower production rates in comparison to fixed automation
- Flexibility to deal with changes in product configuration
- Suitability for batch production

12.3.3 Flexible automation

It is an extension of programmable automation, which is capable of producing a variety of products with no time loss for any changeovers from one style to the other (Sullivan, 2009) i.e., no loss in production time, while reprogramming the system. As the amount of changeovers being very minimal, the system can produce various combinations and schedules of products. Features of flexible automation are as follows:

- High investment involved for custom design
- Production of variable products

- Medium rates of production
- Flexibility to deal with product design variations

In the context of the adoption of apparel automation, a list of strategies identified from the literature is given in Table 12.1.

12.4 Advancements in production planning

In responding to the ever changing globalization, customization, and technology-driven competition, apparel manufacturers and trading firms must shift the traditional operation of production toward a new paradigm of manufacturing. The advanced systems and tools and their mode of operation in production system need to be planned before their installation. In addition the use of enterprise resource planning (ERP) and APS system assists in PPC and MRP.

One of the important operations in garment manufacturing is the PPC (Das and Patnaik, 2015). With the increased demand on quality and productivity, accurate planning is essential to meet the date of shipment and reduce bottlenecking. Hence, the use of advanced tools and computer software can help to achieve these objectives. Production planning is the planning of managing the machineries and people in the due course of production (Figs. 12.3 and 12.4 as shown in the next sections). The activities of employees, material flow, and allocation of production capacity is managed by the PPC department to deliver the products as per the buyer's requirements. In many cases the production of garments is very time consuming and complex process, which can lead to failure in meeting the deadline of buyers. The definitions of some of these terminologies are given below:

ERP: ERP involves the application of advanced software and technology for managing and integrating various operations of core business processes in real-time. ERP software integrates various applications in the design, planning, production, and quality control to automate many functions related to technology, materials, services, and human resources.

APS: Any computer program that uses advanced mathematical algorithms or logic to perform optimization and/or simulation on finite capacity scheduling, sourcing, capital planning, resource planning, forecasting, demand management, and others. These techniques simultaneously consider a range of constraints and business rules to provide real-time planning and scheduling, decision support, available-to-promise, and capable-to-promise capabilities. APS often generates and evaluates multiple scenarios (APICS, 2007).

MRP: This is the application of software for material planning using a bill-of-material and a master production schedule to calculate the optimum quantity of material needed to be stored to avoid any delay in the production. MRP systems ensure that: (1) the raw materials are available on right quantity and quality for production so that the products are delivered to consumers timely, (2) a safety stock of material is maintained in the store to avoid any production delay, and (3) the manufacturing activities are smoothly organized to meet the delivery schedules.

IS: The application of IS assists in obtaining accurate information relating to different operations at different locations by the uses information technology. These information can be related to design, planning, manufacturing, financial, or human resource.

Unlike the traditional production planning approaches, the contemporary PPC uses several software including the ERP, to manage the complex activity. The

Table 12.1 Apparel automation strategies

	Strategy choices	Highlights of the strategies
System operation and implementation	S1: Specialization of operation	To use special purpose equipment/tools designed to perform one operation with the possible efficiency
	S2: Combined operations	To reduce the number of distinct workstation or work-in-process to save time in operations
	S3: Integration of manufacturing operation	To link several department processes and workstations processes into integrated automated handling equipment, thus increasing the overall output of the system
	S4: Improved material handling and storage	To reduce nonproductive time of using automated material handling and storage systems, leading to a shorter manufacturing lead time and transportation time
	S5: Process control and optimization	To reduce the individual process time and to improve product quality
	S6: Factory operation control	To manage and coordinate the aggregate operations in the factory more efficiently
	S7: Increased flexibility	To achieve maximum utilization of equipment for small and medium volume situations by using the same equipment for a variety of products
	S8: Integrated business operation	To achieve integration of business communications with overseas customers to serve a quick response for their global market competitiveness
	S9: Encapsulation and modularity	To encapsulate the information or operations and build a system step-by-step with modules of components including hardware and/or software.
	S10: Mechatronic principle	To exploit the synergistic combination of mechanical, electronic, and computer engineering, leading to an increased flexibility of a workstation and communication

System design and new/change requirement	S11: System improvement	To design a system that has the ability to cope with internal and external changes and flexibly conform to new manufacturing and/or business requirements
	S12: Object-oriented paradigm	To use an object-oriented methods to encapsulate the system building modules with manufacturing and business components
	S13: Hierarchical control concept	To provide a reference factory and coordination center structure for the construction of automated system
	S14: Open system architecture	To employ a computer-integrated manufacturing architecture to describe a project's life cycle and the structure of tasks
	S15: Mass customization	To produce medium to large amount of production volume with varied products through flexibility and quick responsiveness, with respect to short product development cycles and short product life cycle
	S16: Concurrent engineering	To manage the design, manufacturing and business activities concurrently to shorten the manufacturing and business lead time
	S17: Manufacturing enterprise wheel	To provide vision for understanding of marketplace and customer requirements
	S18: Business reengineering	To restructure and reengineer the organization to meet the new requirements and changes

Reproduced from Chin, K.-S., Pun, K.-F., Lau, H., Leung, Y.S., 2004. Adoption of automation systems and strategy choices for Hong Kong apparel practitioners. The International Journal of Advanced Manufacturing Technology 24 (3/4), 229–240.

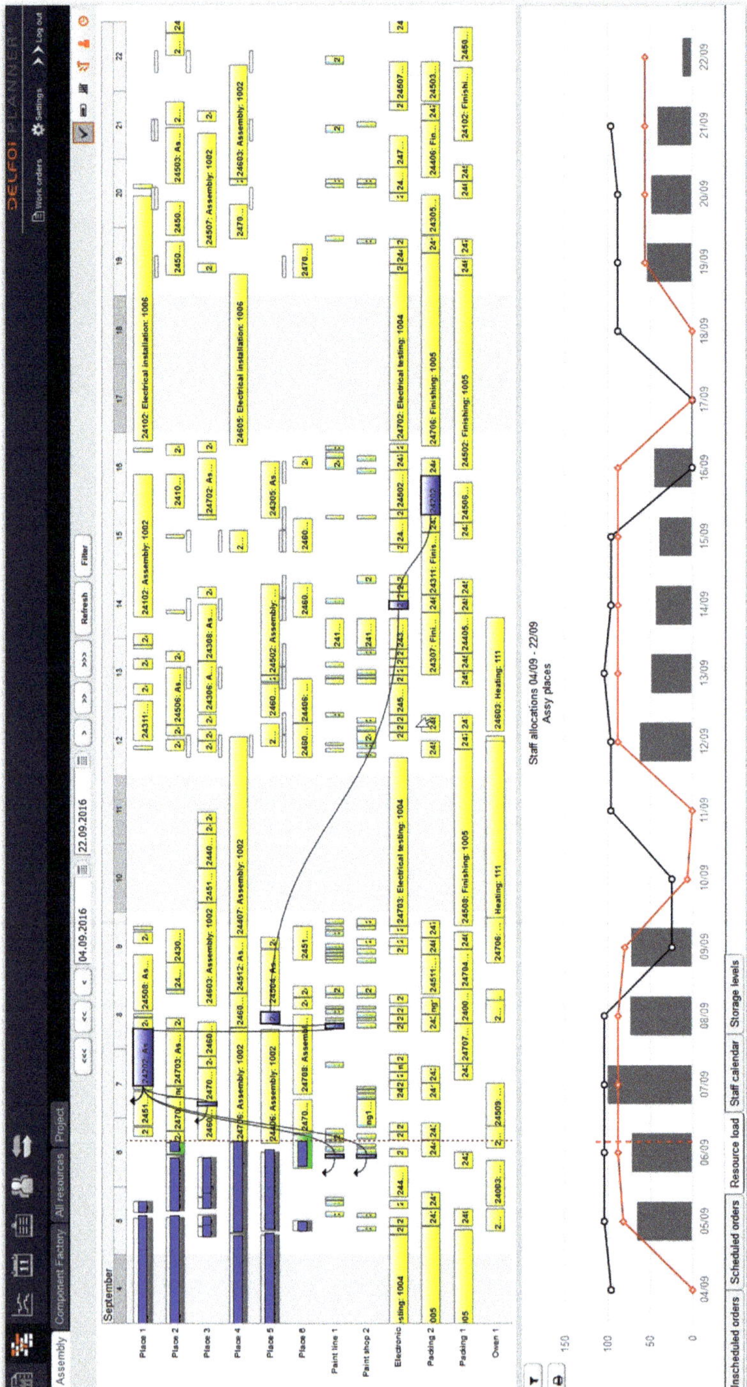

Figure 12.3 Production planning software for advance multisite planning.
Reproduced from Delfoi Planner, 2017. Advanced Planning and Scheduling. https://www.delfoi.com/web/products/delfoi_planner/en_GB/advanced-planning-and-scheduling/.

implementation of ERP takes an integrated approach to manage the business and automate many functions relating to procurement, storage, transportation, production, design, and quality control in garment manufacturing. The following section describes the application of different software in PPC of a garment industry.

12.5 Application of different software and planning tools in production planning and control

Software and planning tools are used as they are powerful, adaptive, and user-friendly. These web-based software and planning tools can be used for APS for cutting, sewing, pressing, packaging, and other operations in the shop floor and automatic data collection from these operations. These software can be accessed by the buyer or other interested parties from remote locations to monitor the work progress. These software and planning tools can be used for advanced planning and production scheduling and reduce the cycle time for planning. Few of the software tools that are used on a daily basis are as follows.

12.5.1 Jobpack production suite

Manufacturing companies with an existing ERP benefit from JobPacks ability to import sales orders, work orders, and/or routings to provide a real-time graphical scheduler without the need to replace or modify an existing ERP. The ability to integrate with other systems keeps cost down and implementation times short so that the company can begin to reap the benefits right away. It helps in gaining control over the shop floor and maximize the amount of work being produced (Jobpack, 2017).

12.5.2 PlanetTogether

PlanetTogether is a scheduling, planning, and analytics platform that makes easier to manage orders, capacity, and inventory in single and multiplant environments. PlanetTogether integrates with SAP, Microsoft Dynamics, Oracle, and other ERP systems (PlanetTogether, 2017).

12.5.3 ProfitFab enterprise resource planning

ProfitFab ERP is designed for the make-to-order manufacturers. ProfitFab is a completely integrated manufacturing software solution designed to implement and streamline management processes, allowing businesses to quickly gain complete control of daily operations. Fully automatic scheduling and management of jobs, inventory, parts, employees, and customers with efficiently integrated ERP software (ProfitFab ERP, 2017).

12.5.4 Resource Manager DB

Resource Manager DB (RMDB) is an affordable, flexible, and quick-to-implement approach to help with the planning, scheduling, and tracking challenges. RMDB features a simple approach for getting results fast. Easily integrates with all systems (Resource Manager DB, 2017).

12.5.5 OmegaCube enterprise resource planning

OmegaCube ERP is delivered and sold as a full suite of integrated modules. Some of the features offered in the system include financial accounting, scheduling, manufacturing, execution, business intelligence, and quality management. OmegaCube ERP also features a proprietary "Developer Studio" that allows users to extend, customize, and build their own applications as well as incorporate their unique business rules into the system (OmegaCube ERP, 2017).

12.5.6 Quantum

Finite Scheduling, part of Quantum Paperless Manufacturing System, offers comprehensive management of workflow processes. Quantum helps to quickly identify bottlenecks, labor shortages, gaps in work, and potential late shipments. Test scenarios for the optimal routing before releasing changes to the shop floor. With dynamic reprioritization and rerouting process control, the manufacturers have an accurate view of all the work and the control necessary to increase profitability and ship the orders on time (CIMx Software, 2017).

12.5.7 SIMATIC IT Preactor advanced planning and scheduling

Preactor International is the world leader in production planning and scheduling software, used by over 4000 organizations of wide ranging types around the world to achieve lean and agile production. Case Studies show that the benefits that can be obtained after installation offer a return on investment measured in months, sometimes in weeks (SIMATIC IT Preactor, 2017).

12.5.8 Delfoi planner

Delfoi Planner is a transparent, web-based APS software, which uniquely includes manufacturing execution functionalities. The APS software is well suited for discrete- and project-type manufacturing, service and maintenance operations. Delfoi Planner can be easily integrated into ERP systems using robust and proven interfaces. The most significant benefits for using Delfoi Planner are drastically reduced lead times (Delfoi Planner, 2017).

12.5.9 Asprova

A factory-oriented production scheduling system for high-speed creation of multi-product multiprocess production schedules (Fig. 12.4). Asprova can be integrated into any ERP system or pull data from any flat files and will optimize processes and track utilization of resources. Asprova gives visibility to daily production plans and

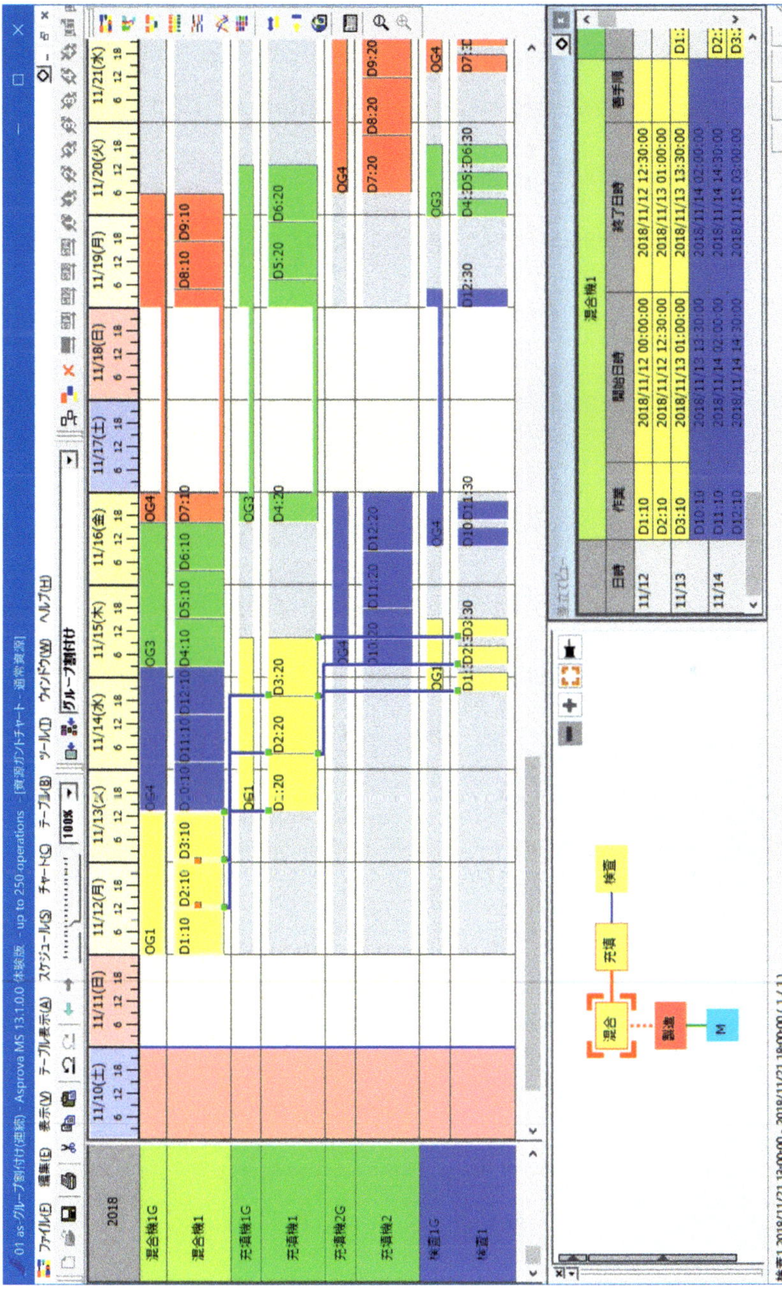

Figure 12.4 Production scheduling system.
Reproduced from Asprova, 2017. Production Scheduling System. https://https://www.asprova.com.

to real-time events that occur. The production schedule drives the shop floor productively and efficiently allowing the manufacturer to increase production and efficiency (Asprova, 2017).

12.6 Computerized manufacturing support systems

Automation is aimed at reducing the manual labor during product design and planning phases and also during the business function phase. In the current scenario, all the manufacturing support systems are implemented using computer systems. Computer technology is used for automation in the factories as well. CIM is used for designing the products, production planning, control of the operations, and perform other business related functions within the firm. CIM involves integration of all the above functions in one system that operates throughout the enterprise. Computer-aided design (CAD)/computer-aided manufacturing (CAM) is the other terms used to identify the CIM system.

12.7 Recent trends

1. The clothing creator is an automated process for apparel manufacturing, which makes loathing without direct human intervention (Stern, 2016) where a garment is finished in a 45 s cycle that combines 3D molding and ultrasonic for cutting/seaming. It can accommodate a variety of knit, woven, and nonwoven materials that contain synthetic fibers. The system has the capacity to produce garments of higher quality, good production lead times, and lower production costs.
2. Gerber Technology—Viet Tien was the first company to utilize Gerber Technology Accumark system—for design, grading, and marker making software for the apparel industry (Gerber Technology, 2016). It resulted in the company having enhanced operations and workforce management. The whole manufacturing procedure works on a CAD system, where they can send patterns and/or markers directly to the workstations via emails saving money by not sending them through couriers.
3. Jeanologia, uses laser as a technology to manufacture denim, which has helped to reshape denim (Donaldson, 2016). It is not a new practice but certain changes have made it possible by replacing human hands in sandblasting, distressing, and even create high-fashion denim patterns.
4. E-flow technology is the current trend where industries are heading to zero discharge machines, e.g., air is transformed into nanobubbles, then finishing products and water create a bubble "skin" of sorts that transports the finishing properties, such as softening or water repellency, to a garment (Donaldson, 2016).
5. An Atlanta-based Softwear Automation Inc. recently introduced a radical new approach to sewing automation (Reddy, 2016). The system that has been developed by the company eliminates issues related to fabric distortion by relying on an advanced computer vision system. The camera tracks the stitching done at the needle and coordinates the precise movement of the fabrics using lightweight robots. This will help to transform today's labor-intensive manufacturing plants into high-tech automation systems.

6. Radio frequency identification technology is also used to capture data that are useful for improving the intelligence associated with the fuzzy logic (Lee et al., 2013; Nayak et al., 2015).
7. Hanging storage systems helps in multitier picking, storage, transport, retail distribution, and sorting, which improves accessibility and visibility (GMS Systems, 2016).
8. During garment production and processing, rail systems and monorail trolley systems increase productivity by reducing supervision (GMS Systems, 2016).
9. Various conveyors are used for transfer between levels such as Rogol conveyor, which is a low maintenance, cost-effective conveyor used during high volume applications (GMS Systems, 2016); similarly Tubular Yoke conveyor is used to transfer cartons to packaging stations; and the Powered Overhead conveyor is used for transfer of trolleys over long distances (GMS Systems, 2016).
10. Vehicle loading is done for transport of garments, where the garments are hanged onto retractable telescopic boom that transfers garments straight or at a 90degree angle.

Therefore, one can adopt and accept various modes of automation and be benefitted in terms of profit, production time, and money.

12.8 Conclusion

As identified since decades as an industry with low technological intensity, the textile and clothing manufacturing can make a major leap in employing more science and technology in the form of automation in its manufacturing processes. Owing to the competition from various sources in the global marketplace, a company needs to introduce a new fashion product in the shortest time possible. The easiest and least expensive way to accomplish such an objective is by opting for the application of advanced tools, software, and automation in production systems in a textile and clothing industry. It will be difficult to develop an automation system without human intervention, but with the current trend, the human component is gradually decreasing in the automation process. Traditional garment manufacturing was labor-intensive and was cost-effective due to less demand on the quality and slow production cycle. Due to increased demands for quality clothing, automation is the approach that offers interesting possibilities and potentials for high-tech and better quality garments with lower cost and quick response to the consumer market in terms of current designs and orders. The future of garment industry will be brighter with the applications of advanced tools and equipment in various departments of garment manufacturing.

References

Anonymous, 2015. Robotics to Rule Textile and Apparel Manufacturing Processes. Available at: http://www.fibre2fashion.com/news/textile-news/robotics-to-rule-textile-apparel-manufacturing-processes-174622-newsdetails.htm.

APICS, 2007. Using information technology to enable supply chain management, APICS certifies supply chain professional learning system. APICS, Alexandria, VA.

Asprova, 2017. Production Scheduling System. https://https://www.asprova.com.

Azmeh, S., Nadvi, K., 2014. Asian firms and the restructuring of global value chains. International Business Review 23 (4), 708–717.

Book, W., Winck, R., Killpack, M., Huggins, J., Dickerson, S., Jayaraman, S., Colin, T., Prado, R., 2010. Automated garment manufacturing system using novel sensing and actuation. In: Proceedings of 2010 International Symposium on Flexible Automation (2010 ISFA), July 12–14, Tokyo, Japan.

Bruno, F.D.S., Pimentel, F., 2016. Apparel Manufacturing 4.0: A Perspective for the Future of the Brazilian Textile and Apparel Industry. Fashion Colloquia, São Paulo, Brazil.

Chin, K.-S., Pun, K.-F., Lau, H., Leung, Y.S., 2004. Adoption of automation systems and strategy choices for Hong Kong apparel practitioners. The International Journal of Advanced Manufacturing Technology 24 (3/4), 229–240.

CIMx Software, 2017. Paperless Manufacturing Solutions. Available at: https://www.cimx.com/.

Das, S., Patnaik, A., 2015. Production Planning in the Apparel Industry. Garment Manufacturing Technology. Woodhead Publishing, pp. 81–108.

Delfoi Planner, 2017. Advanced Planning and Scheduling. https://www.delfoi.com/web/products/delfoi_planner/en_GB/advanced-planning-and-scheduling/.

Distler, J., Fernandez-Seara, J., Gottstein, H., Haemmerle, V., Rasch, S., Rohrhofer, S., November 2014. Apparel at Crossroads: The End of Low Cost Country Sourcing. The Boston Consulting Group.

Donaldson, T., 2016. Automation to Reduce Outsourcing in Apparel Manufacturing. Available at: https://sourcingjournalonline.com/automation-reduce-outsourcing-apparel-manufacturing/.

Eloot, K., Huan, A., Lehnich, M., June 2013. A New Era for Manufacturing in China. McKinsey Quarterly.

Euler Hermes Economic Research Development, January 25, 2013. Reindustrialization of the United States. Special Report. Economic Outlook no. 1187.

Foresight, 2013. The Future of Manufacturing: A New Era of Opportunity and Challenge for the UK. Summary Report. The government office for Science, London, UK.

Gerber Technology, 2016. Vietnamese Garment. Manufacturers Gain Respect from Foreign Trading Partners Because of Manufacturing Automation Systems. Available at: http://www.gerbertechnology.com/fashion-apparel/interior/pattern-design-software/vietnamese-garment-manufacturers.

Gereffi, G., Memedovic, O., 2003. The Global Apparel Value Chain: What Prospects for Upgrading by Developing Countries? United Nations Industrial Development Organisation, Vienna, Austria.

GMS Systems, 2016. Garment on Hanger Systems. Available at: www.gms-systems.com.

Hannam, R., 1996. Computer Integrated Manufacturing; from Concepts to Realization. Addison Wesley, London, pp. 103–125.

Hax, A.C., Majluf, N.S., 1996. The Strategy Concept and Process: A Pragmatic Approach, second ed. Prentice Hall, USA, Upper Saddle River, NJ.

JobPack, 2017. Advanced Planning- Dynamic Graphical Scheduling. Available at: https://jobpack.com.

Johnson, G., Scholes, K., 1997. Exploring Corporate Strategy, second ed. Prentice-Hall, Hertfordshire, UK.

Lee, C.K.H., Choy, K.L., Ho, G.T.S., Law, K.M.Y., 2013. A RFID-based resource allocation system for garment manufacturing. Expert Systems with Applications 40 (2), 784–799.

Maleki, R., 1991. Flexible Manufacturing Systems-Technology and Management. Prentice Hall, New Jersey, USA.

Michelini, R.C., Razzoli, R.P., 2013. Robotics in clothes manufacture. International Journal of Mechanical Engineering and Applications 1 (1), 17–27.

National Research Council, 1991. Improving Engineering Design. National Academy Press, Washington, USA, pp. 15–29.

Nayak, R., Padhye, R., 2015a. Introduction: the apparel industry. In: Nayak, R., Padhye, R. (Eds.), Garment Manufacturing Technology. Elsevier.

Nayak, R., Padhye, R., 2015b. Garment Manufacturing Technology. Elsevier.

Nayak, R., Singh, A., Padhye, R., Wang, L., 2015. RFID in textile and clothing manufacturing: technology and challenges. Fashion and Textiles 2, 1–16.

OmegaCube ERP, 2017. Manufacturing and Distribution ERP Solutions. Available at: https://www.omegacube.com.

Owano, N., June 10, 2012. Darpa Issues Robot Challenge to Clothing Imports. Phys Org. Available at: https://phys.org/news/2012-06-darpa-issues-robot-imports.html.

Owen, P., May-June 2000. Flexible Automation to Increase Productivity. Text Horizons, pp. 21–22.

PlanetTogether, 2017. Available at: https://www.planettogether.com.

Porter, M.E., 1998. Competitive Advantage: Creating and Sustaining Superior Performance, second ed. Free press, New York.

ProfitFab, 2017. Available at: https://www.prfitfab.com.

Reddy, K.P., 2016. The Rise of Robotic Automation in the Sewing Industry. Textile World. April/May/June 2016. Available at: http://textileworldasia.com/textile-world-asia/2016/05/the-rise-of-robotic-automation-in-the-sewing-industry/.

Resource Manager DB, 2017. Planning and Scheduling Solutions. Available at: https://www.usersolutions.com/production-scheduling-products/resource-manager-db/.

Rolstadas, A., Andersen, B., 2000. Enterprise Modeling Improving Global Industrial Competitiveness. Kluwer Academic Publishers, London, UK.

Russell, R., March 1998. Management Information System. World Cloth Manu, pp. 10–17.

Russell, R., June 2000. Manufacturing Software. World Cloth Manu, pp. 13–32.

Scheines, J., 1988. Flexible Apparel Manufacturing. American Apparel Manufacturers Association, Arlington, USA.

Segal-Horn, S. (Ed.), 1998. The Strategy Reader. Open University/Blackwell Publishers, Milton Keynes, UK.

SIMATIC IT Preactor, 2017. Advanced Planning and Scheduling. https://www.siemens.com/mcms/mes/en/mescomponents/preactor/pages/ildefault.aspx.

Sirkin, H.L., Zinser, M., Hohner, D., 2011. Made in America, Again: Why Manufacturing Will Return to the U.S. The Boston Consulting Group.

Stern, B., 2016. An Automated Process for Apparel Manufacturing. Available at: http://contest.techbriefs.com/2016/entries/machinery-automation-robotics/6327.

Sullivan, D.O., May 2009. Industrial Automation. Course Notes. Universidade do Minho.

Sun, H.Y., 1994. Patterns of organisational changes and technological innovations. International Journal of Technology Management 9 (2), 213–226.

Sun, H.Y., Riis, J.O., 1994. Organizational, technological, strategic, and managerial issues along the implementation process of advanced manufacturing technology: a general framework of implementation guide. International Journal of Human Factors in Manufacturing 4 (1), 23–36.

Thakkar, A., 2009. Automation in garment manufacturing-Its likely impact on advanced and emerging economies. In: In International Conference on Steering Mature Business-A Leading Challenge to the Textile Industry, 3rd October 2009, Ahmedabad, India.

The Economist, March 10, 2012. The End of Cheap China: Why Do Soaring Wages Mean for Global Manufacturing. Business Manufacturing.

Use of advanced tools and equipment in industrial engineering

13

Maria-Carmen Loghin, Irina Ionescu, Emil-Constantin Loghin, Ionuţ Dulgheriu
"Gheorghe Asachi" Technical University of Iaşi, Iaşi, Romania

13.1 Introduction

The manufacturing process in the clothing industry is complex, made of subprocesses with different degrees of automatization that are highly segmented and labor intensive (Chen et al., 2014; Nayak and Padhye, 2015a). Clothing manufacturing depends on manual operations for production, handling of materials, and storage. The process incorporates complex systems that involve operators, raw materials, information, machines, and energy; which is in a continuous state of redesign and improvement to be work efficient and cost-effective (Guo et al., 2011). Although there is scope for automation, there are certain problems in applying automation, such as the nature of fabric, which is the major component of a garment. However, other areas are involved in production that can improve the automation process, for example industrial engineering (IE) is one such area (Martin-Vega, 2004). Specific IE techniques that are successfully used for activities of automation in clothing manufacturing systems are:

- workplace layout
- work study analysis
- time study and databases with standard times
- line balancing

In today's clothing manufacturing environment, the processes are characterized by short production cycles and constantly diminishing batch sizes, whereas the variety of product types and models continues to increase (Nayak and Padhye, 2015b). Constant pressure to shorten the lead time adds to these demands and the need to decrease wasted time becomes a challenge. Hence, concepts such as lean manufacturing come into play to address the issues of shorter production cycle and the reduction of waste. According to lean manufacturing principles, the seven causes of waste are:

- overproduction
- inventory
- waiting
- motion
- defects (rework)
- overprocessing
- transportation

Automation in Garment Manufacturing. http://dx.doi.org/10.1016/B978-0-08-101211-6.00013-6

Among these seven types of waste, "motion" is greatly controlled by IE applications (Babu, 2012). The optimum work method, from the point of view of motion, presumes that the motion, which is not necessary can be eliminated, with direct impact on execution time. Another way to optimize time is to redesign the workplace by reducing distances and finding technical solutions to ease motions. Line balancing has effects on diminishing or eliminating waiting time, reducing online stock, and optimizing transport time, which is controlled by work study and motion study.

13.2 Work study

The work study or work method is composed of rules established for work execution by the operator in a standard procedure. In clothing production, there is more than one technological variant to achieve an objective, according to the number and order of technological operations (Nicolaiov et al., 2005). Compared with the work method, the working mode means customizing the work method for the operator's individual execution. The working pedagogy is important, and the way in which a method is taught, practiced, and learned determines the efficiency of an activity. The contents of the work study includes (Loghin et al., 2005):

- formulation of the actions of the work task and splitting of development segments, according to the proceeding steps
- distribution between the operator and the machine
- establishment of a succession in time of the segments
- quantitative analysis of the segments, based on methods-time measurement (MTM) analysis
- planning and structuring of the time rate for operations
- evaluation of alternative working methods for the same task, based on time rate and the operator's effort

The targets of the work study are:

- methods are optimized according to the endowment
- workplaces have a rational configuration and location
- there is optimum task distribution among operators of the technological line
- there is endowment of the workplace with the necessary means for the working tasks
- good working conditions are ensured for the executant, based on ergonomic criteria
- the necessary training programs are identified

As far as specific operations for the clothing manufacturing process are involved, the following sequences are identified as common elements: (1) feeding, (2) execution, and (3) disposal. Work study is composed of elements of movement necessary for the development of these operations, with important implications for operational time and work productivity (Nicolaiov et al., 2010). The necessary time for each operation is objectively influenced by factors such as:

- difficulty in handling fabric
- the number of cut pieces/semifinished pieces involved in the operation
- the size of cut pieces/semifinished pieces
- workplace configuration (distance, basic and supplementary endowment)

The following section describes common elements (feeding, execution, and disposal) involved in general operations in garment manufacturing.

1. Feeding of cut pieces/semifinished pieces:
   In the method study of sewing operations, the beginning of the feeding sequence is considered to be the first reach of the hand toward the cut piece/semifinished piece, and the end is the final position under the presser foot. Feeding consists of grasping the cut piece and moving it to an intermediate or final position. These elements are repeated for all pieces. An example is "attaching a pocket on a shirt," which involves:
   a. front feeding (grasp the front, place on the sewing table, and eventually place under the presser foot)
   b. pocket feeding (grasp the pocket, position on front, and match on the mark)
   c. position under the presser foot
   Grasping methods for the cut pieces/semifinished pieces can vary according to the difficulty in handling:
   a. easy grasping (by closing the fingers: G1A)
   b. difficult grasping (with supplementary fingers move: G1B and G1C)
   Grasping methods for the cut pieces/semifinished pieces, according to the workplace configuration and technological requirements, may involve:
   a. easy grasp
   b. grasp with turn
   c. simultaneous, symmetrical grasp (two pieces with both hands)
   d. successive grasp (two pieces with one hand)
   Position methods for the cut pieces/semifinished pieces vary, according to difficulty:
   a. easy position, with acceptable tolerance
   b. accurate position, with precision
   Position methods of cut pieces/semifinished pieces, according to the destination:
   a. positioning on another piece
   b. positioning under the presser foot
   By combining these steps, different work methods can be designed for the feeding segment and different values for the operational time can be obtained, as explained in the following example:
   Example: Run stitch flap on shirt (feeding step)
   Method 1: Grasp front flap; then grasp back flap with turn and position on front flap; align edges, positioning under the presser foot (time needed: t=0.107 min)
   Method 2: Grasp front and back flap simultaneously (symmetrically); align edges, positioning under the presser foot (t=0.071 min). With Method 2 the time is reduced by 34% compared with Method 1.
2. Execution (sewing of cut pieces/semifinished pieces):
   In the method study, the beginning of the sewing sequence is considered to be the moment when the sewing machine is set into action, and the end is the when the threads are cut. Between these two sequences, the steps involved are usually:
   a. handling of cut pieces/semifinished pieces
   b. sewing/stop sequences
   c. devices or auxiliary materials operation and then thread cutting
   Handling during the sewing sequence is necessary for:
   a. positioning at notches
   b. positioning at the sewing line end (long seams)
   c. maintaining layers in contact (attaching labels)
   d. checking the length of the sewn pieces (applying collar)

e. orientation of the sewing allowances (one side or symmetrically)
f. positioning a supplementary ply (three plies overlapped)
g. moving or straightening the fabric (turning on face/back)
h. pleating, simple or double
i. changing the sewing direction (turning the fabric around the needle, small or big piece)

The sewing sequence is essentially influenced by the operator's qualification and skill set, but there are also objective influencing factors, such as the technical performance of the machine:

a. sewing speed
b. needle position (programming the needle down at intermediate stops eliminates hand movement to the pulley, to insert the needle into the fabric: approximate time, 0.015 min)
c. devices and operation mode (in/out possibility)
d. existence of a control panel, possibility of automatic back-tacking (eliminates hand movement to the back-tack lever: approximate time, 0.02 min for one back-tack)
e. thread cutter on the machine

The sewing time is also influenced by the stitch parameters:

a. stitch density (for a 10-cm seam, at the same speed, the time is 0.015 min for 3-stitch/cm density and 0.020 min for 5-stitch/cm)
b. sewing path (straight, big radius curve, small radius curve, polygon, etc.).
c. back-tacking stitches at the seam's beginning and end, and number and layout of back-tacking stitches.

Regarding the work method, the sewing path steps (the number of stops) have the greatest influence over the operational time, as discussed in the example:

Example: Elbow seam, men's jacket, 50 cm length, one notch, sewing speed 4000 stitches/min, density 4 stitches/cm.

Method 1: Back-tack at start, position at notch, sewing 15 cm, checking notches and end, sewing 15 cm, position at the end, sew 20 cm, end back-tack [time 1 (t_1) = 0.194 min].

Method 2: Back-tack at start, position at notch, sew 25 cm, position at the end, sew 25 cm, end back-tack (t_2 = 0.164 min).

Method 3: Back-tack at start, position at end (fixing the two-ply with two hands), left-hand move with the sleeve to maintain the ply position, sew 50 cm with gradual return of the left hand to the body, end back-tack (t_3 = 0.159 min).

Method 3 has a time economy of 12% compared with Method 1 and 4% compared with Method 2. The time economy results from the decreased number of stops (one stop consumes 0.007 min for deceleration and acceleration).

Thread cutting after sewing can be performed:

a. manually, with classic or thread scissors. This means a hand reach toward the scissors, grasp, cut, and release to the sewing table (t = 0.05 min for one cut). The method can be improved by keeping the thread scissors in hand (t = 0.013 min).
b. manually, by passing the thread allowance over a cutting edge (with or without grasp, t = 0.02 min),
c. mechanically, using the treadle at the sewing machines with thread cutter (t = 0.006 min).

3. By disposing of the cut pieces/semifinished pieces:

In the method study, the beginning of the disposal sequence is considered to be the moment the thread is cut, and the end is the withdrawal of the hand after releasing the semifinished piece. The disposal sequence is influenced by the:

a. size and manufacturing stage of the semifinished piece
b. fabric's characteristics (tendency to wrinkle, flexibility etc.)
c. logistic organization of the technological unit (formation of bundles, transport, etc.)
d. supplementary location at workplace to collect the sewn pieces (boxes, cases, trolleys)

According to these factors, the disposal sequence can be performed:

a. with grasping, moving to the stacking place, and release

b. with grasping, "shaking" to straighten, moving to the stacking place, and release (usually for big pieces)

c. with grasping, "shaking" to straighten, moving to the stacking place, folding, and release (usually for big pieces)

d. without grasping, and sliding on the sewing table (usually small pieces)

e. continually sewing in a chain. In this case, the evacuation time is zero but the operational time is supplemented by cutting and stacking pieces.

13.2.1 Design principles and optimization of work study techniques

The method study is a part of the work study that includes systematic and critical analytic research on existent or programmed work processes, that can help in the design and practice of efficient processes. The method study must ensure good work conditions for the operator. The method study can cover a wider area, if it is referring to the whole company, the departments, and the technological lines, or a limited area if it implies only the working place (Loghin, 2005) (Fig. 13.1).

The workplace configuration is based on the knowledge and analysis of the working procedure, working method, and mode. The method study can be used in various fields such as:

- work planning: activities, means of production, manufacturing time, delivery time, and necessary personnel
- work structure
- labor remuneration;
- establishment of manufacturing/production costs
- check for exploitation costs

The level of data obtained from the method study is:

- very precise for retribution
- estimated and of lower precision for predefined costs or work structure

Factors that influence the use of a specific work method are presented in Fig. 13.2.

Rationally, the work method should be in a continuous process of change and improvement, owing to the model's variability and technological solutions. The staff's attitude toward the change in work method has a significant influence on the work method. The tendency is to resist to change brought about by the new methods. Usually, people involved in work methods want practical confirmation of the usefulness of the change before accepting it. With knowledge of the psychological character of the personnel and local conditions, some measures can be taken to diminish difficulties:

- explanation of the general principles and of the real objectives of the study to the operators
- explanation of the advantages of the improved method at the workplace, especially regarding unwanted operations or operations with high concentration, for which it can be shown that the method study will effectively reduce effort and tiredness
- demonstration of the method and the steps involved in it
- benefits that will result from using the improved method

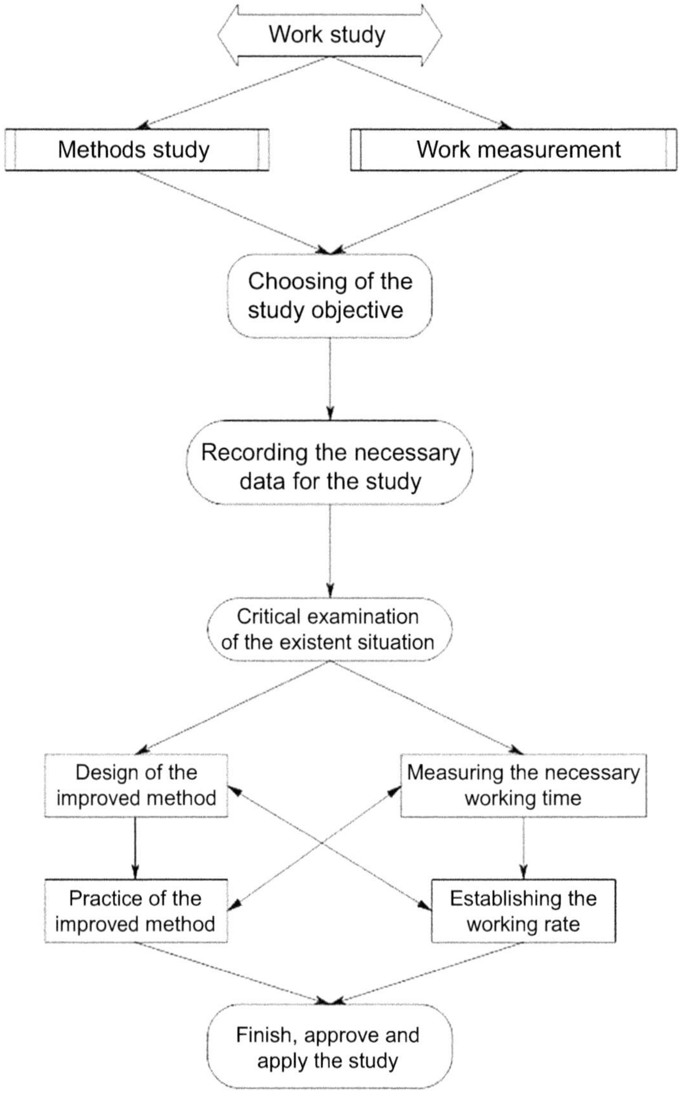

Figure 13.1 The flow diagram of work study.

A work method is characterized as shown in Fig. 13.3.

A very important characteristic of the work method is flexibility. If strict methods are imposed, negative effects may result, such as an increase in execution time or a lack of motivation by the operator. A work method must be changed rationally, not just "change for the sake of change." In practice, only the work result is important. If this

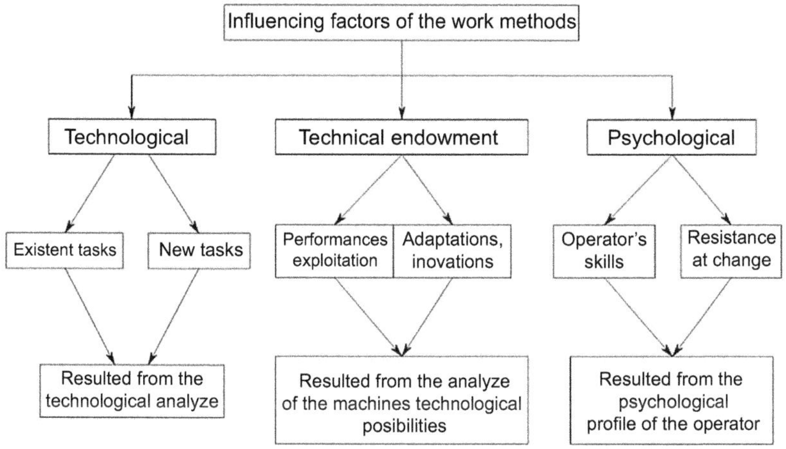

Figure 13.2 Influential factors of work methods.

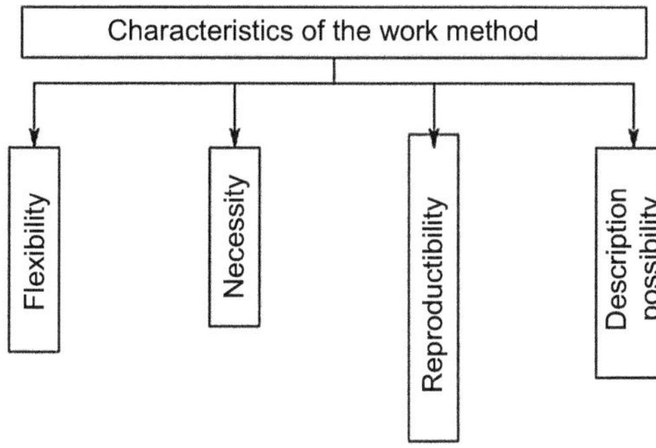

Figure 13.3 Characteristics of the work method.

implies a convenient effort, the method does not have to change. Hence, the method characteristic, necessity, has a role. Sometimes, too rigid a formulated method can lead to bad results. In the meantime, a complex method can be simplified, if the result is the same.

From the reproducibility point of view, a method is considered to be good if it is easily taught, the cost is low, and it has a high degree of repeatability. A method must be precisely described, in simple words, with lots of graphic descriptions, termed the description possibility. A large amount of information must be transmitted with drawings, eventually with a color code for clearer understanding. The method study can be used to solve different problems:

- rationalization of the manufacturing process from a workplace, technological line, or department
- technological flow organization
- improvement of machines and workplace layout

- movement rationalization (means of transportation or operators)
- rationalization of operator's movements at the workplace, etc.

Two types of method studies can be identified:

1. General method study of the manufacturing process, including:
   a. general analysis: the target is to emphasize the changes in work objects and the checking activities needed to confirm these changes, to rationalize their succession and build a base for further studies
   b. detailed analysis: the study deepens the whole process of analysis or a part of it, considering all of the stages followed by the work object during its transformation
   c. movement analysis: refers to all movements made by elements involved in the process, to reduce or eliminate them
   d. layout analysis: the target is to find the most suitable places for the work means, to eliminate unnecessary movements and the time necessary to save the means of transportation or the workspace
2. Work process study at the workplace, including:
   a. analysis of the operator's activity: emphasizes activities developed for productive reasons, but also useless activities; analyzes targets raising the percentage of productive work and its efficiency. For teamwork, the analysis establishes the best action modalities for operators, so that nonproductive time between the different moments of intervention are reduced. In this case, an objective is to reduce the operators' number of activities for one operation, by growing the degree of use.
   b. analysis of operator: machine activity. The target is to reduce the total operation cycle by improving the degree to which the machine and the operator are useful, to establish the best correlation between the operator's interventions and the machine's work.
   c. analysis of the operator's movements: here, the study is more thorough, to rationalize the operator's movements in the working area, based on the economy of movement.

According to the targets of the method study, the following steps explain the justifications for performing a method study:

1. For improvement reasons:
   a. choosing the objective (department, technological line, or work place)
   b. recording data for the study (working means position and situation, time used in the process, execution mode of the operations, etc.)
   c. critical examination of the existing situation in logical order
   d. elaboration on a new, improved method
   e. implementation of the new method
   f. completion, approval, and implementation of the new method
2. For design reasons:
   a. documentation
   b. analysis
   c. design
   d. practical checking
   e. completion and approval

The success of a method study consists of a detailed analysis based on the principle "Nothing is obvious; nothing is absolutely needful."

13.2.2 Work methods analysis and evaluation based on motion study

Within the total time to process an order, only 65% to 70% is represented by manufacturing (Fig. 13.4); the rest is used to supply and deliver the raw material to the recipient. Of the manufacturing time, only 30% is dedicated to value added activities; the rest is represented by logistic activities (transport and storage). Of the value added activities, 30% or more than half are represented by secondary activities (handling); the rest are effective interventions to transform work objects (Nicolaiov et al., 2010).

Looking at Fig. 13.4, the following aspects of manufacturing need to be considered to increase work efficiency:

1. technical aspects:
   a. choosing the proper machines and using all of their technological possibilities
   b. maintaining the performance of the machines
   c. using advanced devices
   d. eliminating all unnecessary process sequences
2. organizational aspects:
   a. reducing transport time between operations by: decreasing the length of routes, using straight routes, and choosing appropriate transport systems
   b. reducing storage time by: preventing or improving bottlenecks in technological flow and reducing the number of pieces simultaneously manufactured
3. work study aspects:
   a. choosing the appropriate working method established by work study
   b. decreasing handling time
   c. ergonomically adjusting the workplace for the operator

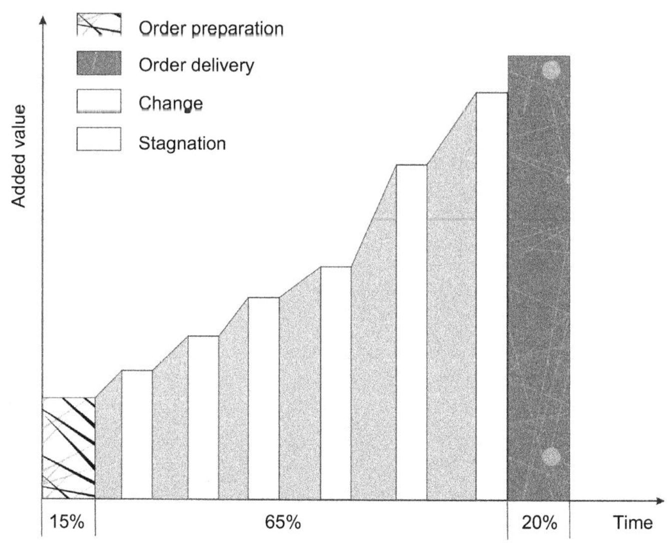

Figure 13.4 Structure of order processing.

Complex manufacturing activities can be described by some elements of motion. By decomposing the processes in elements of motion, possibilities can be identified to increase the work efficiency of an operator, to improve productivity. The elements of motion are the basis of predetermined time standards (Joung and Noh, 2014), made up of instructions for the analysis of motion processes and tables that show time values. Based on these instructions:

- the motion processes are split into standard elements of motion
- at the execution of movement, the process characteristics and parameters are considered
- the elements of motion are associated with time values from tables

Based on this work study, to configure a workplace, different configurations can be compared at the planning stage. Using the motion technique principle, workplace configurations are based on the following principles (Loghin and Nicolaiov, 2013):

- the results of motion are simplified by recognizing and decomposing difficult elements in a movement process
- handling optimization means recognizing and eliminating all unnecessary motions from a process
- automatization and partial mechanization allow work productivity to increase above limits reached by optimizing manual labor.

13.2.3 Elements of motion

Basic elements of motion for work methods involved in garment manufacturing can be of five types: reach, grasp, move, position, and release (Hidoş and Isac, 1972), according to MTM-1, as presented in Fig. 13.5 and described subsequently:

1. reach represents hand motion toward a work object
2. grasp represents fingers bending to pick up a work object

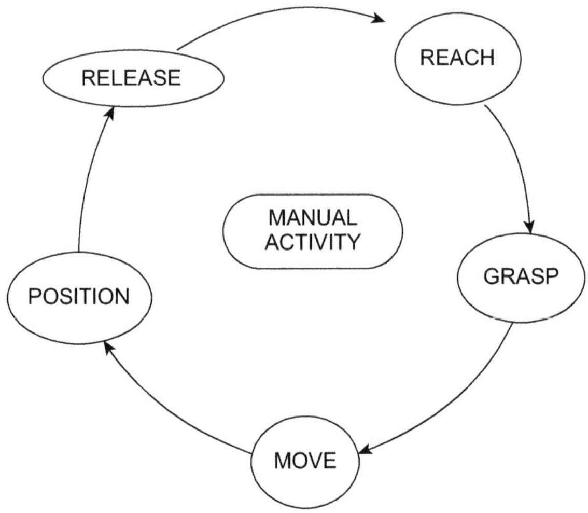

Figure 13.5 Basic elements of motion for manual activity.

3. move represents a work object motion using the hand
4. position represents joining or sliding the work objects
5. release represents opening the fingers containing the work object

13.2.3.1 Reach

Definition: Hand or finger motion toward a precise or indefinite place.
Coding: Example: R40B

Motion cases are determined by the degree of control necessary for motion; they are determined by the location, shape, and size of the object toward which the hand reaches:

- case A – Automatically
- case B – Banal
- case C – Choosing
- case D – Delicate (dangerous)

Case A: the fingers or hand reach toward an object located in a precise, constant, well-known position, or on the other hand. Example: Reaching the hand toward the sewing machine drawer, at 20 cm: R20A.

Case B: the hand or fingers are moving toward an usually remote object. The location can vary from one cycle to another. For Case B, an indirect look is necessary. Example: Reaching the hand toward the scissors placed on the sewing machine table, at 25 cm: R25B.

Recommendations

- It is recommended that workplaces have well established position for objects that are used frequently.
- It is recommended that all other cases be transformed in Case A or B.

Case C: the hand or fingers are moving toward an object mixed with others of the same or different kind, so that at reaching it requires a search and selection.

Case C must meet two conditions:

1. the objects are not hidden, so they can be chosen and selected only by looking
2. the objects are not tangled or hooked onto each other

There must also be maximum of two types of objects mixed together; otherwise the work method must be changed because it is irrational.

Example: Reaching the hand toward a coat button mixed with other buttons (36 cm): R35C.

Recommendations

- find solutions to separate mixed objects (buttons and labels)
- pay attention to the length of the route, which influences time

Case D: the hand or fingers are moving toward an isolated object, a small one that requires caution or precision to be reached. Small objects are considered to be those that need a direct glance. Caution is necessary to avoid damaging the object or hurting the hand.

Example: Reaching the hand toward a blouse button (size: d=7.5 mm, 12L) (pin from a box), at 35 cm: R35D.

13.2.3.2 Grasp

Definition: Grasping is a motion of the hand or fingers to control one or more pieces for the next basic move.

Coding: G1B

Example: Based on the factors of influence, there are five types of motion of grasp:

- G1: grasp for picking
- G2: regrasp
- G3: grasp for transfer
- G4: grasp with selection
- G5: grasp to make contact

The G1 type of grasp consists of ensuring control with an object by taking the object between the fingers. Depending on the shape, size, or location of the object, there are three cases:

- Case G1A
- Case G1B
- Case G1C

G1A: ensuring control over an isolated object regardless of the size, simply by closing the fingers (Fig. 13.6).

Example: Grasp the chalk, which is placed on a vertical position, from a support: G1A.

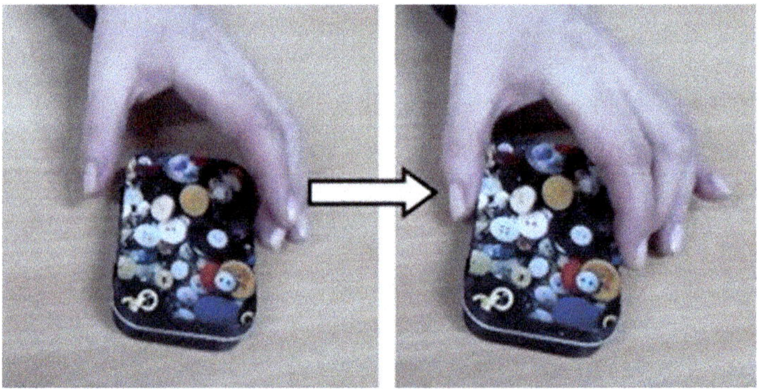

Figure 13.6 Grasp: G1A case.

G1B case: ensuring control over a small or flat object placed on a flat surface (Fig. 13.7).

Example: Grasp a template or sheet of paper from a stack on the worktable: G1B.

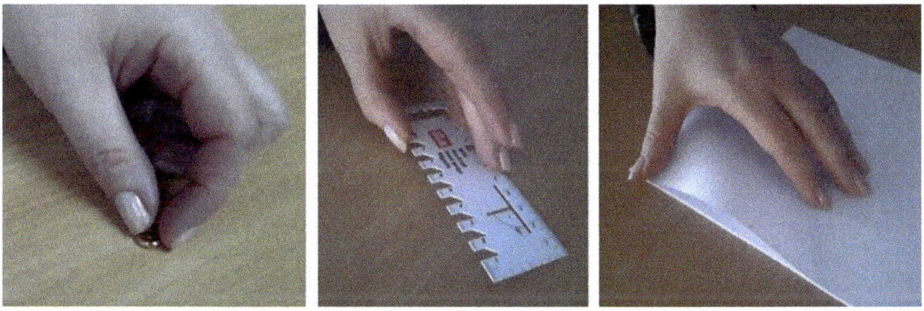

Figure 13.7 Grasp: G1B case.

The G1C case consists of ensuring necessary control over an object with a cylindrical section that is in contact with the surface on which it is placed, and with a side surface (Fig. 13.8).

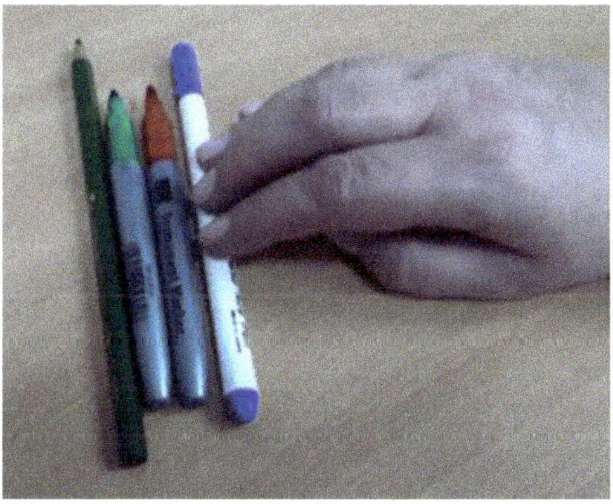

Figure 13.8 Grasp: G1C case.

The motion in the G2 category consists of changing the place from which an object is grasped, to bring it under the fingers' control or to control over it better or to bring it into a position suited for the next move (Fig. 13.9). A classic example is changing the grasping place of a pen to position it for writing.

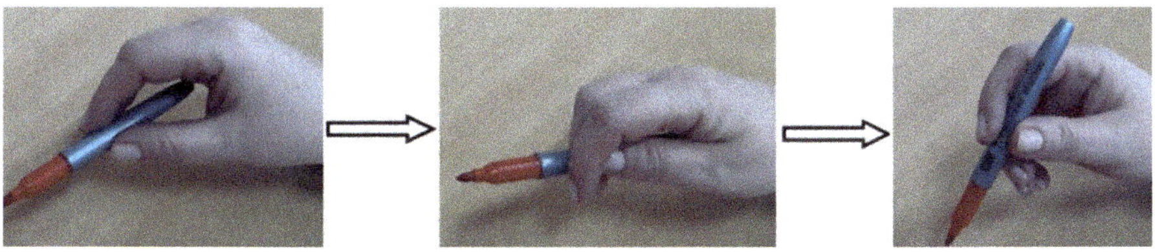

Figure 13.9 Grasp: G2 case.

Motion in the G3 category consists of passing an object from one hand to another (Fig. 13.10).

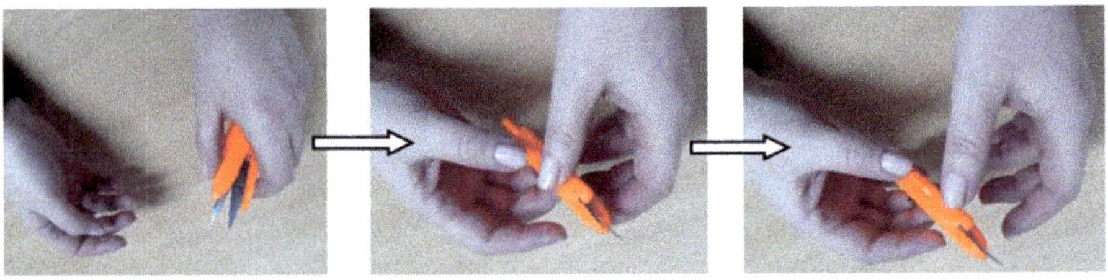

Figure 13.10 Grasp: G3 case.

Motion in the G4 category consists of controlling one object when it is mixed among others so that, before it is grasped, the object is searched for and selected (Fig. 13.11).

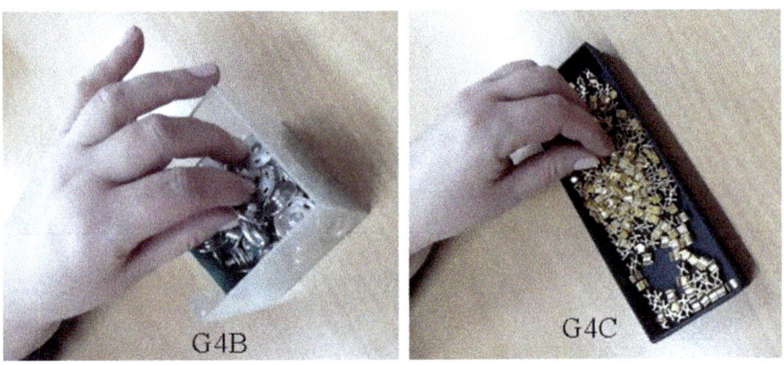

Figure 13.11 Grasp: G4 case.

According to the selected object size, the following categories for G4 movement are established:

- G4A: objects larger than $25 \times 25 \times 25$ mm
- G4B: objects between $6 \times 6 \times 3$ and $25 \times 25 \times 25$ mm
- G4C: objects smaller than $6 \times 6 \times 3$ mm

Examples: Grasp a coat button from a box with other buttons: G4A; grasp a bobbin placed in the sewing machine drawer among other bobbins: G4B; grasp a pin from a box: G4C.

Recommendation

- the complexity of motion gradually increases from G1 to G4, and so does the time for the task. Technical solutions need to be found to decrease the motion case: for example, in case G4, to separate mixed objects.

13.2.3.3 Move

Definition: Move one or more pieces, using the hand or fingers, to a precise or indefinite place.
Code: M40B

Example: According to the destination there are three cases, depending on the control needed to move the object:

- Case A: Automatically
- Case B: Banal
- Case C: Controlled

Case A: The object is moved by hand or fingers to one obstacle or to the other hand (Fig. 13.12).

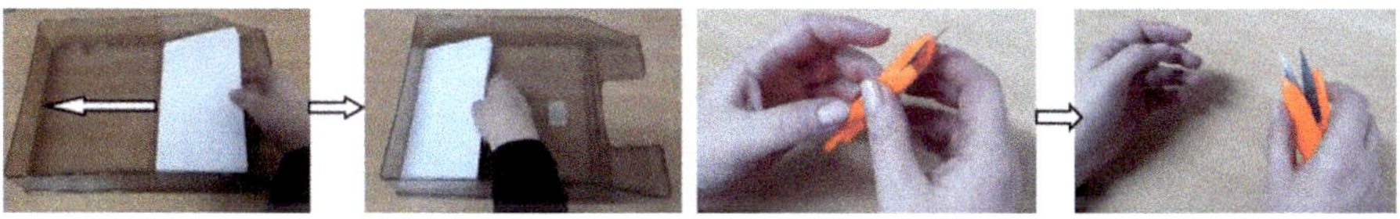

Figure 13.12 Move case A.

Recommendation

- use systems to limit the position and reduce the time for movement

Case B consists of moving an object with the hand or fingers to an approximate or indefinite location (Fig. 13.13). This is the most frequent case and it is recommended that a solution be found to change it in MA case. For example, move the scissors 40 cm to cut threads at the sewing machine: M40B.

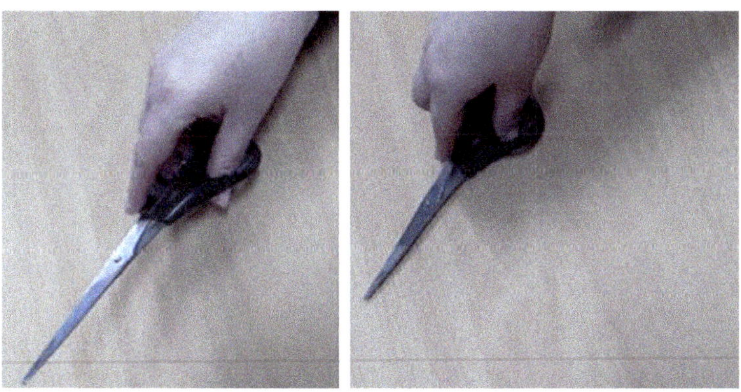

Figure 13.13 Move case B.

Case C consists of moving an object by hand or fingers to a precise location. The object must be placed in a specified position in relation to another object (positioning follows). For example, move the bobbin 20 cm to the bobbin case placed in the other hand (positioning follows): M20C.

13.2.3.4 Position

Definition: A motion of the fingers or hand so that two or more objects are oriented, aligned, and arranged in a precise physical link.

Code: P2SSE

Example: Position is defined as an effort for control, with precision <25 mm of the target; this precision separates position from move.

Adjusting the class gives information about the effort necessary for precision and the pressing force. There are three adjusting classes (Table 13.1).

Table 13.1 Adjusting classes

		Position 1	Position 2	Position 3
Joining each other	Necessary force	No pressure	Easily pushed	Strongly pushed
	Control force	Without special precision ≤±6 mm	With particular precision ≤±1.5 mm	With high precision
Slide one over another			**Overlap tolerance**	
		Up to ±1.5 mm	Up to ±0.4 mm	≤ ±0.4 mm

The symmetry class express the orientation need for positioning according to the geometrical characteristics of the cross-section of the joined pieces and their position before joining. There are three symmetry classes (Table 13.2).

Table 13.2 Symmetry classes

Class	Symbol	Maximum orientation	Symmetry class description
Symmetrical	S	0°	The object does not need orientation (rotation) for joining
Semisymmetrical	SS	90°	The object needs rotation around its axis
Nonsymmetrical	NS	180°	The object can be joined in a single position

The operation mode refers to working conditions and the state of the objects to be positioned. There are two degrees of difficulty:

- E: easy
- D: difficult

To differentiate between the two degrees of difficulty, the next indication list is used (Table 13.3) to determine how difficult something is to handle.

Examples for "difficult" positioning: thread the needle, P2SD; attach cuff at sleeve, P2NSD.

13.2.3.5 Release

Definition: Releasing is the basic motion that is executed when the fingers or hand ceases control over an object.

Code: RL1

Table 13.3 Degree of difficulty

The degree of difficulty is difficult (D) if at least one of the following conditions is fulfilled	
Work conditions • supplementary grasp is necessary • positioning place is hidden • obstruction owing to lack of space at positioning place	Type of piece • piece with sharp edges • fragile piece • slippery piece • flexible piece • piece weight > 1 kg

Figure 13.14 Release RL1.

For example, release is a simple motion with the lowest degree of control. The necessary time it takes is imperceptible. According to the way in which contact with the object is terminated, there are two cases:

- Case 1: RL1
- Case 2: RL2

In Case 1, contact with the object stops simply by opening the fingers (Fig. 13.14).

In Case 2, contact between the fingers or hand and the object stops instantly (Fig. 13.15).

13.2.4 Case study: improvement in work method for a sewing operation

The sewing operation that is analyzed is attaching labels to the lining of the waistband (Fig. 13.16).

The operation of attaching labels to the lining of the waistband was chosen because the operator was new and did not have a proper method. Also, the operation involves a lot of handling and manipulation movements (Nicolaiov et al., 2010).

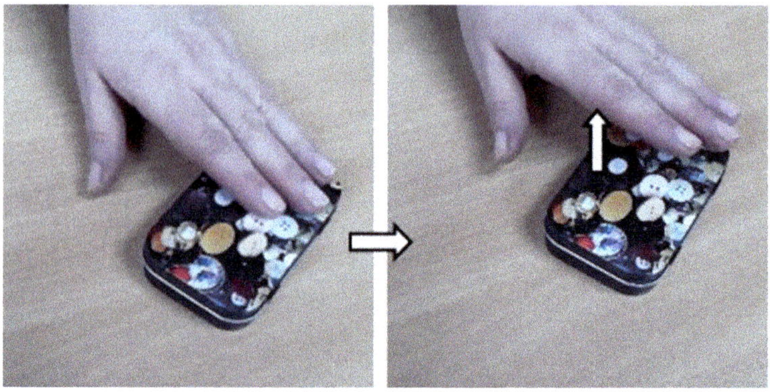

Figure 13.15 Release RL2.

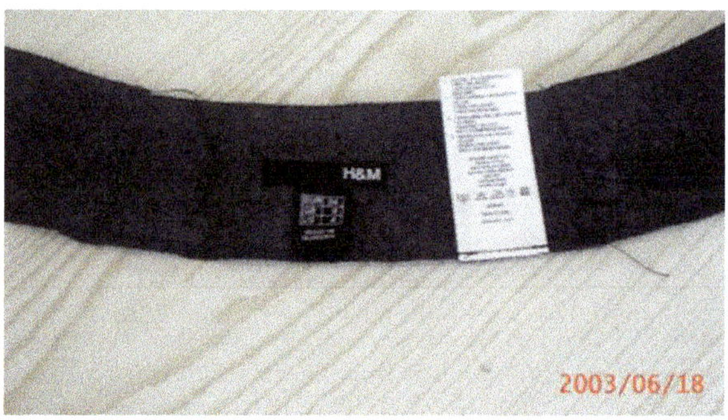

Figure 13.16 Case study: attaching labels to the lining of the waistband.

13.2.4.1 Analysis of existing situation

The workplace layout for attaching labels to the lining of the waistband and its relation to the previous and next operations are shown in Fig. 13.17.

The existent configuration of the workplace is shown in Fig. 13.18. The work system of the analyzed operation is defined by the characteristic elements:

1. work task: sign and attach the brand label onto the lining of the waistband in a rectangular shape, also attaching the size label on one side; mark and attach the care instruction label on one side, cutting the threads; then check the operation.
   Normal time, Nt = 0.95 min
   Number of operators, m = 0.4 operators
   Sewing line production, Np = 377.22 pieces/8 h
   Work productivity, W = 8 pieces/8 h/operator
2. input: bundles with various numbers of waistband linings. The bundles come from the previous operation; some are bound and some are not. Previous operations: sew the waistband lining pieces, inserting the hanger tape into the side seam. All label types (brand, size, and care instructions) are placed randomly onto the machine table.

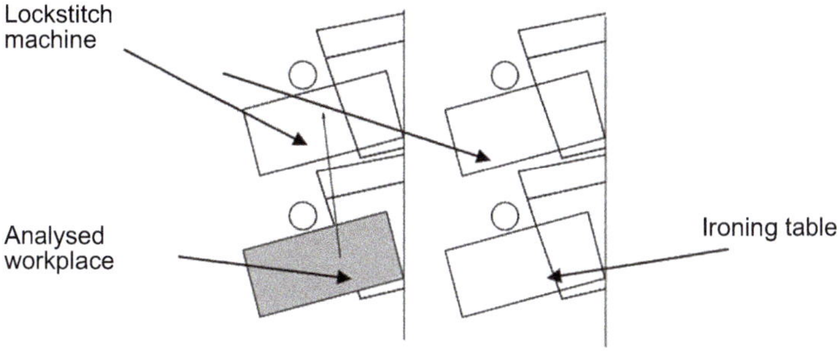

Figure 13.17 Sketch of the analyzed workplace layout.

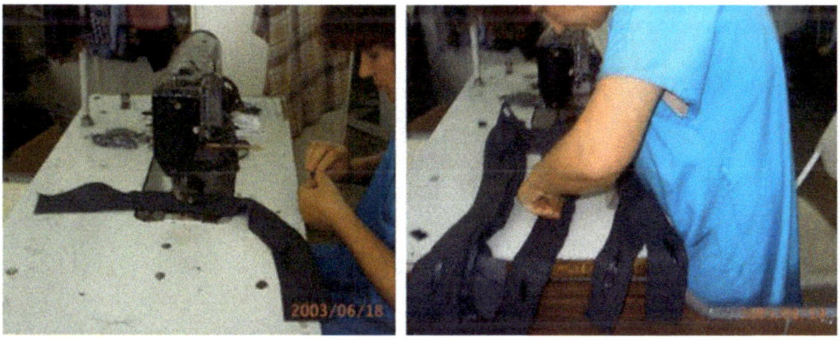

Figure 13.18 Workplace configuration.

3. activity development:

a. Workplace is fed with bundles of waistband lining.

b. Mark the label position. Process sequence:

i. Unfold a bundle of waistband linings. Place it on the machine table.

ii. Take one waistband. Place it on the table.

iii. Take the template from the machine drawer. Place it on the waistband and mark it with a pencil.

iv. Place the waistband lining in the feed box.

c. Attach the brand and size labels. Cut the threads.

i. Feed one waistband lining from the box. Position it under the presser foot.

ii. Take one brand label from the machine table. Position it at the mark.

iii. Back-tack and sew the large side. Turn at the corner and sew the small side (three stitches). Turn at the label corner and sew four stitches (Fig. 13.19).

iv. Take the size label and position it under the brand label. Complete the large side. Turn at the corner and sew the small side. Back-tack the threads and cut (Fig. 13.19).

d. Mark the position of the care instructions label and sew. Process sequence:

i. Mark the label position using the template.

ii. Feed one label. Position at the mark. Sew the short seam with back-tacking at the beginning and end. Cut threads.

iii. Place the waistband on the side of the box.

Observation: the operator is using scissors to cut threads.

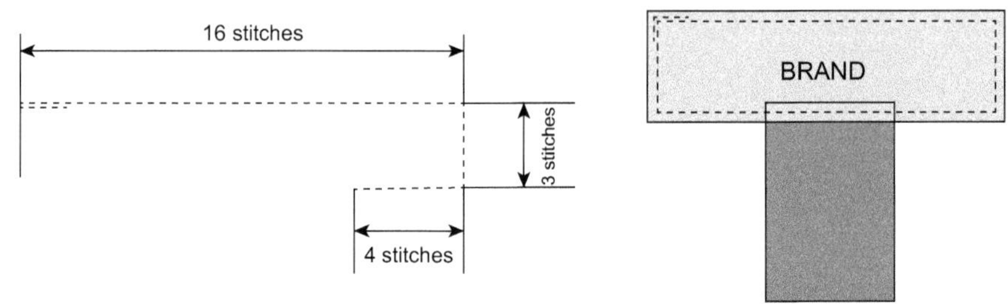

Figure 13.19 Technological conditions.

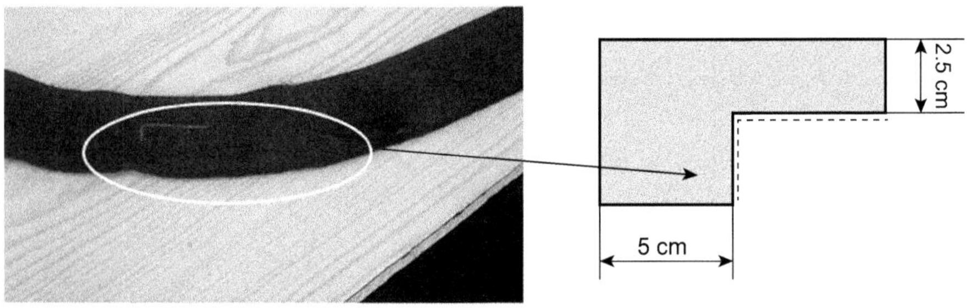

Figure 13.20 Marking template.

4. Working mean:
   Lockstitch machine "Minerva" without a thread cutter.
   Thread scissors.
   Templates for marking the label's position (Fig. 13.20)
   Feed box on the left side
5. Output: waistband lining with labels attached (Fig. 13.16) and bundles with different number of pieces, according to the next operation.

Work method analysis

At this workplace, the operator first marks all pieces from the bundle and then attaches all three labels at every waistband lining, piece by piece. The development of activity and analysis using the MTM method is presented in Table 13.4.

Method observations As shown in Fig. 13.21, from an anthropometric point of view, the configuration of the workplace is not good. The table height is not suited for a standing position, even if this work sequence is not used all of the time.

- The labels are placed out of order, mixed on the machine table. A lot of time is wasted in identification. The size labels are also mixed, and if they are attached incorrectly, it can lead to a defective product. There is no precise size identification of the waistbands from the bundle.
- The waistband is fed from the box. The distance is too large. The operator turns to grab a piece from the bundle. Usually, the waistband linings are tangled and additional time is necessary to separate them.
- The sewing machine has a high speed. For this type of operation, a reduced speed is better. Sewing is performed almost entirely manually owing to the small number of stitches.

Table 13.4 **Methods-time measurement analysis**

No.	Development	Methods-time measurement	Time measurement unit	Time (min)
1.	Grasp one waistband with left hand, move to right hand; position under the pressure foot, approximately at mark	R30C G4A G3 M30C P2NSD RL1	65.1	0.039
2.	Grasp a brand label, from the machine table; position at mark	R30C G4A M30C P2NSD RL1	65.1	0.039
3.	Choose and grasp a size label from the machine table; position at mark	R25C G4A M25C P2NSD RL1	62.6	0.038
4.	Grasp the scissors from the machine table cut threads; release scissors on the machine table	R30B G1A M30C P1SSE M30B RL1	54.3	0.033
5.	Grasp the template for the instruction care label; position; grasp the pencil for marks; mark the position; release the pencil on the machine table	R20B G1B M20C P1SE R25B G1C1 M25C P1SE M25B RL1	83.4	0.05
6.	Grasp an instruction label; position at mark	R25C G4A M25C P2NSD RL1	62.6	0.038
7.	Grasp the scissors from the machine table; cut the threads; release the scissors to the table	R30B G1A M30C P1SSD M30B RL1	54.3	0.033
8.	Place the waistband on the side of the box	G1B M30B P1SE RL1	24.4	0.015
	Total		471.8	0.283

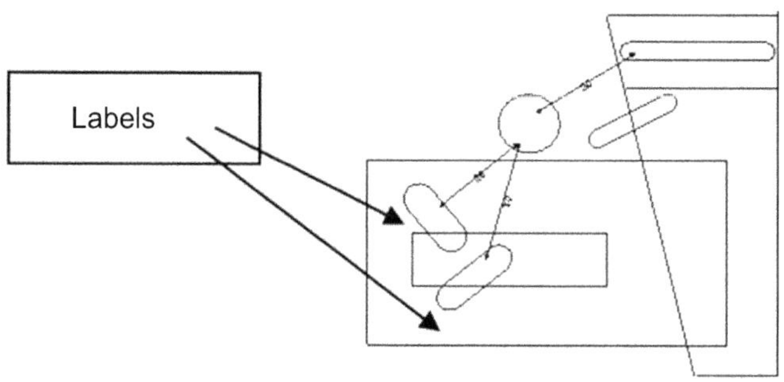

Figure 13.21 Sketch of the analyzed workplace layout.

Identification of possible improvements and solutions

After a thorough analysis of the existing situation, the following possible improvements were identified:

- Marking the label's position
- New working method for the "attach labels" operation;
- Workplace reconfiguration;

Improvement objectives

1. Removing the label's marking operation using guide marks on the machine table in the sewing area.
2. Using modular boxes for the labels
3. Changing the way labels are fed

Elimination of the marking operation

Because the label's position is not influenced by size, the marking operation can be eliminated using templates. Proposal:

- Placing a mark at 5 cm on the machine table in the sewing area, parallel to the side seam of the waistband.
- Use a guide to feed the waistband at 2.5 cm (Fig. 13.22).
- A spotlight can be used to increase productivity. This will mark the beginning of the seam.
- Placing a mark 2 cm from the middle back seam to sew the care instruction label.

Changing the possibility for label feeding

Devices to feed labels can be adapted for all types or models of garments. The following variations are proposed for feeding labels (Fig. 13.23):

- Multilevel box with jointed stand, placed on the machine table on the right side or under the sewing head (Fig. 13.23(a)). Each compartment is used for a specific label type, e.g., the top for brand labels, the middle for care instruction labels, and the bottom for size labels. Compartments can have different colors, which helps develop automatic moves.

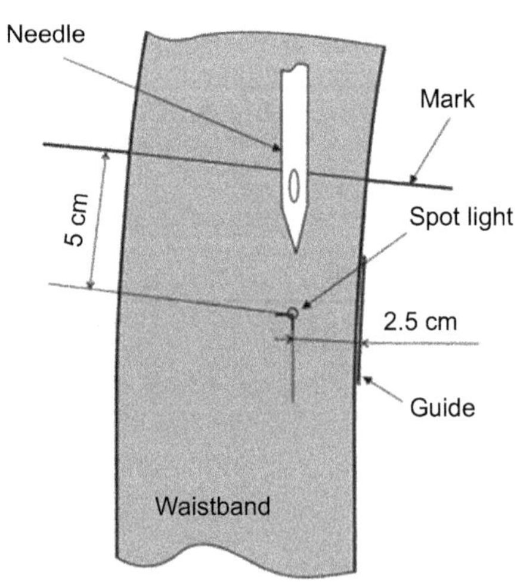

Figure 13.22 Technical solution.

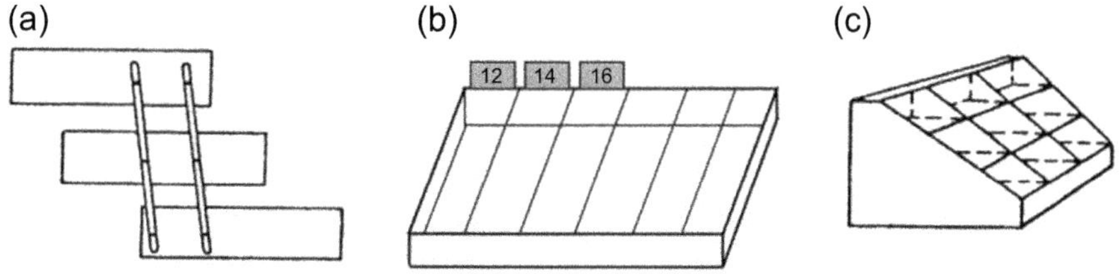

Figure 13.23 Types of boxes.

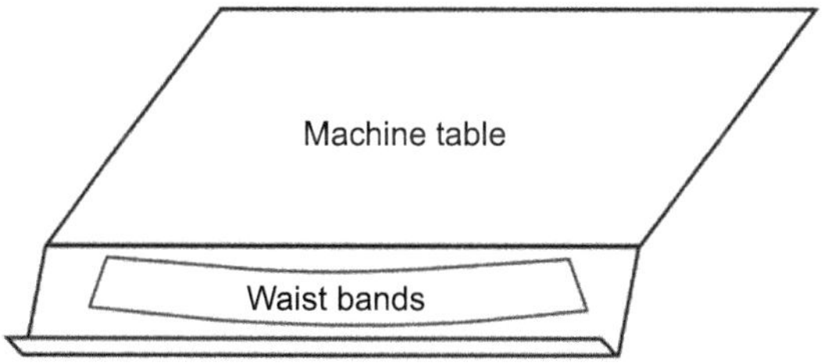

Figure 13.24 Stand for waistbands.

- The bottom compartment can be divided into cells for each size. Each cell is labeled according to size (Fig. 13.23(b)). In this way, time spent in searching and identifying labels will be eliminated.
- For a box with fixed compartments (Fig. 13.23(c)), the principle is the same.

Changing the way to feed waistbands

The proposal is to feed the waistbands from the lap, instead of the box. Another solution can be attaching a stand in front of machine table (Fig. 13.24).

Changing the work method

After the changes, the new work method is:

Workplace preparation

1. Place the waistband bundle on the lap or the stand at the front of the machine table.

Work mode

1. Grasp a waistband with both hands; position it with the side seam at the 5-cm mark and with the upper side at the 2.5-cm guide.
2. Grasp a brand label from the box. With the right hand, position it in front of the spotlight or needle tip.
3. Back-tack two stitches at the beginning. Sew the right side at the lower speed (place the hand on the pulley). Turn. Check the label position. Sew three stitches by manually rotating the pulley with the right hand. Turn. Sew four stitches in the same way. Stop the machine with the needle in the lowered position.
4. Grasp a size label from the right compartment with the right hand. Position it near the needle. Insert it under the brand label. Resume sewing the large side. Turn. Sew the small side. Back-tack two stitches at the end. Remove the waistband from under the presser foot. Cut the threads.

Table 13.5 Methods-time measurement analysis of improved method

No.	Development	Methods-time measurement	Time measurement unit	Time (min)
1.	Grasp a waistband with both hands from the lap; position at machine table at mark and guide	R14A G1B M20A RL1	21.9	0.013
2.	Grasp a brand label from box; position at mark	R20B G4A M20C P2NSD RL1	57.6	0.035
3.	Grasp a size label from box; position at mark	R20B G4A M20C P2NSD RL1	57.6	0.035
4.	Grasp the scissors from the machine table; cut the threads, place the scissors on the table	R30B G1A M30C P1SSE M30B RL1	54.3	0.033
5.	Grasp a care instruction label; position at mark	R20B G4A M20C P2NSD RL1	57.6	0.035
6.	Grasp the scissors from the machine table; cut the threads, place the scissors on the table	R30B G1A M30C P1SSD M30B RL1	54.3	0.033
7.	Place the waistband on the side of the box	G1B M30B P1SE RL1	24.4	0.015
	Total		327.7	0.1966
	Old method total (min)		0.283	
	Reduction (%)		30.53	

5. Position the middle back seam over the 2-cm mark. Grasp a care instruction label from the box. With the right hand, position at the needle or spotlight. Sew with back-tack at the beginning and end. Cut the threads. Place the waistband on the feed box side.

The MTM analysis of the improved method emphasizes decreasing time (Table 13.5). After the case study, 13 rules can be established to increase work efficiency:

1. less use of hands to grasp and hold objects
2. use both hands equally and ensure symmetrical movements (equal on right and left sides)
3. overlap left hand and right hand movements
4. keep hands close to body
5. use fingers movement instead of arm or body movements
6. if possible, avoid sudden movements at beginning and end of work task
7. use the treadle when possible
8. reduce distances
9. use "strategic" position for work objects or tools
10. use free fall, sliding, or gravity ("water flow" principle) whenever possible
11. employ technological interventions without pulling pieces out of the feeding system
12. create and use obstacles to limit movement
13. use the machines and all of their technological possibilities at full capacity

13.3 Motion study and standard time setting

Two techniques are widely used to establish operational or standard time:

- time study, based on time registration of work sequences of an operation
- method development and time setting, based on MTM principles

13.3.1 Time registration systems

Time registration (or time study) is developed according to a standard methodology that splits an operation into process sections, records the time for every sequence using a stopwatch, evaluates the operator's efficiency for every sequence, statistically evaluates the recording, etc. A widely used procedure is the standard program for time study established by Reichsausschuß für Arbeitsstudium (REFA), Germany.

A synthesis of the methods of time measurement is presented as follows (REFA, Production data management):

1. manual registration: establishes the sequences. Measuring points and time recordings are made by an operator using a stopwatch (mechanical or electronic); transcriptions and data statistical checks are made manually in a time study form
2. electronic registration: establishes the sequences. Measuring points are made by an operator and the recording, data statistical processing, and storage are performed automatically. The following variations on recording devices can be used:
    a. mobile recording board with integrated microcomputer (tablet or recording board) and dedicated software (e.g., Multidata.6 System from Drigus, REFA Chronos 4.0 by REFA Consulting AG) (Fig. 13.25(a) and (b))
    b. Mobile one-hand system with integrated microcomputer without a recording board (e.g., ORTIM b11 from ORTIM, SYMODAT DE 64 from UBR).
3. Autoregistration using video systems:
    a. video recording with efficiency evaluation by the operator
    b. video processing by digitalization and compression
    c. separation of process sequences by segmenting the video recording and generating time building blocks

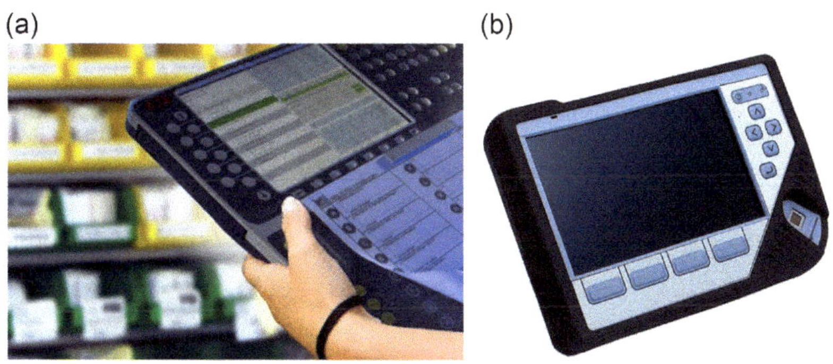

Figure 13.25 Mobile recorder boards: (a) Multidata 6, (b) Chronos 4.0.
(a) Drigus. (b) REFA Consulting.

Figure 13.26 Multimedia Video Task Analysis (https://mvta.engr.wisc.edu/).

Recording and evaluation can be separated if the process is recorded with a video camera. The registration process can be evaluated with an image-evaluating device in which the individual process sections are marked and times are allocated. Thus, it is possible to analyze the process by indicating the work operation. Frozen images are shown and measuring points are set. It is possible to structure the work process into process sections even after the time study because measuring points can be set during evaluation.

An example of dedicated software for a time and motion study is Multimedia Video Task Analysis (MVTA) (Fig. 13.26), developed by Robert G. Radwin and Thomas Y. Yen.

The MVTA system analyzes the video recording of the work process through image processing. Motions are transposed in graphic form (Fig. 13.26) and process segments are identified, delimited by break points. A break point is an arbitrary event in the process recording, assimilated with a large amplitude executed by the operator for activity development. Time allocation for each individual process section is followed by time between the first and last images determining the section. In this way, the analyzed process is split into process steps and the time for every process, and the analysis of motions can be performed (real time, slow motion, fast motion, frame by frame, etc.). Data can be stored and used to establish the standard time for specific processes. Nevertheless, time recording methods do not eliminate the stage of evaluating the operators' efficiency.

13.3.2 General presentation of SSD4Pro

The second technique for setting standard time uses the MTM principles based on the analysis of the elements of motion and composition of time. The software SSD4Pro [Standard Sewing Data (SSD)] produced by AJ Consultants from Finland is dedicated to clothing

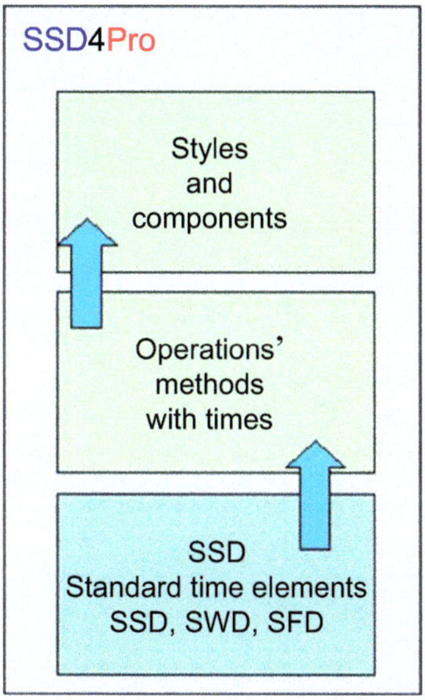

Figure 13.27 Basic Standard Sewing Data.

manufacturing; the application includes predefined macros specific to the clothing manufacturing process, such as sewing, ironing, and special operations and movements.

Basic SSD4Pro is a tool that offers a solution to developing methods and setting accurate time standards for operations. The system manages operations and their methods and times as well as styles and components with operation lists and other product specifications (Fig. 13.27).

The primary function of the SSD Basic system is the methods (operations) with times and descriptions. A method is built of SSD elements. The method also has a descriptive name and other title data to separate it from other, similar methods. Each method has a unique operation (method) code. The method is created using SSD elements (Fig. 13.28). Once the elements are added, the SSD system generates the total time for the operation.

In the first step, the operation is fully described in the method description window, the Title data tab (Fig. 13.29).

The method Title data include different information:

- description of the operation, both brief and detailed
- machine type and code
- sewing speed for the sewing machines
- type of device
- number of pieces from the bundle
- sewing line and group of operation
- required skills and work category of operator
- codes for product, model, order, and client
- technical drawings, photos, or uploaded videos
- allowance factor established at the company level [different values depending on the operation type (manual, lock stitch, chain stitch, automated, etc.) or constant value]

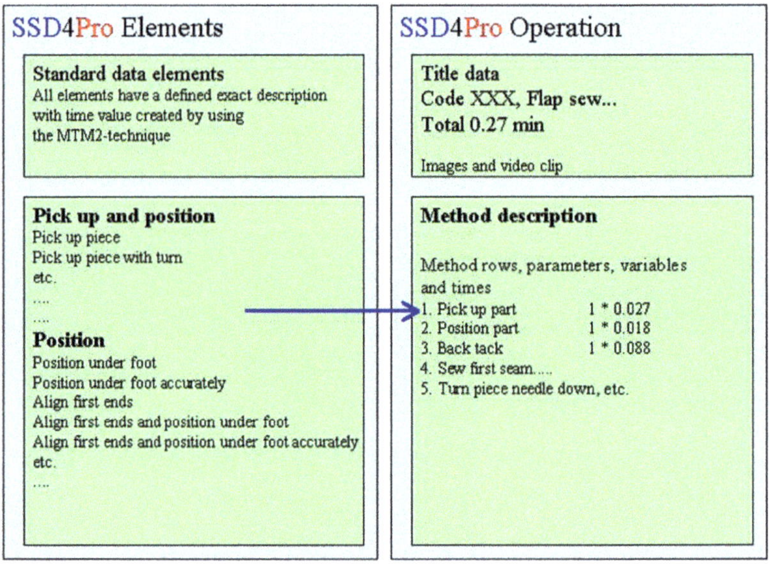

Figure 13.28 Creation of method from elements.

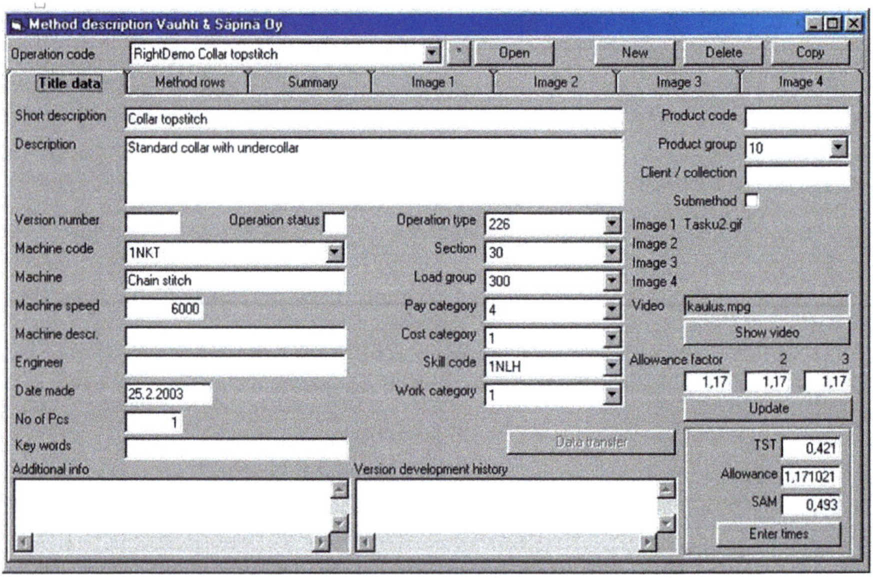

Figure 13.29 The Title data tab in the method description window.

Time values [time study time (TST)] and SAM (standard allowed minutes) are displayed in the Title data window. SAM is the result of multiplying TST time by the allowance factor.

The Enter elements window (Fig. 13.30) is automatically opened when the SSD program is started.

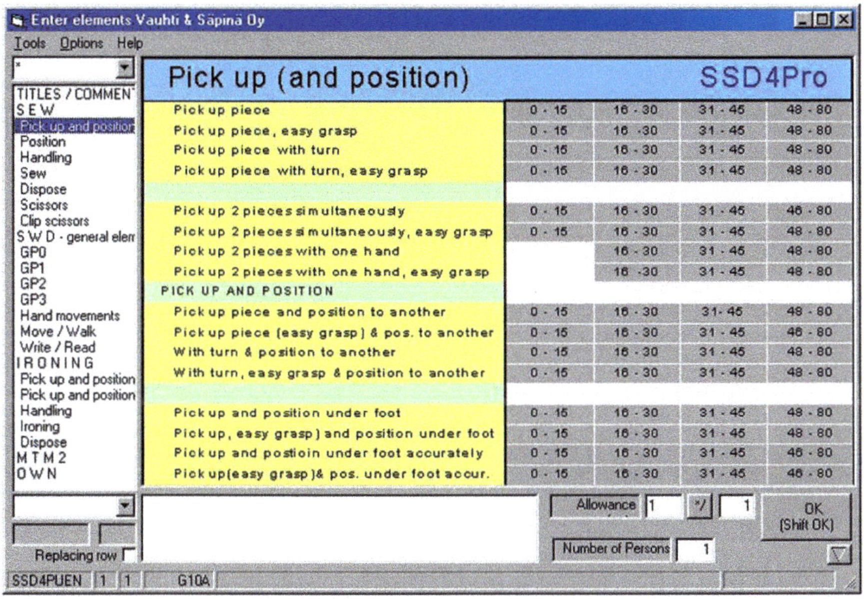

Figure 13.30 Enter elements window.

A work method for a sewing operation is effectively composed by selecting and successively adding predefined elements of movement. For example, for the sew module (for sewing), the following groups of movement are defined:

- pick up (and position) for cut pieces and semifinished pieces; pick up and move, with or without positioning over another piece, mark, or device
- position, for precise position sequences or intermediate position during the sewing operation
- handling for different intermediate movements of pieces during sewing (turning, straightening, pleating, etc.)
- sew for the sewing sequences, indicating the number of back-tack stitches, sewing speed, and length of the seam
- disposal for evacuation sequence
- scissors and clip scissors for different manual cutting sequences using scissors

For every movement group, possible cases are defined, according to the accuracy, complexity, and precision of movement or the fabric type (soft, normal, or heavy). For sequences that include movements, the operator must select the distance of movement from the distance gap matrix attached to every movement case (0–15, 16–30, 31–45, or 46–80). The distances are in centimeters.

13.3.3 *Case study: analysis of a sewing operation*

The case study consists of using the SSD program to calculating the SAM for one sewing operation. The study took place in a Romanian clothing company specializing in men's and women's jackets.

Figure 13.31 Workplace for attaching sleeves.

Operation: Attach sleeves on a woman's jacket (Fig. 13.31)

Machine: Sleeve setting machine by Juki

Short description of the operation: The body of the jacket is placed on a hanger on a trolley and the sleeves are clipped on. The operator takes one jacket, places it on his or her lap, takes the sleeves, and separates the right sleeve. The left sleeve is put aside. First the right sleeve is sewn, and then the left sleeve. In the end, the jacket is placed on the hanger.

The TST time calculation is presented in Table 13.6 (method report from SSD4Pro).

Table 13.6 **Operation "sleeve setting" time calculation**

Method description: subtotals		SSD4Pro AJ-Consultants 11.11.		
BL65000	650 **Sleeves set**			
SL	300INLV			
BL	248055			
09.11.2007 Mrs. Carmen Loghin				
#	**Code**	**Element**	**Freq.**	**TST**
1.	*	Feeding components		
2.	G11D	Pick up piece, easy to grasp, 46–80 cm	1*	0.026
3.	GP11A	Easy to get, position inaccurately, 6–15 cm	1*	0.007
4.	MTA	Apply pressure	2*	0.017
5.	MTR	Regrasp (MTM2)	1*	0.004
6.	G11B	Pick up piece, easy grasp, 16–30 cm sleeves on knees	1*	0.015
	01=			**0.068**
7.	–	Pick up right sleeve		
8.	G50C	Pick up piece and position to another 31–45 cm	1*	0.038
9.	H10B	Move or straighten piece 16–30 cm	2*	0.030
10.	P10	Align first ends side seam set (difficult fabric)	2*	0.046
11.	P35B	Position under foot accurately 16–30 cm	1*	0.018
	02=			**0.131**

Table 13.6 Operation "sleeve setting" time calculation—Continued

Method description: subtotals		SSD4Pro AJ-Consultants 11.11.
BL65000	650 **Sleeves set**	
SL	300INLV	
BL	248055	

09.11.2007 Mrs. Carmen Loghin

#	Code	Element	Freq.	TST
12.	–	Attach right sleeve		
13.	MA100	Back-tack 3+3	1*	0.011
14.	P50A	Reposition 0–15 cm	1*	0.026
15.	MA35	Sew 3.5 stitches/cm, speed? cm: 4, RPM: 500	2*	0.070
16.	P50A	Reposition 0–15 cm	2*	0.053
17.	H10B	Move or straighten piece 16–30 cm	1*	0.015
18.	MA35	Sew 3.5 stitches/cm, speed? cm: 4, RPM: 500	2*	0.070
19.	P50A	Reposition 0–15 cm	4*	0.106
20.	H10B	Move or straighten piece 16–30 cm	1*	0.015
21.	MA35	Sew 3.5 stitches/cm, speed? cm: 4, RPM:500	1*	0.035
22.	P50A	Reposition 0–15 cm	1*	0.026
23.	MA35	Sew 3.5 stitches/cm, speed? cm: 4, RPM:500	1*	0.035
24.	P50A	Reposition 0–15 cm	1*	0.026
25.	H10B	Move or straighten piece 16–30 cm	1*	0.015
26.	MA35	Sew 3.5 stitches/cm, speed? cm: 4, RPM:500	1*	0.035
27.	P50A	Reposition 0–15 cm	1*	0.026
28.	MA35	Sew 3.5 stitches/cm, speed? cm: 4, RPM:500	1*	0.035
29.	P50A	Reposition 0–15 cm	1*	0.026
30.	H10B	Move or straighten piece 16–30 cm	1*	0.015
31.	MA35	Sew 3.5 stitches/cm, speed? cm: 4, RPM:500	2*	0.070
32.	P50A	Reposition 0–15 cm	2*	0.053
33.	MA35	Sew 3.5 stitches/cm, speed? cm: 4, RPM:500	2*	0.075
34.	P51A	Reposition, easy grasp 0–15 cm	1*	0.016
35.	H10B	Move or straighten piece 16–30 cm	1*	0.015
36.	HS2C	Pick up scissors, cut within 30 cm, dispose of scissors, 31–45 cm cuts: 4	1*	0.081
	03=			**0.951**
37.	–	Pick up left sleeve		
38.	G50C	Pick up piece and position another 31–45 cm	1*	0.038
39.	H10B	Move or straighten piece 16–30 cm	1*	0.015
40.	P10	Align first ends side seam set (difficult fabric)	2*	0.046
41.	P35A	Position under foot, accurately 0–15 cm	1*	0.016
	04=			**0.114**
42.	–	Attach left sleeve		
43.	MA100	Back-tack 3+3	1*	0.011
44.	P50A	Reposition 0–15 cm	1*	0.026
45.	MA35	Sew 3.5 stitches/cm, speed? cm: 4, RPM: 500	2*	0.070

Continued

Table 13.6 **Operation "sleeve setting" time calculation—Continued**

Method description: subtotals		SSD4Pro AJ-Consultants 11.11.
BL65000	650 **Sleeves set**	
SL	300INLV	
BL	248055	

09.11.2007 Mrs. Carmen Loghin

#	Code	Element	Freq.	TST
46.	P50A	Reposition 0–15 cm	2*	0.053
47.	H10B	Move or straighten piece 16–30 cm	1*	0.015
48.	MA35	Sew 3.5 stitches/cm, speed? cm: 4, RPM: 500	2*	0.070
49.	P50A	Reposition 0–15 cm	4*	0.106
50.	H10B	Move or straighten piece 16–30 cm	1*	0.015
51.	MA35	Sew 3.5 stitches/cm, speed? cm: 4, RPM: 500	1*	0.035
52.	P50A	Reposition 0–15 cm	1*	0.026
53.	MA35	Sew 3.5 stitches/cm, speed? cm: 4, RPM: 500	1*	0.035
54.	P50A	Reposition 0–15 cm	1*	0.026
55.	H10B	Move or straighten piece 16–30 cm	1*	0.015
56.	MA35	Sew 3.5 stitches/cm, speed? cm: 4, RPM: 500	1*	0.035
57.	P50A	Reposition 0–15 cm	1*	0.026
58.	MA35	Sew 3.5 stitches/cm, speed? cm: 4, RPM: 500	1*	0.035
59.	P50A	Reposition 0–15 cm	1*	0.026
60.	H10B	Move or straighten piece 16–30 cm	1*	0.015
61.	MA35	Sew 3.5 stitches/cm, speed? cm: 4, RPM: 500	2*	0.070
62.	P50A	Reposition 0–15 cm	2*	0.053
63.	MA35	Sew 3.5 stitches/cm, speed? cm: 4, RPM: 500	2*	0.075
64.	P50A	Reposition 0–15 cm	1*	0.016
65.	H10B	Move or straighten piece 16–30 cm	1*	0.05
66.	HS2C	Pick up scissors, cut within 30 cm, dispose of scissors, 31–45 cm cuts: 4	1*	0.081
	05=			**0.951**
67.	–	Control right armhole		
68.	H10C	Move or straighten piece 31–45 cm	1*	0.020
69.	H15B	Move or straighten piece without grasp 16–30 cm	1*	0.014
70.	H10B	Move or straighten piece 16–30 cm	1*	0.015
71.	P65A	Regrasp with hand movements 0–15 cm	1*	0.005
72.	MT2EL	Eye motion	1*	0.004
	06=			**0.058**
73.	–	Control left armhole		
74.	H10C	Move or straighten piece 31–45 cm	1*	0.020
75.	H15B	Move or straighten piece without grasp 16–30 cm	1*	0.014
76.	H10B	Move or straighten piece 16–30 cm	1*	0.015
77.	P65A	Regrasp with hand movements 0–15 cm	1*	0.005
78.	MT2EL	Eye motion	1*	0.004
	07=			**0.058**
79.	–	Control symmetry		
80.	H10D	Move or straighten piece 46–80 cm	1*	0.026

Table 13.6 **Operation "sleeve setting" time calculation—Continued**

Method description: subtotals			SSD4Pro AJ-Consultants 11.11.	
BL65000	650 **Sleeves set**			
SL	300INLV			
BL	248055			
09.11.2007 Mrs. Carmen Loghin				
#	**Code**	**Element**	**Freq.**	**TST**
81.	HR3	Read and compare words: 2	2*	0.084
82.	D10D	Dispose of 80–>cm	1*	0.026
	08=			**0.136**
83.	BUNDLE	Bundle time, bundle size: 10	1*	0.08
	09=			**0.08**
		TST		**2548**
		Allow		1.12
		SAM		**2.854**

RPM, revolutions per minute.

Using the SSDPro software application leads to the following advantages:

- all times calculated can be stored and used again for minor changes in the method
- these elements represent a deep analysis of the work method
- solutions for improvement are easily suggested after the movement analysis
- the ratio between the effective sewing time and the movement time can be emphasized
- it is not necessary to evaluate the operator's efficiency; the resulting SAM is issued for 100% efficiency

13.4 Line balancing and work efficiency in clothing manufacturing

The productivity of a clothing company is based not only on the efficiency of every operator but also on the possibility of groups working (Nicolaiov et al., 2013). In this field, in the sewing department, it is defined by the sewing line. It is possible for a firm to have good workers, but if they are not trained to work like a group, the results will not be good. On the other hand, a good line organization will reflect the best part of every operator. The efficiency of a sewing line organization depends on: (1) the ability of the chief of the line to balance the line; (2) the means of transportation between workplaces; (3) the skill level of every operator (the number of operation he or she knows); (4) the production follow-up system; and (5) the organizational type. To appreciate the efficiency of a sewing line, in the following section some specific indicators are proposed for the evaluation of line efficiency.

13.4.1 *Specific efficiency indicators*

To evaluate the efficiency of a sewing line, the following data need to collected (Morshed and Palash, 2014): (1) product type (model, order, and number of pieces), (2) the number of operators (N_o), (3) the working time (e.g., T = 480 min), (4) the time per garment (T_e, the sum of the operational time for a garment model), and (5) the maximum operation time [T_b, the maximum operation time at a workplace (in min)]. All of these data are usually present in the technological process of the garment (Adeppa, 2015). Using these data, some parameters can be calculated to indicate the line efficiency:

$$\text{Basic pitch time (p): } p = T_e/N_o \text{ (min)} \tag{13.1}$$

$$\text{Minimum/maximum pitch } (p_{min}/p_{max})\text{: } p_{min/max} = \pm 10\%\ p \text{ (min)} \tag{13.2}$$

$$\text{Theoretical daily output } (W_p)\text{: } W_p = T \times N_o/T_e \text{ (pieces/day)} \tag{13.3}$$

$$\text{Line efficiency } (L_e)\text{: } L_e = p/T_b \times 100\ (\%) \tag{13.4}$$

$$\text{Real output } (W_r)\text{: } W_r = W_p \times L_e/100 \text{ (pieces/day)} \tag{13.5}$$

$$\text{Bottleneck productivity } (W_b)\text{: } W_b = W_r/N_o \text{ (pieces/operator)} \tag{13.6}$$

For a more accurate line balancing evaluation, a pitch time diagram should be drawn. This diagram consists of a chart of each operator's load compared with the basic pitch (middle horizontal line). The ideal situation is considered when all points are placed between the P_{min} and P_{max} horizontal lines. Using the diagram, the bottlenecks and "easy places" are identified, so the chief line can be measured to rebalance the operations or conduct a work study for some operations.

13.4.2 Case study: evaluation of efficiency of a technological line

The activity in this case study took place at a clothing company from Romania that specialized in women's jackets. To evaluate the efficiency of the technological sewing line, these steps were followed:

1. gathered necessary data (technological process in Table 13.7) and directly surveyed the technological line
2. calculated the efficiency indicators of the technological line (Table 13.8)
3. drew the pitch diagram
4. observed

13.4.2.1 Analysis of the technological line

The analysis focused on Line E15, Model 320302. The distribution of operations (as the line leader made it) is given in Table 13.7.

13.4.2.2 Direct observation

The technologists and time study experts have excellent technological knowledge and rich experience. At the planning department a huge technological database exists on hard copy, with times according to different technological variants, based on practice. The operational time is correct according to the technological complexity and

Table 13.7 **E15: Line balancing**

Workplace no.	Operation no.	Operation	Time (min)	Total time/ operator	Operator no.	Machine
1	1	Feed 1	3.45		O1	Worktable (WT), lockstitch machine (LM)
	26	Sew hanger	0.45			
	31	Sew seams and shoulders	2.45			
	48	Pick up the adhesive dots	1.575			
	30	Feed 2	1.65			
		Total		**9.575**		
2	2	Mark under-collar	1.8		O2	WT
	3	Mark front edges	2.75			
	4	Mark facing	1.2			
		Total		**5.75**		
3	5	Topstitch flaps	2.2		O3	Hand stitch machine (AMF)
	36	Hand stitch front edges	10			
		Total		**12.2**		
4	6	Mark under-collar	0.725		O4	LM
	37	Mark front edges	6.4			
	43	Run stitch facing to front	1.5			
		Total		**8.625**		
5	6	Topstitch flaps	0.725		O5	LM, sleeve setting machine
	37	Topstitch front edges	6.4			
	38	Sew in sleeves	4.9			
		Total		**12.025**		
6	7	Run stitch flaps	0.6		O6	LM
	9	Sew breast darts	1.65			
	10	Sew tape to sleeve head	0.45			
	12	Mark the flaps	0.425			
	13	Finish the pocket	5			
		Total		**8.125**		

Continued

Table 13.7 **E15: Line balancing—Continued**

Workplace no.	Operation no.	Operation	Time (min)	Total time/ operator	Operator no.	Machine
7	7	Run stitch flaps	0.6		O7	LM
	9	Sew breast darts	1.65			
	10	Sew tape at sleeve head	0.45			
	12	Mark the flaps	0.425			
	13	Finish the pocket	5			
		Total		**8.125**		
8	8	Iron flaps	1.5		O8	IT
	11	Iron darts	3.05			
	14	Iron pocket	1.15			
	15	Iron pocket inside	0.5			
		Total		**6.2**		
9	16	Sew center back seam	1.5		O9	LM, overlock machine, button-hole machine
	17	Overlock center back	2			
	19	Sew tape on back	0.45			
	20	Cut and sew band	0.275			
	22	Overlock shoulder seam	1.55			
	49	Sew buttonhole	1.5			
		Total		**7.275**		
10	16	Sew center back seam	1.5		O10	LM
	19	Sew tape on back	0.45			
	20	Cut and sew band	0.275			
	27	Mark and attach label	0.9			
	33	Sew under-collar	1.7			
	35	Sew facing with back neck hole facing	2.75			
	43	Run stitch facing to front	1.5			
		Total		**9.075**		

11	21	Sew sleeve seam	6.55		O11	LM
		Total		**6.55**		
12	18	Iron back	1.25		O12	Ironing table (IT)
	23	Iron sleeve hem	2.575			
	29	Iron lining	1.1			
	32	Iron side seam	1.35			
	34	Iron armhole seam	0.8			
	39	Iron sleeve head	0.7			
	44	Turn and iron collar	2.75			
	45	Trim front edges seam	0.85			
	50	Iron facing- lining seam	1.225			
		Total		**12.6**		
13	18	Iron back	1.25		O13	IT
	23	Iron sleeve hem	2.575			
	29	Iron lining	1.1			
	32	Iron side seam	1.35			
	34	Iron armhole seam	0.8			
	39	Iron sleeve head	0.7			
	44	Turn and iron collar	2.75			
	45	Trim front edges seam	0.85			
	50	Iron facing- lining seam	1.225			
		Total		**12.6**		
14	24	Sew darts on lining	0.4		O14	Safety stitch machine (SSM)
	25	Sew front and back lining	1.4			
	28	Sew body lining	4.025			
		Total		**5.825**		
15	24	Sew darts on lining	0.4		O15	SSM
	25	Sew front and back lining	1.4			
	28	Sew body lining	4.025			
	48	Pick up adhesive dots	1.575			
		Total		**7.4**		

Continued

Table 13.7 **E15: Line balancing—Continued**

Workplace no.	Operation no.	Operation	Time (min)	Total time/ operator	Operator no.	Machine
16	43	Run stitch facing to front	3		O16	LM
	51	Secure facing in front	5.2			
		Total		**8.2**		
17	40	Sew fullness on sleeve head	2		O17	LM, WT
	41	Attach shoulders pads	1.55			
	42	Secure shoulder pads	1			
	52	Trim threads	1			
	53	Finish garment	2			
	54	Final steaming on hanger	0.3			
		Total		**7.85**		
18	46	Secure fusible tape	0.95		O18	LM
	47	Attach lining	7.325			
		Total	8.275			
19	46	Secure fusible tape	0.95		O19	LM
	47	Attach lining	7.325			
		Total		**8.275**		
		$\mathbf{T_e}$		**164.55 min**	$N_o = 19$	

Table 13.8 Efficiency of the technological line

Indicators of E15 sewing line	Results	Remark
Basic pitch time (min)	8.66	
Maximum operation time (bottleneck) (min)	12.60	Ironing (O12, O13)
Theoretical daily output (pieces/day)	55.42	Just for calculation purposes
Line efficiency (%)	68.73	
Bottleneck daily output (pieces/day)	38.10	Just for calculation purposes
Bottleneck productivity (pieces/operator)	2	

difficulty level of the fabric, but these do not take into consideration the work method specificity and workplace configuration.

Recommendations

1. There is a need to update techniques for the registration and evaluation of operational time. It is heartily recommended to review operational times, especially for complex operations with too high values (e.g., 6, 5, 12 min).
2. There is a need to make the work method a priority. The activity of implementing a new model by the technologist must be associated with implementing the methods. The technologist becomes a method trainer.
3. There is a need to organize databases for electronic support using specialized software (with graphics and descriptive options).
4. It is strongly recommended to establish the actual level of operational times, to have a true image of efficiency (for example, the existing 120% efficiency is in real terms only 70%).

13.4.2.3 Efficiency indicators of the technological line

Using the data from the technological process (Table 13.7) and the formula presented in Section 13.4.1, the effective values for the E15 sewing line efficiency are calculated in Table 13.8.

From the analyzed data (processing product time, operational time, and planned/realized production norm) related to the type of product to the level of endowment level and by personal evaluation, the observation is that the actual efficiency level is low; it drops to approximately 75%.

13.4.2.4 Pitch diagram

For a thorough investigation of the sewing line, a pitch diagram was drawn (Fig. 13.32) using basic data from the technological process presented in Table 13.7.

13.4.2.5 Observations

1. The real situation is that operators have different levels of efficiency well known by the line leader. Under these conditions, the line leader takes into consideration operators' individual efficiency for optimum line balancing. For example, Operator O2, with an average efficiency of 70%, was assigned to an operation with an operational time placed below the p_{min} line. On the other hand, if one operator has an average individual efficiency of 130%, that operator will be assigned an operation with an operational time situated above the p_{max} line.

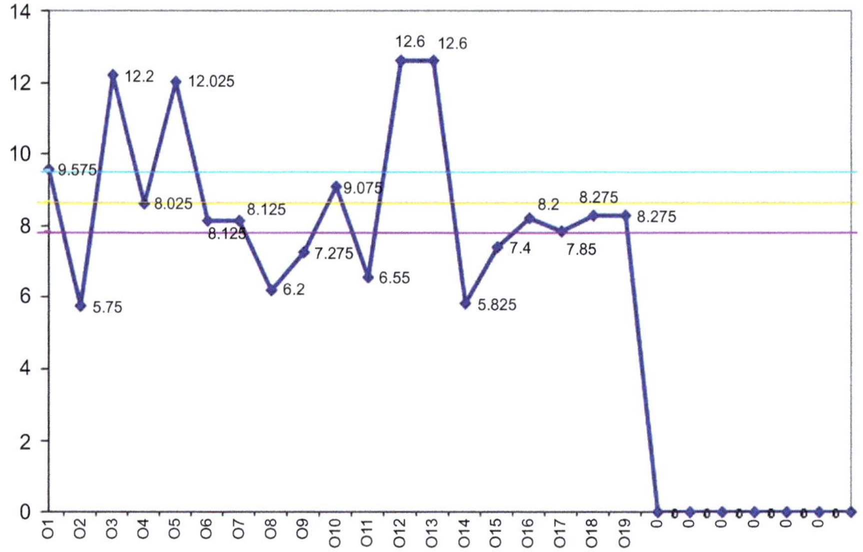

Figure 13.32 Pitch diagram for E15 sewing line.

2. The balancing of operator loading was noticed (11 of 19 operators for E15). The line leader well knew the operators' capacity and realized line balancing by redistributing operations during order processing.
3. It is possible to reduce the number of operations in a technological process by adding them in microprocesses. Some operators executed more than three operations, and from line leader's remarks, they were always executed.
4. The line feeder should not execute technological operations. He or she should be paid according to the line's efficiency. The duty must be continuous feeding with bundles for each operator. He or she must "pull" the product on the technological line.
5. All technical possibilities of equipment should be surveyed to be effective (for example, the thread cutter of a sleeve machine was broken, and the operator had to use scissors; this was unplanned and decreased efficiency).
6. The complete lack of attachments (liners, lasers, etc.) was observed, and the author noted a diminution of about 10% individual efficiency, mainly for long sewing. The marks were not sufficiently near the needle.
7. There were important areas in which efficiency could be increased by improving work methods (standard methods for long sewing, for small pieces, and for less signing and ironing). Some operators improved their operations by themselves. These improvements must be analyzed by technicians and extended to the rest of the lines. Every improvement must be recompensed.

13.5 Conclusion

The clothing manufacturing process is characterized by a great percentage of manual mechanical work. This is why an investment in state-of-the-art machines is not enough. To increase productivity, emphasis must be given on the operator, the work method, the workplace configuration, and the technological line organization. As demonstrated in this chapter, with minimal investment, great results can be achieved. The first step is to have the necessary knowledge (work study, method study, MTM techniques, and line balancing). The second step is to observe the manufacturing process carefully, starting with each operator and finishing with the technological line system.

Based on the MTM, every workplace can be analyzed from both the method and workplace organization points of view, so that a better solution can be found, as shown in the case study. The most important thing is that at a low cost, the time necessary for every operation can be reduced. In the meantime, a better method will reduce the operator's effort. For better results, software programs can help the person in charge with daily activities. As shown in the case study, the final results are thorough and can be used for further purposes. It can be seen that technological solutions for work study are increasing.

At the level of the technological line, the results of the work study at every workplace are adding up. Line balancing is the final step for achieving high productivity in the company. Partial data can be used to calculate specific indicators and take necessary measures to improve the manufacturing process. As can be seen, all of the elements presented in this chapter are correlated. Finally, human knowledge and skills are the most important factors for the success of industrial engineering in garment manufacturing.

References

Adeppa, A., 2015. A study on basics of assembly line balancing. International Journal on Emerging Technologies 6 (2), 294–297 (Special Issue on NCRIET-2015).

Anonymous, 2005a. REFA Seminar Documentation. REFA Consulting AG.

Anonymous, 2005b. SSD4Pro Users' Guide. AJ Consultants, Finland.

Babu, V.R., 2012. Industrial Engineering in Apparel Production. Woodhead Publishing Limited.

Chen, J., Chen, C.-C., Lin, Y.-J., Lin, C.-J., Chen, T., 2014. Assembly line balancing problem of sewing lines in garment industry. In: International Conference on Industrial Engineering and Operations Management, Bali, Indonesia, January 7–9 2014, pp. 1215–1225.

Guo, Z.X., Wong, W.K., Leung, S.Y.S., Li, M., 2011. Applications of artificial intelligence in the apparel industry: a review. Textile Research Journal 81 (18), 1871–1892.

Hidoş, C., Isac, P., 1972. Work study. In: Predetermined Time Systems Based on Movement Study, vol. 6. Tehnica Publishing House, Bucureşti.

Joung, Y.-K., Noh, S.D., 2014. Integrated modeling and simulation with in-line motion captures for automated ergonomic analysis in product lifecycle management. Concurrent Engineering: Research and Applications 22 (3), 218–233.

Loghin, C., Nicolaiov, P., 2013. Workplaces Configuration in Garment Manufacturing. Case Studies. Performantica Publishing House, Iaşi. ISBN: 978-606-685-055-1, pp. 15–36.

Loghin, C., Nicolaiov, P., Ionescu, I., 2005. Work method - decisional tool for technological changes. In: Proceedings of the International Conference Management of Technological Changes, vol. 2. ISBN: 978-960-8475-05-2, pp. 53–58.

Loghin, C., 2005. MTM analysis - a tool for the management of technological changes. Romanian Textiles and Leather Journal (2), 22–34.

Martin-Vega, L., 2004. The purpose and evolution of industrial engineering, chapter 1. In: Maynard's Industrial Engineering Handbook, Section 1 - Industrial Engineering: Past, Present, and Future.

Morshed, N., Palash, K.S., 2014. Assembly line balancing to improve productivity using work sharing method in apparel industry. Global Journal of Researches in Engineering: Industrial Engineering 14 (3), 39–47.

Nayak, R., Padhye, R., 2015a. Introduction: the apparel industry. In: Nayak, R., Padhye, R. (Eds.), Garment Manufacturing Technology. Elsevier.

Nayak, R., Padhye, R., 2015b. Garment Manufacturing Technology. Elsevier.

Nicolaiov, P., Loghin, C., Ionescu, I., 2005. The technological reconfiguration, an instrument for increasing the sewing work productivity. In: Proceedings of the International Conference Management of Technological Changes, vol. 1. ISBN: 978-960-8475-04-5, pp. 99–104.

Nicolaiov, P., Florea, A., Loghin, C., 2010. Innovative Methods for Working Tasks in Garment Manufacturing (Ch. 1–6). Certex Publishing House, Bucureşti. ISBN: 978-973-1716-61-9.

Nicolaiov, P., Florea, A., Loghin, C., 2013. Specific Innovation Approaches in Garment Manufacturing Companies. AGIR Bulletin. ISSN: 1224-7928 1, pp.168–172.

Automation in quality monitoring of fabrics and garment seams

14

Thomas Gries, Volker Lutz, Volker Niebel, Marco Saggiomo, Kristina Simonis
Institut fuer Textiltechnik of RWTH Aachen University, Aachen, Germany

14.1 Introduction

The major raw material for making a garment is the fabric. Garment manufacturers source the fabric from external suppliers and inspect them for quality before the production. During the inspection, various fabric defects are marked and points are scored on the basis of the severity of faults (Nayak and Padhye, 2015). The fabric manufacturers also inspect the fabrics before sending them to garment manufacturers. Fabric inspection is done in a lighted environment by skilled workers using a table or a machine. The faults present in the fabric are marked, and some points are given for each type of fault. If the total points exceed the prescribed value, the fabric is rejected. Various systems such as 4-point, 10-point, and Dallas system are used for fabric inspection (Vijayan and Jadhav, 2015). Fabric inspection ensures the amount of garments with fabric faults are minimized by providing fabric without any fault. Fabric inspection not only ensures good finished garment quality but also reduces rejection and timely delivery.

In the context of automation, machine vision is playing an important role for fabric inspection, which is an imaging-based tool. In a garment industry, this is helpful for automatic inspection, process control, and monitoring. For proper working of machine vision, a large number of technologies, software, and hardware are integrated in a real-time environment of a computer. The existing technologies are used to solve the existing problems. Machine vision consists of an automatic inspection system, a camera or number of cameras, lighting system, or other imager, a processor, software, and output device such as the monitor. The following section highlights the use of machine vision for automatic fabric inspection.

14.2 Quality monitoring of woven fabrics

Machine vision is already being applied in textile production, and the methods used are still undergoing constant development. Several surveys and reviews on the topic have been conducted before, some of which are presented in this chapter. Besides the defect detection in fabric production, methods for classifying these defects have been developed as well as applications for quality control via the

Automation in Garment Manufacturing. http://dx.doi.org/10.1016/B978-0-08-101211-6.00014-8

analysis of fabric structure and properties. Furthermore, systems for textile quality assurance featuring selected methods for detection and classification have been implemented and tested to verify real-time applicability.

A significant amount of work on machine vision in textile applications has been done on the topic of defect detection and classification. Indeed, because of the nature of fabric and its production process, images are suited very well for fabric quality assurance. In the following sections, an overview of the quality monitoring of woven fabrics has been discussed.

14.2.1 Detection methods

Several surveys and reviews on the topic of fabric defect detection have been conducted. For a better overview, different methods have been categorized into several types. Kumar defines three categories of methods: statistical, spectral, and model-based used for fabric inspection (Kumar, 2008). In his survey, he also suggests that a combination of methods from different categories can yield better results. However, another researcher, Xie categorized differently dividing the methods into statistical, structural, filter-based, and model-based (Xie, 2008). Ngan et al. took a different and more generalized approach while classifying the methods (Ngan et al., 2011). Starting from patterned texture defect detection and distinguishing between different groups of texture, they defined two classes: motif- and nonmotif-based approaches. The nonmotif-based class contains two further subclasses, one of which contains the same three categories of approaches used for categorization in (Kumar, 2008; Ngan et al., 2011) in addition to the other two—learning and structural approaches.

The other nonmotif-based subclass is defined for different groups of texture and contains the category of hybrid approaches. The class of motif-based approaches only contains the motif-based approach category used for any group of texture. In another, more recent survey, texture analysis is divided into four categories: structural, statistical, spectral, and model-based. The survey gives an overview of detection and classification methods. Being the greatest common measure, the set consisting of statistical, spectral, and model-based will be used at the end of the section to categorize the referenced approaches (Loonkar, 2015).

In the study conducted by Zhou et al. (2011), autoregressive spectral analysis (AR) was introduced. As the name suggests, AR is a spectral approach. The feature extraction in this method takes place in horizontal and vertical projections of image windows. The approach in Sun and Long (2011) used an adaptive pulse-coupled neural network (PCNN) and ridgelet transformation, which is a combination of Radon and wavelet transformations. In the paper the proposed method is compared to a regular PCNN method and the Otsu thresholding method, which is a clustering-based image thresholding tool. Using a combined measure for comparison of the segmentation results, the novel method outperforms the others significantly.

The local contrast deviations-based method in Shi et al. (2011) utilized reference images of the fabric being examined. For the segmentation, a bilevel threshold function is used. The results showed an improvement in robustness compared to the use of modified local binary patterns (LBP) (Tajeripour et al., 2008).

Ananthavaram et al. proposed a method based on the use of regular bands (RB) and independent component analysis (ICA) (Ananthavaram et al., 2012). The shift-invariant RB method was based on periodicity and has therefore been chosen for textile texture analysis. ICA is used in different other fields, which is a statistical analysis method detecting directions not depending on one another, and that corresponds to irregularities and therefore defects in texture.

A new optimal objective function for the improvement of Gabor wavelet defect detection was proposed by Li et al. (2012b). The method used a measure based on mean and variance to detect especially warp and weft defects. It was of low computational complexity, and reported to show a detection rate of 93%.

The method presented in Li et al. (2012a) used Log-Gabor filters, the L*a*b* color space, and LBP. The LBP-based detection is conducted on energy feature images containing combined brightness and color information. Operating with nongray-scale images, the method can not only detect structural defects but also identify color defects in yarn-dyed fabric. The broader variety of the detected flaws leads the method to a comparatively low reported detection rate of 90% retaining the performance to be used in real time.

In Wan et al. (2013), a defect segmentation method based on PCNN and symmetric cross Tsallis entropy has been proposed. Using the latter to select the iteration number and threshold, better results have been achieved than with another PCNN method and Otsu segmentation.

Singular value decomposition (SVD) has been used for fault detection and segmentation in Chandra and Datta (2013). The images are successively partitioned into sub-images of different scales on which SVD is then performed. To increase efficiency the method used regions of interest considering pixel intensities to only perform analysis on potential parts of the image, where defects are present. The method detected 95% of the defects in the used textile defect image database.

For allowing the use in fabric defect detection and segmentation the bidimensional empirical mode decomposition has been adapted in Li et al. (2014). Images with several types of defects have been used to test the method's accuracy, which was determined to reach 96.1% using the same image database as the one used in Chandra and Datta (2013). A disadvantage of this approach is the high computational complexity, which prohibits real-time use of the algorithm as presented in the paper. In Table 14.1, a sourcewise overview of the reviewed methods recently developed for fabric defect detection has been discussed.

14.2.2 Defect classification methods

As described before, the defect detection step may be followed by a classification. The methods used for classification can be divided in several method families, among which the important ones are support vector machines (SVM), clustering methods, statistical inference, and artificial neural networks (ANN). In Habib et al. (2014) a survey focused on defect detection classifiers has been conducted, and an overview is given in Kumar (2011). In Ortiz-Jaramillo et al. (2014) different types of features for textile defect classification purposes are reviewed and evaluated.

Table 14.1 **Recently developed fabric defect detection methods**

Type	Method	Detection rate
Spectral	AR spectral analysis	–
Spectral	PCNN, ridgelet transform	–
Statistical	Local contrast deviation, bilevel thresholding	–
Statistical	RB, ICA	–
Statistical	Gabor wavelets, mean/var.	93%
Hybrid	Log-Gabor filter, L*a*b* color space, LBP	90%
Statistical	PCNN, Tsallis entropy	–
Model-based	Subimage SVD	95%
Statistical	Gabor filter, double thresholding	–
Spectral	BEMD	96.1%
Hybrid	Gabor filters, double thresholding, morphological operations	99.6%

AR, autoregressive; *BEMD*, bidimensional empirical mode decomposition; *ICA*, independent component analysis; *LBP*, local binary patterns; *PCNN*, pulse-coupled neural network; *RB*, regular bands; *SVD*, singular value decomposition.

In Habib et al. (2011), novel features have been introduced and used in conjunction with known/established features to comprise a set of features for robust defect classification. Using a total of 100 images in 6 categories (color yarn, vertical missing yarn, horizontal missing yarn, hole, spot, defect free) a classification accuracy of 98.99% has been reported, with only the spot as a defect not classified correctly in 100% of the cases. For the classification, an NN with backpropagation was used; the defect detection was done with a double thresholding method.

Li et al. (2013) combined a spectral estimation method with rough set theory for the detection and classification of fabric defects. In a first step the defect-free fabric pattern is extracted via parameter estimation and then the image is compared to match the pattern to detect flaws. With the tested method a recognition of oil warp and weft defects with ~96% accuracy was achieved.

The method proposed in Li and Cheng (2014) used an SVM to recognize six types of defects in yarn-dyed woven fabric (cracked ends, reed mark, broken picks, weft crackiness, hole and stain). To classify the defects, nine features are used leading an overall classification rate of ~91%. For the defect detection, the image is converted to a different color space and Log-Gabor filtering is applied. Table 14.2 contains an overview of the reviewed methods.

14.2.3 Example of machine vision system for fabric production

The weft insertion in air-jet weaving machines is not reliable in 100% of the cases. So-called "faulty pickings" can occur because of:

- inappropriate jet ending time,
- inappropriate air pressure settings, and
- unsynchronized weft yarn unwinding and picking fluid jetting.

Table 14.2 **Recently developed fabric defect classification methods**

Method	Classification rate (%)
Backpropagation NN	98.99
Rough set	96
SVM	91
Bayesian NN	95
Fuzzy logic	96.8
Cooccurrence matrix, feedforward NN	96.3

NN, neural networks; *SVM*, support vector machines.

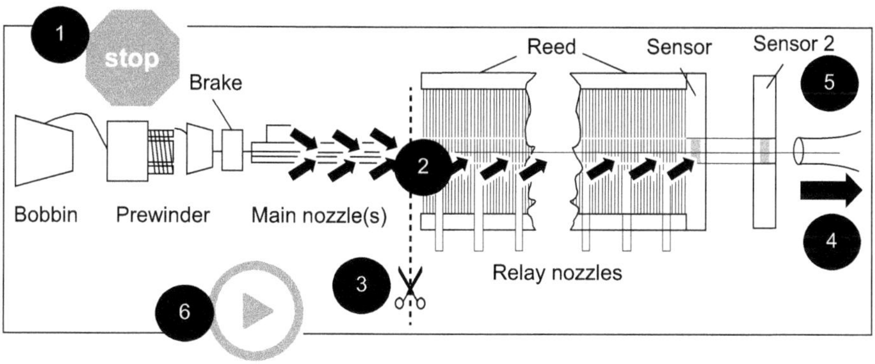

Figure 14.1 DORNIER's automatic filling repair principle.

The term "faulty picking" is defined as a weft thread launched by a main nozzle that does not reach the opposite side of the shed within a predetermined time interval (Mangold, 2000). An add-on for weaving machines enables the process to detect faulty pickings, to automatically repair them and to restart the machine (Wahhoud and Hehle). The weaving machine manufacturers respectively provide their own systems (extract):

- Lindauer DORNIER GmbH, Lindau, Germany: "automatic filling repair" (AFR)
- Picanol nv., Ieper, Belgium: "pick repair automation"
- Toyota Textile Machinery Europe AG, Uster, Switzerland: "Toyota automatic pick operator"
- TSUDAKOMA Corp., Ishikawa, Japan: "automatic defective pick remover"

The working principle of the automated removal of faulty pickings is shown in Fig. 14.1 (the example of DORNIER's AFR).

The automatic filling repair (AFR) removes faulty weft threads along the following steps:

1. Sensor 1 detects faulty weft thread—machine stops
2. Faulty weft thread is blown out by nozzles
3. Faulty weft thread is cut by scissors

4. Faulty weft thread is suctioned off
5. Sensor 2 checks if thread was suctioned off

After checking if a faulty weft thread was suctioned off successfully, a signal sent by sensor 2 indicates if the machine can restart or not. The checking of sensor 2 is so far the only possibility to verify the functionality of the AFR. In 80% of the cases, sensor 2 indicates a successful AFR procedure and therefore, initiates a machine restart although there are fragments of the faulty weft thread remaining in the shed. The AFR procedure is supposed to restart the weaving machine automatically after the faulty thread is removed (see step 6 in Fig. 14.1). The automatic restart of the machine makes the process independent from the operator who usually has to remove faulty threads and restart the machine manually. If the AFR restarts the weaving machine automatically, despite, the weft thread remaining in the shed, the reed beats up the thread, which results in fabric defects, see Fig. 14.2.

The depicted malfunction of the automatic filling repair (AFR) can occur because of the following reasons:

- Weft thread twists around one of the relay nozzles during the AFR procedure (especially step 2 in Fig. 14.1) and breaks.
- Weft thread is too long and activates sensor 2 during the AFR procedure.
- Weft thread breaks while it is suctioned off; a part of the thread activates sensor 2, while the rest of the thread remains in the shed (step 4 in Fig. 14.1).
- Sensor 2 is soiled and wrongly detects a successfully removed weft thread.

Because of the lack of reliability, the AFR with automatic machine restart is not used by weaving mills producing high quality technical textiles. Weaving mills that do not apply systems for the automated removal of faulty weft threads have to forego a 10% increase of productivity.

An approach for a vision-based support for the automated removal of faulty weft threads in air-jet weaving machines is undertaken in the German project WeftAlert (IGF funded) (Jansen, 2016). A machine vision system carries out an optical inspection of the shed after a supposedly successful AFR procedure. A positive result of the digital image processing means there are no yarn remaining in the shed. A negative result is achieved if there are yarn remaining in the shed after sensor 2 indicates a successful AFR procedure. Through signal feedback of the digital image processing to the machine's control unit, the machine automatically restarts only in case the optical

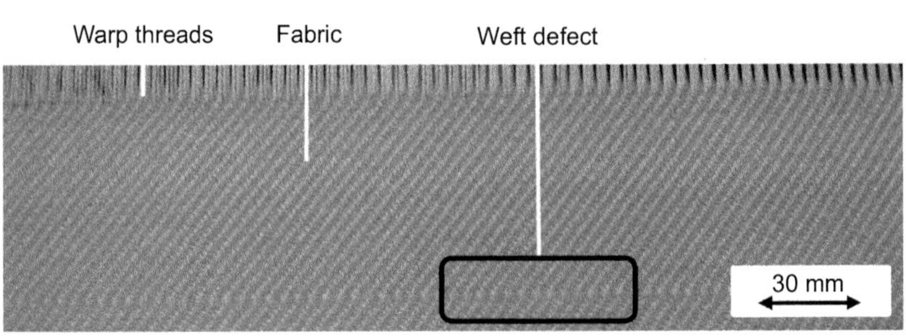

Figure 14.2 Example of a recognizable weft defect.

inspection approves the starting signal from sensor 2. The integration of the machine vision system into the AFR procedure increases the reliability of the system and makes it applicable for weaving mills producing high quality technical textiles.

At first, a defect catalogue is created. The defect catalogue contains pictures of weft remains in the shed. The photos are taken after sensor 2 indicates a successful AFR procedure although there are weft remains in the shed. The defect catalogue is the basis for the design of the machine vision system.

Secondly, a list of requirements is created based on surveys of weaving mills and machine manufacturers and expert interviews. The list of requirements contains necessary information to develop the machine vision system (hardware and image processing algorithms) and to design the integration of the system into an exemplary weaving machine (support, signal feedback, etc.).

Next, proper hardware (camera, lens, illumination, etc.) for the machine vision is chosen, and the support of the machine vision system is designed. Moreover, an image processing algorithm to detect weft remains in pictures of the shed is developed.

Finally, the machine vision system is validated in lab scale and in weaving mills.

The defect catalogue contains pictures of malfunctions of the AFR, which is the basis for the design of the machine vision system. Before the defect catalogue can be created, a suitable camera position needs to be determined. Fig. 14.3 shows the examined camera positions.

As shown in Fig. 14.3, three different camera positions are examined to detect weft remains in the shed. The camera is positioned perpendicularly and parallel as well as in a 10 degrees angle to the shed. For the determination of a suitable camera position and for creation of the defect catalogue, a digital reflex camera EOS 6D with the lens EF 17–40 mm from Canon Kabushikigaisha, Tokyo, Japan is used. Settings and relevant specifications of camera and lens are summarized in Table 14.3.

All experimental results of this work base on trials performed on an air-jet weaving machine A1 from Lindauer DORNIER GmbH, Lindau. Pictures taken from the mentioned camera positions are shown in Fig. 14.3.

It becomes obvious from Fig. 14.3 (left) that the weft remains is barely visible because of moire patterns. Moire patterns are based on the so-called moire effect.

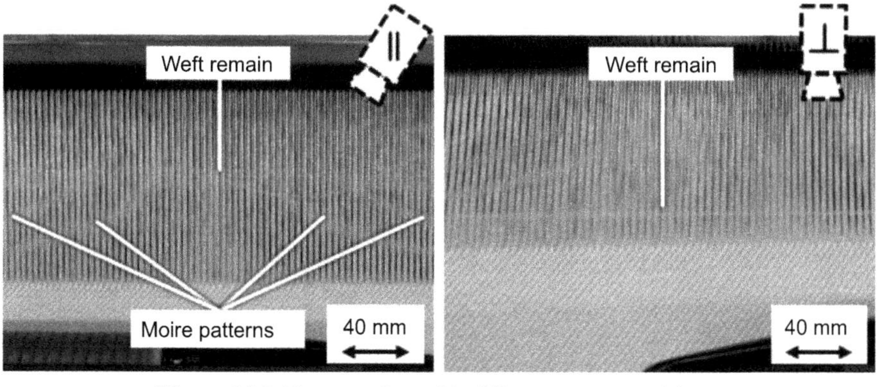

Figure 14.3 Pictures taken with different camera positions.

Table 14.3 Settings and specifications of camera and lens

Setting/specification	Value
Camera type	Canon EOS 6D
Lens type	Canon EF 17-40
Resolution	2736×1824 px
Sensor size	35.8×23.9 mm
Aperture	F/11
ISO	3200
Exposure duration	1/250 s

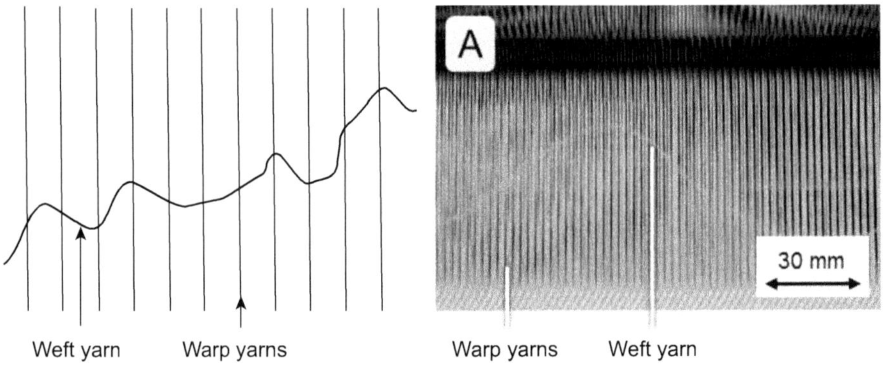

Figure 14.4 Defect category A, principle and example of a weft remain between upper and lower shed.

The patterns are similar to interference patterns. Moire patterns are formed when repetitive structures (in this case warp yarns) are overlaid with another structure. Camera-based influences such as subsampling favors the occurence of moire patterns. As can be gathered from Fig. 14.3 (right), the perpendicular camera position allows a better recognition of weft remains. A picture of the remaining camera position (10 degrees angle to shed) is not shown here because from this angle a view through the warp yarns was not possible.

Based on the results in Fig. 14.3, a perpendicular camera position is chosen to take pictures for the defect catalogue. The pictures for the defect catalogue are taken with a preliminary experimental setup in the ITA laboratory (Italab Private Limited) as well as in two weaving mills. At first, 100 pictures of malfunctions of the AFR were taken in each case. Afterward, the pictures were analyzed and clustered into four different defect categories. Examples for the determined defect categories are shown in the following Figs. 14.4 and 14.5. Fig. 14.4 shows the defect category A ("Standard") describing a weft remain that lies between the upper and lower shed.

Fig. 14.5 shows defect category B ("whirls"). A weft defect fits into category B if the weft yarn shows whirls.

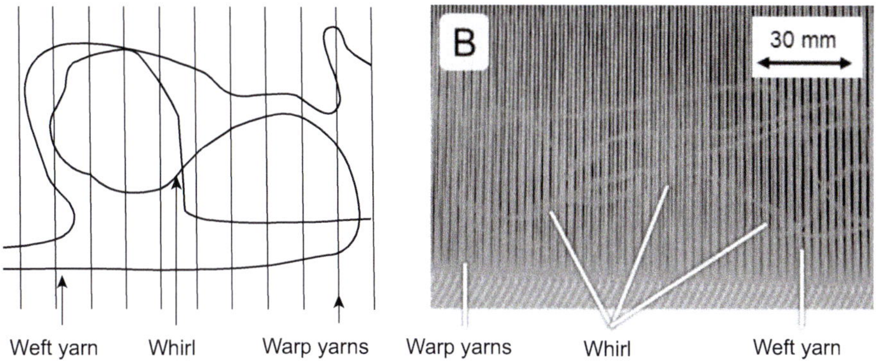

Figure 14.5 Defect category B, principle and example of a weft yarn showing whirls.

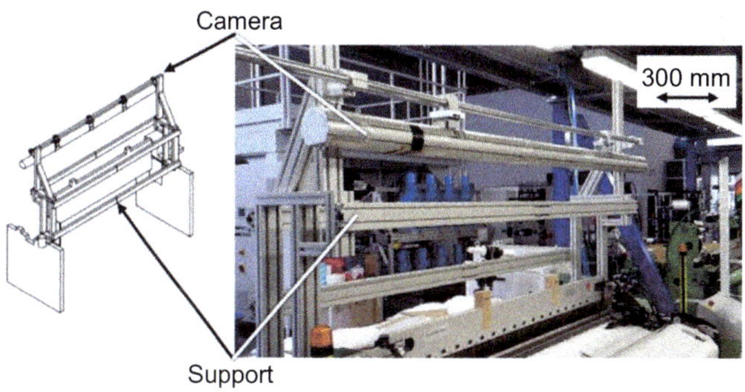

Figure 14.6 Mounted prototype on an A1 air-jet weaving machine.

Based on the defect catalogue and a requirement list, the machine vision system is designed. Furthermore, a first prototype of the machine vision system was set up in the laboratory of ITA, see Fig. 14.6. The machine vision system prototype is mounted on an A1 air-jet weaving machine from Lindauer DORNIER GmbH, Lindau.

14.3 Quality monitoring of seams

An overview of seam quality aspect and measurement methods is given in this section. Monitoring technologies are divided into machine parameter measurements and optical seam quality measuring principles. The result of the sewing process is influenced by the operating materials, the characteristics of the material to be sewn, the sewing thread, and the sewing machine needle (Nayak et al. 2013a). Therefore, an optimum setting of all sewing machine parameters is necessary. The seam balance (perfect position of thread entanglement points) is a main quality feature of a seam. It is accomplished by a balanced ratio of the restraining forces applied to the top and bottom thread. The interaction of the presser foot force with the material feed also influences the position of the thread entangling points. The needle sticking force and the needle temperature resulting from sewing speed and friction are further influencing factors (Amirbayat and Alagha, 1993; Ferreira et al., 1994; Anon., 1993; Anon., 1997; Rödel, 1996).

In addition to the sewing machine parameters, the seam is also influenced by the characteristics of the sewing material, the sewing thread, the needle and the processing technique. An overview of the different seam affecting parameters is shown in Fig. 14.7.

The company Dürkopp Adler AG, Bielefeld, Germany has developed a commercial system for the documentation of a sewing process with a thread-based sewing machine (see Fig. 14.11). The system records the data of the upper thread tension during the sewing process and thus allows conclusions to be drawn on the seam quality. An approximation sensor measures the distance toward a bending element and converts the measured values into the thread tension.

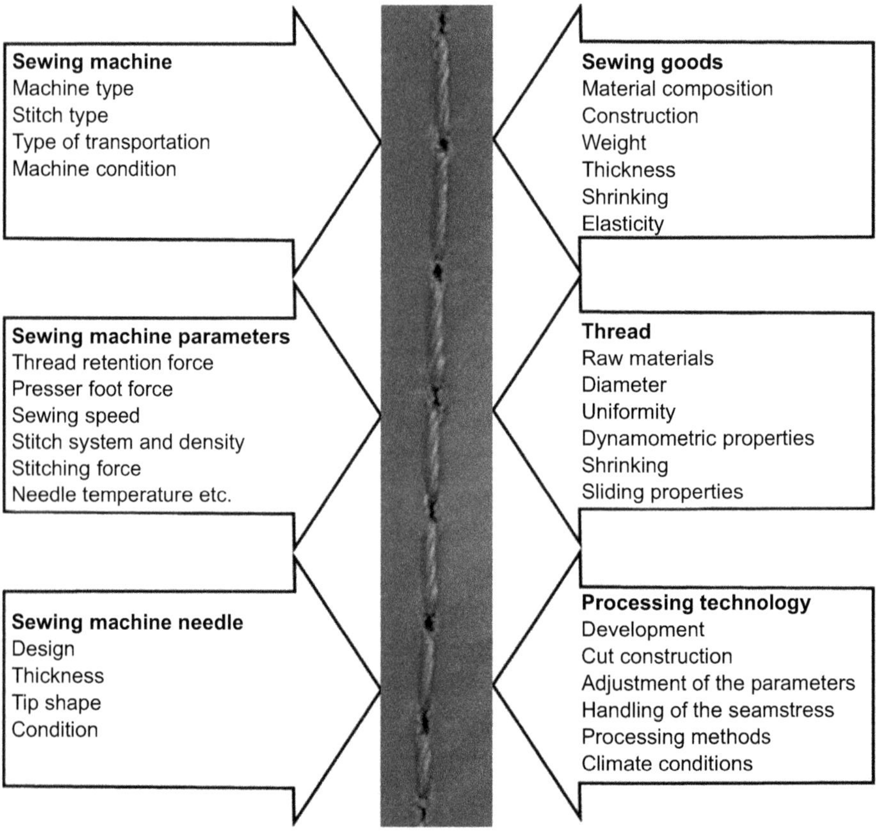

Figure 14.7 Seam influencing parameters.
Based on Klopp, K., Jussen, B., Gries, T., 2004. Einfluss der Nähparameter auf textile Endprodukte. In: Knecht, P. (Hrsg.), Jahrbuch 2004 – Textil, Bekleidung, Handel. Schiele u. Schön, Berlin, pp. S.173–S.181.

14.3.1 Current quality problem in the sewing technology

Today, the main problem in the production of clothes consists of faulty seams. These show up in seam crimping, seam shifts, or stitch length fluctuations, which cause additional cost of reworking. It also lacks flexibility of the joining processes. The main reason for this is the crafty character of the sewing process. The classy sewing machine is still used as main resource in the stitching. The worker is indispensable as operator on this machine. However, the operator is a potential error source. The product quality is heavily affected by the worker and depends on his/her ability, skill, and well-being (Zöll, 2002).

Various measuring systems based on sensors for the monitoring of processes have already been developed in the textile technology. Fig. 14.8 shows all existing sensor applications in various textile manufacturing processes. Yet, in garment manufacturing, only the measuring of tensions in the upper thread and different image analyzes on sewing machines have been applied. In many other areas of textile technology, a lot more online and off-line applications of sensor systems have been established (Klopp, 2013).

The following section describes the current measurement methods and projects that dealt with the automation and the development of measuring systems in the sewing process. This part of the work is an overview of some research works, but it does not claim to be complete coverage of the broad range of existing literature.

14.3.2 Previous solution concepts

Today, the seams of joined textiles are mostly controlled by off-line monitoring and 100% visual inspection. There are also online monitoring, which are used to determine

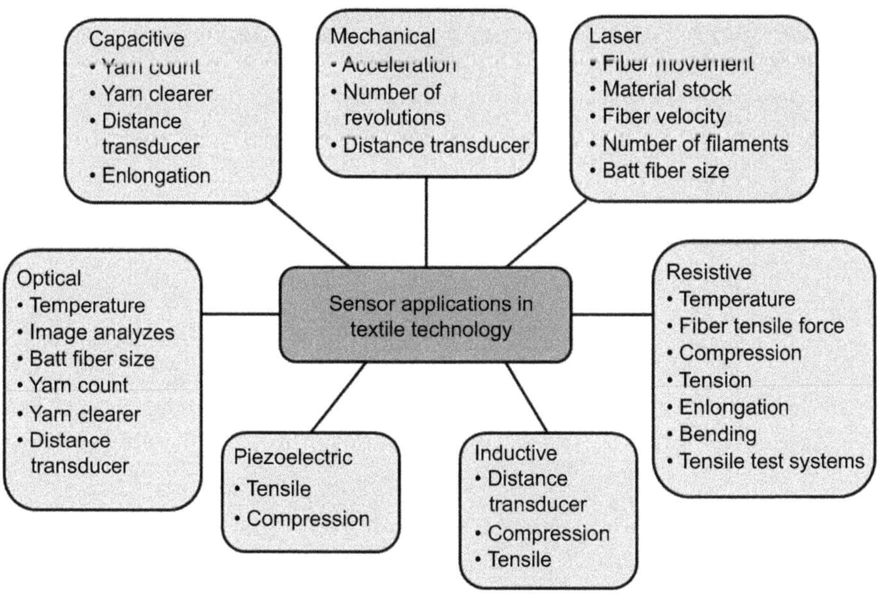

Figure 14.8 Sensor applications in textile technology (Gries and Klopp, 2008).

process and product characteristics, and they are integrated into the process. Seam tensile strength and seam transverse tear strength are determined by means of off-line testing procedures to assess the seam quality. Factors such as seam curling, seam height profile, and the seam sling behavior are also used for seam control. For the seam curling, measurements are made to determine the tendency to displace curling. A further approach is provided by an ultrasonic measuring method, whereby the seam can be recorded and objectively assessed without any contact (Anon., 1987; Reumann, 2000).

The online comprehensible machine parameters include, for example, upper and lower thread tension, as well as sewing speeds and the presser foot force. There are also some developed sensors, which are used for material detection and the determination of the sewing material handling and single transport operations. Sewing material detection can be carried out by miniaturized fibers in the presser foot or can be detected by infrared and microwave sensors. Optoelectrical and optical sensors are used to detect movements or changes in length directly at the sewing thread or the sewing machine control components. In optoelectronic methods, the movements and positional changes can be detected by shifts in the sensor, based on the reflected laser light. An example of this is a measuring system for determining the filling level of the lower filament coil (Reumann, 2000; Bäckmann, 1998; Jussen et al., 2002; Profos and Pfeifer, 1994; Rogale and Dragcevic, 1998; Wulfhorst, 1996).

A further optically measurable parameter is the seam balance. This describes the optimum position of the entanglement points and is characterized by a balanced ratio of the restraining forces applied to the upper and lower threads.

Fig. 14.9 shows two types of entanglement points. In the balanced seam (1), the entanglement points of the upper and lower thread lie centrally between the sewing material layers. In contrast, the seam entanglement (2) represents an unbalanced seam. Here, the ratio of the yarn restraining forces does not match, and the entanglement point shifts out of the centre of the textile layers.

A laser triangulation sensor has been developed by Inui and Shibuya in 1992 for the optical control of the crimping intensities. The sensor is used off-line. The sample is traversed under a fixed laser. The laser line triangulation consists of a laser line and a camera. These determine a 3D image through height profiles of the object, as the object is moving. This allows the shape of the object, consisting of height, diameter,

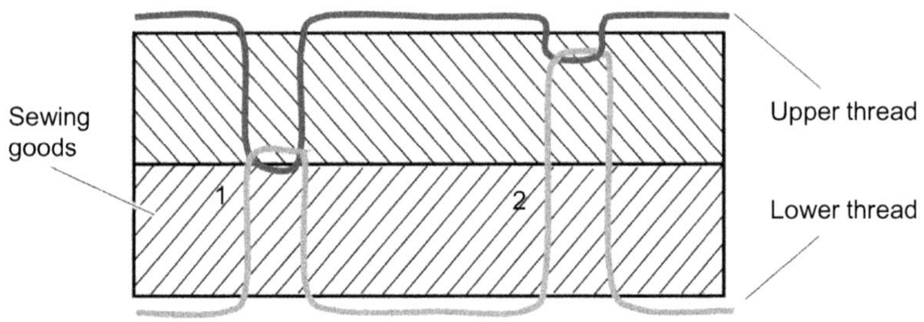

Figure 14.9 Seam balance.

and volume, to be recorded and analyzed (Amirbayat and Alagha, 1993; Heckner, 1995; Inui and Shibuya, 1992).

The principle of laser triangulation is illustrated in Fig. 14.10. The sensor of the camera detects the individual elevation profiles generated by the laser on the surface of the object. The individual profiles can be joined by subsequent data processing. This creates the 3D profile (Anon., 2013).

In addition to the above, researches have been done on in processes inspection, such as fabric production, solutions for the evaluation of textile surfaces that are based on digital image processing. For example, an optoelectrical system has been developed, which detects structural deviations, changes in the surface texture, and stiffness (Wulfhorst, 1998).

Thread friction and upper thread tensile forces can be measured, e.g., using bending cantilever principle. For this purpose, a strain gauge is glued onto the bending beam. This can convert the deflection signal into an electrical signal. The lower thread tensile forces can also be determined with this principle. However, this arrangement is restricted in scope and permits only evaluations at specific points in the process. In a project by Ferreira, Harlock, and Grosberg, strain gauges were used to record various machine parameters. These serve for the combined online detection of upper thread retention force, presser foot force, and stitch plate detection force (Ferreira et al., 1994; Jussen et al., 2002).

As a further parameter, the needle temperature can also be detected by means of infrared sensors or thermosensors. An increased needle temperature can damage the thread and the fabric. In this measuring system, thermocouples detect the occurring temperature change by a change of voltage. Infrared sensors determine the temperature over the intensity of the absorbed heat radiation (Grote and Feldhusen, 2007; Luenenschloss and Gerundt, 1978). The upper thread movement can be detected, e.g., by piezoelectric sensors. This type of sensor reacts under pressure or tension with an electrical charge, which generates electrical signals via an amplifier system (Grote and Feldhusen, 2007; Dorrity, 1995; Dorrity and Olson, 1996).

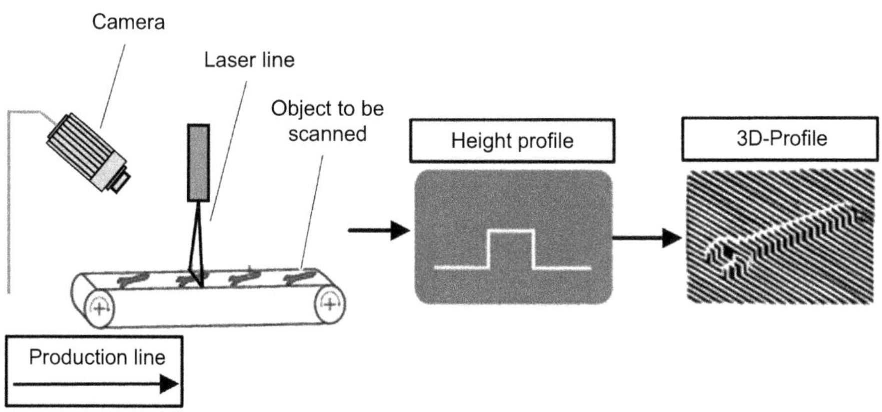

Figure 14.10 Principle of laser line triangulation, in accordance with Anon. (2013).

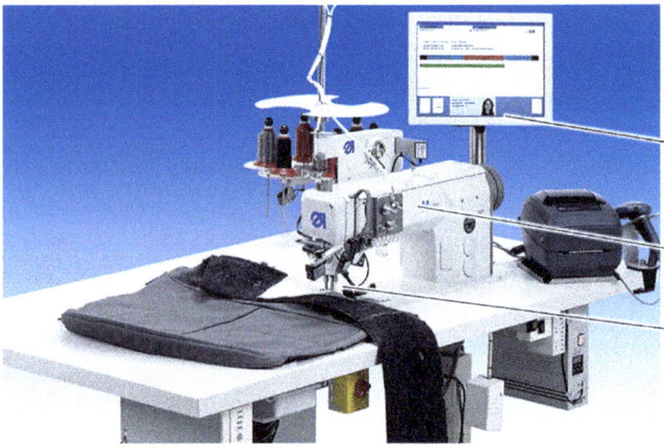

Figure 14.11 Dürkopp Adler seam documentation system for creation of airbag seam.

Today, especially for quality and safety critical sewing processes, such as in airbag production, there is an online measuring system, which measures the yarn retention force of the upper thread. This was developed to meet the requirement for a "tear seam," which is supposed to guarantee the smooth exit of the airbag at the automobile seat (Nayak et al., 2013b). At the same time, the documentation of the creation of the seam was improved, which is an important factor for the reproducibility of the seam. Fig. 14.11 shows the car interior and the airbags seam used to join the fabrics. In the car seats, the safety relevant nominal tear seams are necessary for the side airbags (Wauer, 1997; Dürkopp Adler, 2013).

Several projects were carried out to automate the sewing process. These are mainly concerned with the total or partial automation (Zöll, 2002). For example, Robotex is a sample project in this field. This project belongs to the third major BRITE project, a research project of the European Community. In this project, the joining process is monitored by a camera system and marking aids. As a further element, a controlled air flow is used, which is required for the perfectly balanced sewing together with the joining parts. However, the project could not be developed up to production maturity (Anon., 1998).

The parameters needle penetration force and needle temperature have also been researched worldwide for many years. Although special measuring technology has already been developed for this but no commercial products are available. A laser triangulation sensor was developed for the optical control of the entanglement points of the upper and lower thread in the work piece during sewing. This sensor is integrated into the presser foot and enables the detection of possible misalignment of the intertwining points with double stitch depending on the application at machine speeds of 3000 and 6000 stitches min^{-1} (Lünenschloß et al., 1973; Nestler and Arnold, 1980).

Because many variants of combinations of the parameter setting are possible to achieve similar sewing results, a separate detection of various parameters, relevant for the quality of the seam, is absolutely necessary regarding reproducible sewing machine settings. From the measured results obtained in this way, manipulated variables can be derived after corresponding evaluation. The condition described above

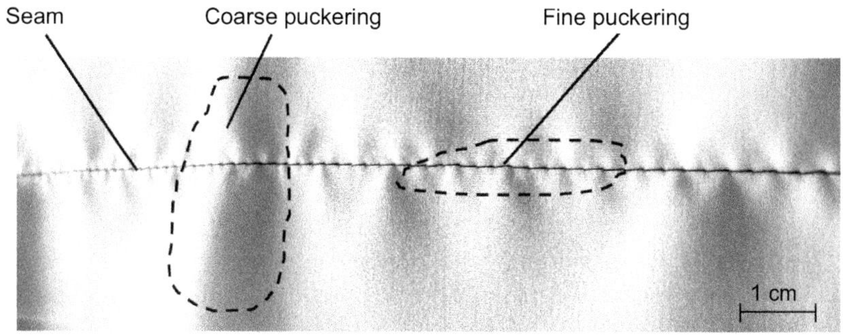

Figure 14.12 Seam puckering appearing on a fabric sewn with a double lock stitch.

shows that this has not yet been realized for reasons of sewing technology. The general development of the control technology with the corresponding sensors and actuators has proceeded in large steps, which has led to remarkable miniaturizations and cost reductions. In the area of sewing technology, however, various devices (sensors and actuators) are missing for reproducible adjustment of the setting values influencing the seam quality. The settings are still subjectively assumed, although the importance of the optimal setting of the sewing machine parameters is known. This means that errors in production are preprogrammed.

14.3.3 Seam puckering

Seam puckering is mainly caused by incorrectly set machine parameters. Incorrect sewing parameters produce a wavy appearance of the fabric along the seam. Many technical textiles such as car seats do not allow any wrinkles of the fabric near the seam. At first glance, these wrinkles are a sign of a poor seam appearance. In addition, the appearance of wrinkles may cause a risk in terms of the expected seam strength.

So far, the degree of seam puckering is determined visually with a sewn sample using a five-set photo standard of the American Association of Textile Chemists and Colorists. Fig. 14.12 shows two different characteristics of seam puckering such as coarse and fine puckering. Frequently fine wrinkles do occur in the immediate vicinity of the seam, whereas more chunky wrinkles have a higher width around the seam zone. Although several options exist, devices based on optical measuring principles are widely used for the evaluation seam puckering, which are explained below (Lutz et al., 2014).

14.3.4 Two-dimensional process—pattern recognition

In pattern recognition, the principle of the shadow is used. When there is a three-dimensional change on the surface, a clear shadow can be detected there by a flat angle of illumination, which is produced by a conventional light source. This type of illumination is referred to as dark-field illumination. If for example, a seam causes wrinkles on the fabric, lighting at an oblique angle identifies them via shadow occurring. The resulting shadow on the textile is then captured by a camera (Hati, 2011).

For monochrome cameras, a gray value is stored, which indicates brightness and intensity values of recorded pixels. To divide these gray values, a threshold operation is performed. Therefore, the gray values are divided into two classes of a binary image according to a fixed threshold. The threshold is placed near the minimum. All values above the threshold value are attached to the gray value of 1 (white) and all the rest is attached to the gray value of 0 (black). The gray value analysis is an established method in the 2D area. It is used mainly for online control in the packaging industry. The shadows must be recognizable with sufficient contrast; therefore, this method provides only satisfactory results relating to bright textiles and minimal ambient light influences (Lutz et al., 2014).

14.3.5 Photogrammetry

Photogrammetry is a triangulation method, in which the object is recorded by two different cameras. Fig. 14.13 shows the structure of a photogrammetry unit schematically. A firmly measured distance "d" to each other and to the object is covered by the two cameras. The processing of generated images of both the cameras is referred to as digital stereo image analysis. Therefore, the distance "h" between the level sensor and the object is calculated from the ratio of the position of the object on the sensor.

Photogrammetry is not commonly used in industrial quality control so far. Nevertheless, it is a system that is quite useful for image analysis and 3D recognition. However, reference marks are required to calibrate the cameras to a common coordinate system. Therefore, labels should be attached to the sewing table, or the seam has to be chosen as a common reference point. The accuracy is declared in the submillimeter range (Lutz et al., 2014).

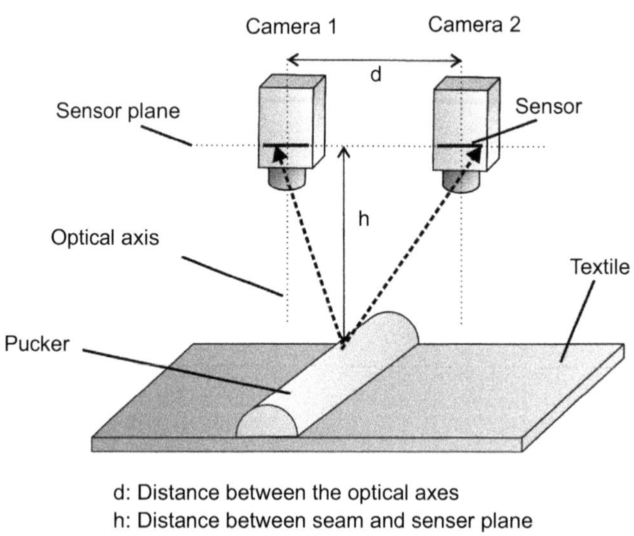

Figure 14.13 Schematic structure of a photogrammetry unit for detecting a wrinkle on a textile surface.

14.3.6 Laser triangulation and light-section method

Within the laser triangulation process emitted laser light is reflected from the target surface to be measured and detected by a sensor. This sensor is rotated by a certain angle to the emergent laser beam. Depending on the height of the measured surface, the laser light impacts at a different angle and is therefore locally shifted to the sensor. The pure laser triangulation is an online procedure in which the height profile of a line can be determined by a linear movement. To capture a height profile of the seam zone, wrinkles on the surface of the fabric surrounding the seam must be detected by a plurality of sensors, which are arranged side by side.

A further development of the punctual laser triangulation process is the light-section method. The light-section method can either be taken in operation with finished encapsulated sensors and laser detector or can be integrated on the sewing machine from the individual components of laser and detector. In the last mentioned method, no further changes may be conducted after the calibration. In contrast, the encapsulated light section sensor is a very robust and fast method to determine a height profile within the process. The method is suitable for 3D surface detection, when the roughness of the material is significantly higher than the wavelength of the laser light. This requirement is given for example in woven materials, when the yarn diameter is significantly higher than the thread diameter of the webbing (Bauer, 2007).

14.3.7 Comparison of measurement methods

In principle, all of the methods described above are suitable for the continuous detection of the seam height profile. The pointlike laser triangulation only provides height information of a point of the surface and is therefore not suitable for the detection of wrinkles in a defined seam width. All methods work smoothly with straight seams. Table 14.4 shows a comparison of different measurement principles based on evaluation criteria.

Once a curved seam is detected, it must be guaranteed that the measured seam lies within the measuring range of the sensor. Such a shift must be considered and balanced during evaluation. To detect a seam height profile, especially the light section procedure serves as a possible method, which was carried out by evaluation. Calibration is not necessary because of the fact that laser and sensor are connected in a compact unit. Furthermore, the laser triangulation is less susceptible to ambient light influences. Modern light section sensors allow a measurement frequency up to 1000 height profiles per minute. An installation of such a sensor unit on an industrial sewing machine is possible because of the compact dimensions without essential restrictions of the operating range (Lutz et al., 2014).

14.4 Quality monitoring of welded seams

Textile welding processes are used to create sealed seams, flat surfaces, or optical effects on seams. Outdoor jackets, marquees, airbags, filters, sportswear, etc., are typical examples for applications of these seams. To create a welded seam, temperature

Table 14.4 **Comparison of different measurement principles based on evaluation criteria (exclusion criteria are highlighted in gray)**

Principle of measurement / Evaluation criteria	Pattern recognition	Photogrammetry	Laser triangulation	Light-section method
Online capability	+	+	+	+
Contactless measuring principle	+	+	+	+
Speed	+	0	+	+
Accuracy	0	0	+	+
Integrability on the sewing machine	+	0	+	+
Covering of seam width	+	+	0	+
Independent of fabric	0	0	+	+
Vibration sensitivity	?	–	0	–
Cost	+	–	–	–
Decision	–	–	–	+

+, appropriate; –, inappropriate; 0, neutral; ?, not known.

and pressure need to be applied to the textiles. Therefore, only thermoplastic textile materials or thermoplastic-coated textiles can be welded. In 2009, in Germany alone textile welding processes were worth ~€100 million (Niebel, 2013; Anon., 2010).

For textile applications, the following industrial welding processes are generally used (Niebel, 2013).

- Hot wedge welding
- Hot air welding
- Tape welding
- High-frequency welding
- Ultrasonic welding

In comparison with seams with a sewing thread, quality control for welded seams is still limited. The optical appearance of the sewing thread on the textile surface is a direct indicator of seam quality. Welded seams should have a flat or regular pattern surface. The welded seam lies between two textile layers and cannot be seen by the machine operator. Possible seam defects can only be detected by process control systems so far. A schematic display of a lockstitch seam (left) and a welded seam (right) are given in Fig. 14.14.

PFAFF Industriesysteme und Maschinen AG, Kaiserslautern, Germany, owns the patent "DE 198 54 259 C2" for temperature control of the textiles at tape welding machines. Tapes are used to seal stitched seams. The surface temperature of the tape and the textile is measured before pressure is applied. Furthermore, the surface temperature of the tape after the sealing is measured. According to the measured data, the welding

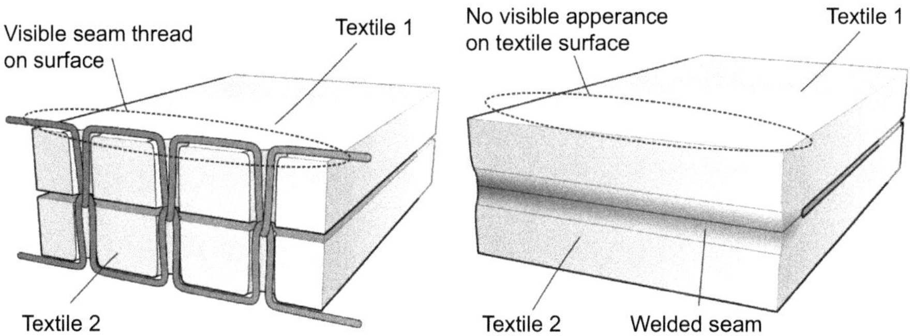

Figure 14.14 Schematic display of a lockstitch seam (left) and a welded seam (right).

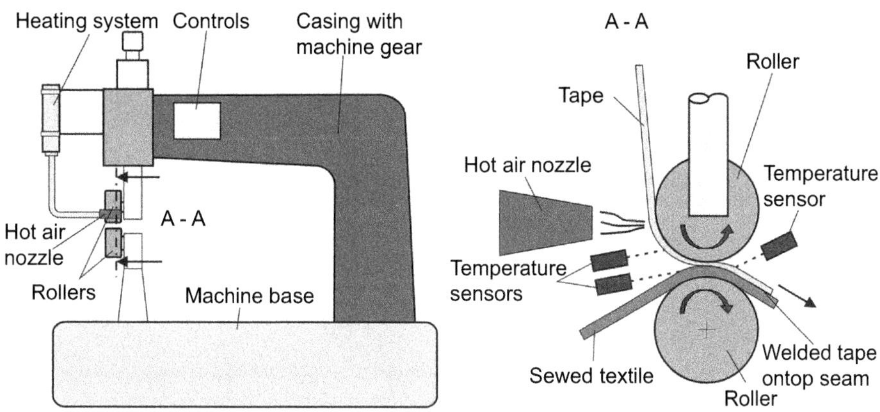

Figure 14.15 Hot air welding machine (left) and details of sensors around the welding area (right) according to Ellenberger et al. (2001).

temperature of the hot air is set. In Fig. 14.15 a hot air welding machine (left) and details of sensors around the welding area (right) are shown (Ellenberger et al., 2001).

A closed loop control system to enhance the seam quality of welded seams was developed at the Institut für Textiltechnik, RWTH Aachen University, Germany in cooperation with PFAFF Industriesysteme und Maschinen AG, Kaiserslautern, Germany. In this work, a temperature measurement system to determine the welding temperature within the weld zone between two fabrics, as well as a control unit was designed. For this purpose, the surface temperature (T_A) of the textile was measured and related to the welding temperature (T_{liquid}) between the two fabric layers, with the help of a mathematical model. Based on the mathematical model, a control and documentation system for the improvement of weld quality was created. With the help of this system, seam tensile strength (F) and reproducibility of these seams were improved. The system is suitable for welding processes with a processing speed of at least 10 m/min. Because of the increase in reproducibility, the rejection rate during production and the number of complaints can be reduced.

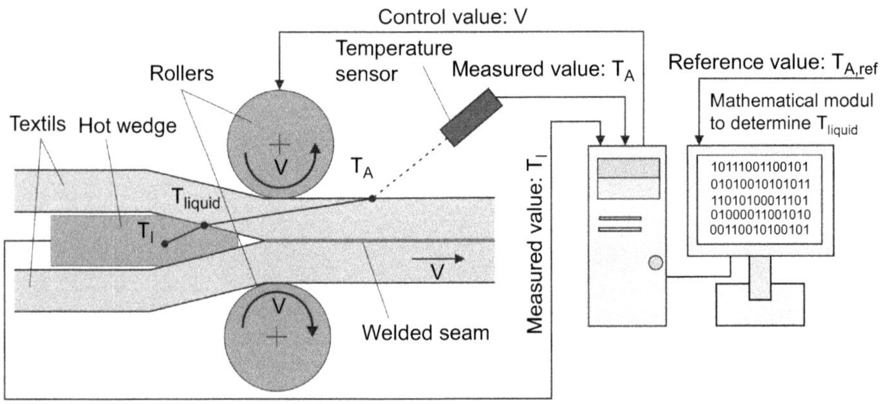

Figure 14.16 Schematic display of the developed quality control system for textile welding processes.

In Fig. 14.16 a schematic display of the developed quality control system for textile welding process is shown.

The surface temperature (T_A) can be measured by the developed system. With the preset internal temperature (T_I) and the mathematical model the relevant welding temperature (T_{liquid}) can be predicted. Based on the mathematical model, a closed loop control system for the relevant welding temperature (T_{liquid}) is subsequently implemented. The regulation is demonstrated by adjusting the welding speed (v), which is set by the transportation rollers.

The temperature sensor detects the surface temperature of the textile (T_A) and transfers it to the control software. The control software reviews if the measured surface temperature (T_A) is within the range of a preset threshold. This threshold is the reference value for the temperature ($T_{A,ref.}$) during control. The size of the temperature range determines the control barrier (R). With the control barrier, the maximum deviation from the mean temperature of the surface $T_{A,max}$ and $T_{A,min}$ is determined. For this study 2°C (R2) and 5°C (R5) were set as control barriers. The schematic display of the control threshold is shown in Fig. 14.17.

If the measured surface temperature (T_A) is not within the set range, the transport velocity (v) is changed until the surface temperature (T_A) returns within the range. Whether the transport velocity (v) is increased or decreased depends on whether the surface temperature (T_A) is higher or lower than the threshold. In the case that the surface temperature (T_A) is too high, the transport velocity (v) increases, and therefore less heat energy from the hot wedge is delivered to the textile. If the surface temperature (T_A) is too low, the transport velocity (v) is reduced, meaning more heat energy is released into the fabric.

Result analysis indicated a significant effect of the regulation of temperature (T_I) and textile speed (v) on the seam tensile strength (F). Depending on the material, an increase of the seam tensile strength (F) can be determined. In two of the three materials—polyamide 6,6 fabric coated with polyurethane and polyester—the maximum seam tensile strength (F_{max}) could be increased by 10% in comparison to uncontrolled samples. Furthermore, the variation of the maximum seam tensile strength (F_{max}) was reduced significantly by the control system for two of the three materials

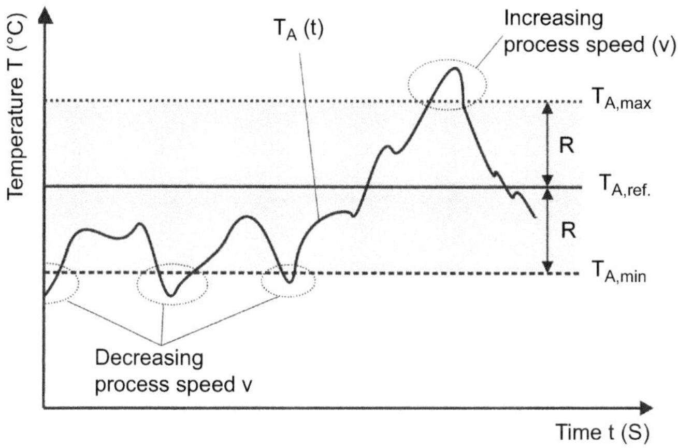

Figure 14.17 Schematic display of control threshold.

(polyester and polyester coated with polyvinylchloride). However, these results all indicated a very high dependency on the material.

Similar systems can be adapted to other welding processes (e.g., ultrasonic welding, high frequency welding etc.). Depending on the welding process, the welded product and the economic circumstances, such quality control systems need to be adopted.

14.5 Conclusion

This chapter presented the current approaches for machine vision systems, in general, and related to textile and clothing production. The solutions outlined in this chapter support textile producers to quickly react on issues occurring in their processes and to increase the quality of the products. All the solutions lead to the vision of smart factories and intelligent machinery. Many online and off-line principles to monitor the relevant aspects of sewn and welded seams have been evaluated. Because of the relatively high investments for vision systems, it is assumed that quality monitoring by measuring relevant machinery parameters is most likely to be implemented in future sewing and welding machinery. Further developments in automation technologies with a lower fraction of manual work will demand a holistic quality control of the automated production steps. The high variation of materials and seam design will be the biggest challenge in development of technical solutions.

Acknowledgements

The IGF project 16,793N of the research association Forschungskuratorium Textil e.V., Reinhardtstraβe 14–16, 10,117 Berlin (Germany) was funded on behalf of AiF in the framework of the programme for funding of the industrielle Gemeinschaftsforschung (IGF). Funding is provided by German Federal Ministry for Economic Affairs and Energy (BMWi).

References

Amirbayat, J., Alagha, M.J., 1993. Further studies on balance and thread consumptions of lock-stitch seams. International Journal of Clothing Science and Technology 5 (2), 26–31.

Anon., 1987. Prüfung der Neigung zum Verdrängungskräuseln (PFB 104). (Prüfempfehlung PFB 104).

Anon., 1993. Nähempfehlungen. Amann & Söhne GmbH, Bönnigheim, (Information).

Anon., 1997. The World of Sewing. Ferd. Schmetz GmbH, Herzogenrath, (Produktinformation).

Anon., 1998. Robotex Projekt von BRITE/Euram Bekleidung/Wear Fashion Technics 9.

Anon., 2010. Konjunkturberichte 01.2009–12.2009. – Berlin: Gesamtverband der deutschen Textil- und Modeindustrie e.V. Gesamtverband textil+mode (Hrsg.). Retrieved from: http://www.textil-mode.de/deutsch/Themen/Konjunktur-Statistik/K291.htm.

Anon., 2013. Sensor Intelligence – Vision Product Catalog 2014/2015. Sick (Hrsg.), Waldkirch, (Firmenschrift).

Bäckmann, R., 1998. Basis für Nähprozess-Controlling- Unterfadenüberwachung an Doppel steppstich-Nähmaschinen. Teil II Bekleidung/Wear 23 (12), S.12–S.14.

Bauer, N., 2007. Handbuch zur industriellen Bildverarbeitung, 1st Aufl. Fraunhofer IRB Verlag, Stuttgart.

Chandra, J.K., Datta, A.K., 2013. Detection of defects in fabrics using subimage-based singular value decomposition. The Journal of the Textile Institute 104, 295–304.

Dorrity, J.L., Olson, L.H., 1996. Thread motion ratio used to monitor sewing machines. International Journal of Clothing Science and Technology 8 (1/2), S.24–S.32.

Dorrity, J.L., 3–4. Mai 1995. New development for seam quality monitoring in sewing applications. In: IEEE Annual Textile Fibre and Film Industry Technical Conference, Charlotte, USA, pp. S.1–S.9.

Ellenberger, B., Herzer, K., Kutscher, K., Neurohr, M., 2001. Heißsiegelmaschine Deutsches Patent DE 198 54 259 C2.

Ferreira, F.B.N., Harlock, S.C., Grosberg, P., 1994. A study of thread tension on a lockstitch sewing machine (part II). International Journal of Clothing Science and Technology 6 (5), 26–29.

Gries, T., Klopp, K., 2008. Füge- und Oberflächentechnologien für Textilien. Springer Verlag, Berlin, Heidelberg.

Grote, K.-H., Feldhusen, J., 2007. Dubbel, Taschenbuch für den Maschinenbau, 22nd Aufl. Springer Verlag, Berlin, Heidelberg.

Habib, M., Faisal, Rokonuzzaman, M., 2011. Distinguishing feature selection for fabric defect classification using neural network. Journal of Multimedia 6, 416–424.

Habib, M., Faisal, R.H., Rokonuzzaman, M., Ahmed, F., et al., 2014. Automated fabric defect inspection: a survey of classifiers. International Journal in Foundations of Computer Science and Technology 4 (1), 17–25.

Hati, S., 2011. Seam pucker in apparels: a critical review of evaluation methods. Asian Journal of Textile 1, S.60–S.73.

Heckner, R., 1995. Tipps und Tricks zur optimalen Nahtqualität – Ausarbeitung der Firma Gütermann & CO. Bekleidungstechnische Schriftenreihe der Forschungsgemeinschaft Bekleidungsindustrie e. V. Anhang zum Band 112.

Inui, S., Shibuya, A., 1992. Objective evaluation of seam pucker using automated contactless measurement technology. International Journal of Clothing Science and Technology 4, 24–33.

Jansen, K., 2016. Forschungsbericht 2016, Bericht 63. Forschungskuratorium Textil e.V., Berlin. http://www.textilforschung.de/uploads/Forschungsbericht-2016.pdf.

Jussen, B., Klopp, K., Diesinger, D., 2002. Reproduzierbarkeit von Nähergebnissen Aif-12134. Forschungsgemeinschaft Bekleidungsindustrie e.V., Köln, (Abschlussbericht).

Klopp, K., Jussen, B., Gries, T., 2004. Einfluss der Nähparameter auf textile Endprodukte. In: Knecht, P. (Ed.), Jahrbuch 2004 – Textil, Bekleidung, Handel. Schiele u. Schön, Berlin, pp. S.173–S.181.

Klopp, K., 2013. Vorlesung Oberflächen- und Fügeverfahren in der Textiltechnik Aachen: RWTH Aachen. (Vorlesungsunterlagen).

Kumar, A., 2008. Computer-vision-based fabric defect detection: a survey. Industrial Electronics 55, 348–363.

Kumar, A., 2011. Computer vision-based fabric defect analysis and measurement. In: Hu, J. (Ed.), Computer Technology for Textiles and Apparel. Woodhead Publishing Limited, Cambridge, pp. 45–65.

Li, W., Cheng, L., 2014. Yarn-dyed woven defect characterization and classification using combined features and support vector machine. The Journal of the Textile Institute 105, 163–174.

Li, W.Y., Di Cheng, L., Xue, W.L., 2012a. Automatic defect detection of yarn-dyed fabrics based on energy fusion and local binary patterns. Advanced Materials Research 472–475, 3039–3042.

Li, Y., Lu, Z., Li, J., Li, T., Cui, L., 2012b. New optimal Gabor wavelets detection algorithm and application. Computer Engineering and Applications 48, 4–6, 11.

Li, M., Deng, Z.-M., Wang, L., 2013. Defect detection of patterned fabric by spectral estimation technique and rough set classifier. In: Fourth Global Congress on Intelligent Systems, pp. 190–194 Piscataway, NJ, USA.

Li, Z., Liu, J., Liu, J., Gao, W., 2014. Fabric defect segmentation by bidimensional empirical mode decomposition. Textile Research Journal 84, 704–713.

Loonkar, S., 2015. A survey-defect detection and classification for fabric texture defects in textile industry. International Journal of Computer Science and Information Security 13.

Luenenschloss, J., Gerundt, S., 1978. Einfluss der Paraffinierung auf die Nadelfadentemperatur beim Nähen von Maschenwaren. Melliand Textilberichte (9), S.211–S.213.

Lünenschloß, J., Krößwang, K., Modi, Z., 1973. Die berührungslose Messung der Nadeltemperatur an laufenden Nähmaschinen Wirkerei- und Stickerei-Technik 7, pp. S.462–S.468.

Lutz, V., Niebel, V., Seidler, C., Gloy, Y.-S., Gries, T., 2014. Optische Messsysteme zur Erfassung der Nahtkräuselung (Optical measurement system for detecting the seam puckering). Technische Textilien 57 (4), S.157–S.159.

Mangold, S., 2000. Problematik unvollständig eingetragener Schüsse beim Luft-weben. International Textile Bulletini 5.

N.N.: Bielefeld Dürkopp Adler, A.G., 2013. Retrived from: http://www.duerkopp-adler.com/de/funct/search.html.

Nayak, R., Padhye, R., 2015. Garment Manufacturing Technology. Elsevier.

Nayak, R., Padhye, R., Dhamija, S., Kumar, V., 2013a. Sewability of air-jet textured sewing threads in denim. Journal of Textile and Apparel, Technology and Management 8 (1), 1–11.

Nayak, R., Padhye, R., Sinnappoo, K., Arnold, L., Behera, B.K., 2013b. Airbags. Textile Progress 45 (4), 209–301.

Nestler, R., Arnold, J., 1980. Beitrag zur Ermittlung der Zusammenhänge zwischen Nadeltemperatur und Nadeldurchstechkraft, während des Stichbildungsprozesses Textiltechnik 3, pp. S.179–S.183.

Ngan, H.Y., Pang, G.K., Yung, N.H., 2011. Automated fabric defect detection—A review. Image and Vision Computing 29, 441–458.

Niebel, V., 2013. Systematische Entwicklung sensorbasierter Online-Qualitätsüberwachung für Schweißverfahren von Textilien Dissertationsschrift. RWTH Aachen, Aachen. Zugl. Aachen: Shaker, 2013.

Ortiz-Jaramillo, B., Orjuela-Vargas, S.A., Van-Langenhove, L., Castellanos-Dominguez, C.G., Philips, W., 2014. Reviewing, selecting and evaluating features in distinguishing fine changes of global texture. Pattern Analysis and Applications 17, 1–15.

Profos, P., Pfeifer, T., 1994. Handbuch der industriellen Messtechnik, 6th Aufl. Oldenburg, München, Wien.

Rao Ananthavaram, R.K., Srinivasa Rao, O., Krishna Prasad, 2012. Automatic defect detection of patterned fabric by using RB method and independent component analysis. International Journal of Computer Applications 39, 52–56.

Reumann, R.-D., 2000. Prüfverfahren in der Textil-und Bekleidungstechnik. Springer-Verlag, Berlin, Heidelberg, pp. S.563–S.565. S.547 ff.

Rödel, H., 1996. Analyse des Standes der Konfektionstechnik in der Praxis und Forschung sowie Beiträge zur Prozesssimulierung Habilitationsschrift. Technische Univ., Dresden. Zugl. Aachen: Shaker, 1997.

Rogale, D., Dragcevic, Z., 1998. Portable computer measuring systems for automatic process parameter acquisition in garment sewing processes. International Journal of Clothing and Technology 10 (4/4), S.283–S.292.

Shi, M., Fu, R., Guo, Y., Bai, S., Xu, B., 2011. Fabric defect detection using local contrast deviations. Multimedia Tools and Applications 52, 147–157.

Sun, Y., Long, H.R., 2011. Adaptive detection of weft-knitted fabric defects based on machine vision system. Journal of the Textile Institute 102, 823–836.

Tajeripour, F., Kabir, E., Sheikhi, A., 2008. Fabric defect detection using local contrast deviations EURASIP. Journal on Advances in Signal Processing 1155, 1–12.

Vijayan, A., Jadhav, A., 2015. Fabric sourcing and selection. Garment Manufacturing Technology 109.

Wahhoud, A., Hehle, J., 1996. Kontrollierte Absaugung an Webmaschinen, Deutsches Patent DE 195 21 100 C1. Veröffentlichungsdatum November 07, 1996.

Wan, H., Wu, Y.Q., Cao, Z.Q., Ye, Z.L., 2013. Segmentation of fabric defect images based on PCNN model and symmetric Tsallis cross entropy. Advanced Materials Research 760–762, 1472–1476.

Wauer, G., Mai 13, 1997. Aktuelle Nähtechnologien für technische Textilien in der Praxis 8. Internationalen Textil-Symposium, Frankfurt am Main.

Wulfhorst, B., 1996. Qualitätssicherung in der Textilindustrie – Methoden und Strategien. Hanser Verlag, Wien, München, p. S.143.

Wulfhorst, B., 1998. Textile Fertigungsverfahren – Eine Einführung. Hanser Verlag, Wien, München, p. S.198. ff.

Xie, X., 2008. A review of recent advances in surface defect detection using texture analysis techniques. Electronic Letters on Computer Vision and Image Analysis 7.

Zhou, J., Bu, H.G., Wang, J., 2011. Feature extraction using auto-regression spectral analysis for fabric defect detection. Advanced Materials Research 175–176, 366–370.

Zöll, K., 2002. Nähtechnik zur Fertigung textiler Hüllen Dissertation. Technische Hochschule., Aachen, pp. 11–16. Zugl. Aachen: Shaker Verlag, 2002.

Recent developments in the garment supply chain

15

Amanpreet Singh, Kanwalpreet Nijhar
RMIT University, Melbourne, VIC, Australia

15.1 Introduction

As the garment industry is huge, heterogeneous, and global in nature, it has seen many transformations in terms of changing dynamics from the simple word "fashion" to the modern and widely accepted concept of "fast fashion." Due to globalization, the garment industries compete not only domestically but also internationally, which drives this industry to stay young and fast. Garment industry is diversified and extends across various subsectors from apparels to dyes, synthetic fibers, and performance fibers, to suit more niche markets such as performance garments, and health care garments (Bruce et al., 2004).

Textile and apparel sectors are predominantly important for the economic growth of a country. These sectors provide a large proportion of employment in manufacturing, distribution, and retail, in both developed and developing countries. A massive shift in garment manufacturing to developing countries from developed countries has been observed because of the cheap labor and low production cost in the former. Stengg (2001) suggested that the impact of globalization can be seen through the localization of manufacturing in the third-world countries; which has changed the competitive direction of companies and nations. The increasing trend of globalization poses a threat to developed countries because they are not able to compete with the prices from lower-cost manufacturing countries (Nayak et al., 2015a).

According to Şen (2008), the garment industries can fundamentally be classified into three categories on the basis of the life cycle of the garments, which includes "basic," "seasonal," and "fashion" products. Basic products have the longest life cycle compared with the other two, which normally run throughout the year. This is followed by seasonal products, which have a product life cycle of around 20 weeks, whereas fashion products have the shortest product life of only 10 weeks. Abernathy et al. (1995) refined this classification: where the "basic" and "seasonal" products primarily target the men's and children's wear, and the women's wear falls into the "fashion" products with more designs, variety, and options.

Bhardwaj and Fairhurst (2010) defined "fashion" as an expression that changes over time and is different for different people. The concept of fast fashion took off with the growth in international trade, increased competition, and clustered manufacturing in developing countries and the ever-changing tastes of customers (Djelic and Ainamo, 1999). According to Sparks and Fernie (2004), factors such as low predictability, high variety, volatile demand, a shorter life cycle, and impulse buying characterize fashion.

Automation in Garment Manufacturing. http://dx.doi.org/10.1016/B978-0-08-101211-6.00015-X

Many of the transnational companies such as Zara, H&M, and Benetton thrive on competing for fashionable products that offer a larger audience to whom to sell their products (Christopher et al., 2004).

According to Porter (1998), it is not the nations but the supply chains that tend to compete in a cutting-edge environment, which is why the dynamics of supply chains have changed dramatically. Mangan et al. (2008) depicted the change as going from a fragmented to a totally integrated supply chain. They suggested that in 1960s, the supply chain functions such as transportation, warehousing, purchasing, and manufacturing used to work in isolation; this was followed by an evolving integrated period in 1980s, in which some functions reflected the importance of integration and the exchange of information. Currently, supply chains are totally integrated and all functions upstream and downstream are virtually linked, sharing real-time information to perform their respective functions.

Until the late 1980s, fashion retailers used to forecast trends and customer demands to place orders and they used to work on the concept of ready-to-wear garments (Guercini, 2001). Doyle (Barnes et al., 2006a,b,c) showed that the contemporary fashion industry totally integrated by changing the structural format from ready-to-wear garments to customized products. This forced the companies to become flexible and quicker, and to offer more varied products to the market, which was possible as the supply chain functions were integrated and the information was shared among all the players in supplier chain. Apart from variety, speed, and flexibility, marketing and capital investment is a major driving force behind companies' competitiveness (Sinha et al., 2001). According to Taplin (1997), today's retailers adopt the concept of "quick fashion" with the underline principle of speed to the marketplace and the customization of products to reduce lead time between the inception of design to consumption. Barnes et al. (2006a,b,c) explained that the fashion industry adopted the correct strategy and developed an infrastructure to promote a quick response by reducing the lead time to offer garments at the lowest possible cost in the late 1980s.

Retailers have started putting less effort into wanting customers to buy in bulk, as they want customers to visit stores frequently to increase their footprint and underpin the idea of "buy today or forget it." This is the strategy that Zara follows to stay ahead in the competitive fashion market (Bhardwaj and Fairhurst, 2010). Low-cost and reduced lead time are the concepts the garment industries today use to compete with each other. Therefore, shifting production processes to low-cost locations overseas is a trend in the garment industry, whether by outsourcing or sending the work offshore. According to Bruce et al. (2004), retailers prefer to source garments from low-cost countries to save on costs as frequent additions to infrastructure to meet customers' changing demands would be impossible in developed countries. This not only allows retailers and companies to concentrate on their core competencies, but also invest their resources and capital elsewhere to offer more return on investment. Although this strategy of outsourcing and offshoring brings substantial cost savings, it also has some downsides, including increased lead time owing to geographical distances, a longer supply chain, less control, and higher indirect costs (Birtwistle et al., 2003). It might also defeat the purpose of cost savings if the cost of carrying inventory and forced markdowns increase (Christopher et al., 2004).

Controlling supply chain activities and becoming flexible is a big challenge in global outsourcing because it might be difficult for companies to replenish fast-selling products in the middle of the season if they run out or if the demand increases. Thus it is imperative to manage logistics and supply chain activities efficiently between suppliers and retailers and to link all of the parties in a global supply chain by sharing real-time information and making collaborative decisions. Most importantly, this trend has created a large number of employment in industrialized countries, but opened up more opportunities in third-world countries.

15.2 Garment supply chain activities

The term "supply chain" has become prominent and has been used extensively since the late 1980s. There are many reasons for the increased importance of supply chain; the main reason being the increased emphasis on global sourcing. Globalization has forced companies to connect virtually. It also pushes them to find alternative ways to become more productive and capable in synchronizing the flow of goods and information. Tyndall et al. (1998) showed that the application of supply chain management is different from perspectives such as operational in which the emphasis is on product and information flow, some believe it be a *-management philosophy* and some as *-management process*.

Jones (1995) stated that the supply chain can be described as the flow of goods and services from the supplier to the end customer. According to Stevens (1989), supply chain management can be defined as the synchronization of activities to conform to the customer's requirements by focusing on a high level of customer service, controlling inventory and the logistics processes. The supply chain also means integrating interorganizational and intraorganizational functions using information flow upstream and downstream to offer the right product at the right price in the right place and at the right quantity and quality, and to the right customer. It can also be viewed as a network of organizations working together virtually, to produce a product or offering services to the end customer.

Porter (1998) showed that in the contemporary world, competition does not lie between companies but rather between their supply chains, and supply chains can work effectively only by integrating activities within and across companies. Because of the involvement of many parties, the supply chain of the textile industry is large, multidimensional, and complex (Jones, 1995). The textile supply chain, like others, involves three major types of flow: the flow of goods, the flow of information, and the flow of funds, that flow upstream and downstream.

The garment supply chain starts with the inception of a design. The first tangible process is fiber production, which is then converted into yarn. The yarn then is woven or knitted to make fabric that is finally converted into a garment after many operations in between. It is assumed that garment manufacturing processes are labor intensive and that the upstream processes such as yarn and fabric manufacturing are capital intensive. The detailed supply chain processes of the garment industry are discussed in the following section.

15.2.1 Design and sample production

The first process in the garment supply chain is the inception of the design. The designs are made in-house or specialist designers may be outsourced for the season. After extensive research and input from the buyers, the designs are constructed to reflect consumers' needs. Designs are also influenced by other brands and designer collections presented in trade shows in cities such as London, Paris, and New York. The designs are made with several factors in mind such as the final customers and their demands, and the end use of the garment.

Consumer demands vary among brands, countries, and seasons, so it is a crucial aspect of the supply chain to determine them properly. Information from the end customer moves upstream to the first contact point, which is the retailer. The retailer passes this information to the head office and then on to the manufacturers. Thus collaboration between the retailer and the manufacturer is important so that the right information is forwarded. On the basis of the information obtained from the customers, the correct types of yarn and fabric are selected, but the design is the first step of the supply chain toward meeting the customers' demand.

Prototype garments are made for internal approval; the whole process takes a substantial amount of time. Some companies such as JC Penny start the design process well in advance, some 40 weeks before the product is to be displayed for sale. Most companies thrive on cutting down this time as close as possible to be more agile and flexible and to reduce the cycle time. Fast fashion companies such as Zara and H&M usually start this process around 17 weeks before the season (Şen, 2008).

To be more responsive in terms of speed, modern technology such as computer-aided design (CAD) is used to make designs. The design process also takes into consideration the fabrication and suppliers required to create the desired garment. The designs are drawn into trade sketches known as "flat drawings." These are black-and-white format technical drawings that show a garment as if it were laid flat, to display all seams, topstitching details, and hardware, and any other design details. The sketches are sent to manufacturers to convey the idea of the design.

15.2.2 Merchandise planning

Once garment designs are finalized, the merchandise team then plans stock allocation; based on previous data, it decides when and where production and sourcing of materials will take place, to ensure it meets the budgets and customer's demands and that the new products will arrive in the store on time.

15.2.3 Production and manufacturing strategies

Many factors influence production and the production strategy to be used. These include the quality of the product, the retail cost, speed, and capital involvement. Production strategies have changed over time with changing demands. Specialized types of products in earlier times needed skilled people because products were handmade; this production strategy is called "craft production." This is now almost obsoleted because of the longer lead time it requires for production and because of the inability to reproduce a similar-quality product. Another issue is the need for a skilled person, if the product needs to be repaired or produced. This type of strategy has a low environmental impact because of the almost negligible use of machinery. This strategy is used in handicrafts, which is extremely expensive. India is a country that specializes in handmade garments.

The early 20th century saw a high demand for production at a fast rate. Because of the labor-intensive approach, craft production was unable to produce at a high volume. Thus the most excited advancement of the time was mass production. Henry Ford (1926) started using this concept in automobile industry and gained significant results. The strategy works on the economies of scale principle, according to which with increased production the cost to produce each unit is lowered as a result of the fixed cost spread over more products. With this strategy, individual activity is divided into labor so that every person continues to perform the same activity over a period of time, which makes the person skilled, and the production level increases. The use of machines also makes this strategy more attractive and widely recognized.

According to Mangan et al. (2008), the downside of this strategy is low product variety. The next prominent era in manufacturing history was lean production, which can be traced back to Toyota's production system principles. Toyota emphasized that importance should not be given to an individual process or person, but the entire production system should be viewed holistically and the efficiency of the process should thus increase (Ohno, 1988). The principle of lean manufacturing is a pull production system, which works on eliminating any type of waste from the organization's supply chain.

The modern strategy of production system that is widely accepted and used in the apparel industry is mass customization with the underlying principle of responsiveness and flexibility (Pine, 1993; Nayak et al., 2015b). According to Giovani Da Silveira et al. (2001), this strategy is facilitated by postponement, in which value-added activities are delayed to the far downstream as much as possible until a company receives a confirmed order. Companies have to trade between responsiveness and cost, depending on their supply chain priority and customer requirements. Another important decision regards where to produce the product, whether in-house or outsourced. Companies that want better control of their supply chain prefer to produce in-house using "vertical integration," whereas cost-competitive brands outsource their raw materials and production activities to stay competitive in pricing.

15.2.3.1 Raw materials (fiber and yarn manufacturing)

Fibers are commonly classified into two types: natural (cotton, jute, wool, etc.) and man-made (polyester, nylon, acrylic, etc.). The production of synthetic or man-made fibers is a capital-intensive process that requires significant skills and knowledge. Examples of synthetic fiber manufacturing companies are DuPont and DAK. Required characteristics can be embedded in man-made fibers while they are produced whereas natural properties are inherent in natural fibers. The use of the fibers depends on the customers' requirements and the end use of the garments.

This chapter mainly covers cotton fiber processing. Cotton yarn manufacturing involves many processes, beginning with ginning, in which cotton fibers are separated from seeds and then converted into cotton bales usually weighing around 700–800 kg. The conventional process of yarn manufacturing is ring spinning whereas the modern process includes open-end spinning. The preliminary process in ring-spun yarn is opening the bale, called the blow room; then the opened fibers are sent to a carding process, where sliver is made and fibers are laid parallel to each other after being cleaned. The sliver then goes to the draw frame to become more even, and then to the roving frame, which is the input process for the ring spinning machine. The outcome of the final process is finished yarn wrapped on bobbins for use in subsequent processes.

15.2.3.2 Fabric manufacturing

Most companies outsource the production of yarn and fabric from various factories across Asia and Europe. At this stage, yarn is transformed into fabric with the use of high-speed weaving machines and delivered to supplier factories. The yarn is first transformed into a big beam and fed into the weaving machines. Before going to weaving machine, this beam undergoes sizing to withstand the chaffing action of heald wires and heald shafts on the weaving machine. The sizing process includes coating of starch or other material on the yarn to give a minimum threshold of strength to prevent yarn breakage. The yarn may also be converted into fabric by knitting or nonwoven processes. The process of dyeing can be carried out at different stages during garment manufacturing, which can be of the fiber, yarn, fabric and even garment stages. To achieve customization and be more responsive, many retailers dye at the garment stage.

15.2.3.3 Garment manufacturing

As discussed, garment manufacturing is a labor-intensive process, so it is easy for newcomers to compete in this area (Abernathy et al., 1995). The final finished fabric (in the case of fabric dyeing) is converted into garment on the basis of the design and samples created and finalized earlier. The fabric is cut into pattern pieces and then assembled using sewing machines, ironed, inspected and finally packed for shipment. The production strategy used depends on the type of the product being manufactured: for basic and seasonal products, for which the demand is comparatively consistent and huge, mass production is employed to save on costs, whereas for fashion products, mass customization may be used.

A large amount of waste may be generated if the fabric is not laid down and cut properly and efficiently. Modern technology such as CAD is used to achieve efficiency and reduce fabric waste. A huge inventory is accumulated during the processes of cutting, sewing, tagging, and packing because there is less automation and more labor-intensive work. Most companies try to implement lean manufacturing tools such as the Kanban system, at this stage to reduce wasteful activities, which is a signaling system between two internal processes of supplier and buyer relationships. It works on the pull principle, in which a visual or an audio signal is sent by the buyer to the feeding process when inventory is nearly exhausted and needs replenishment (Hodge et al., 2011). Kanban is used to reduce work in process inventories and helps the process to discover faults and defects, if any.

15.2.4 Intermediaries or logistics service providers

Finished garments need to be transported to countries where they will be sold; therefore logistics services are required. Masson et al. (2007) put forward that generally many retailers and companies do not perform logistics activities by themselves because they do not have a competitive advantage in this operation. In a globally diverse garment operation, a retailer uses a logistics service provider (LSP) that not only can provide logistics services but also can offer integrated, end-to-end services. These intermediaries may not have physical infrastructure on their own but they usually have a network of suppliers in different countries.

The advantage of using intermediaries or LSPs is that retailers can focus on their own competitive advantages rather than expanding their business portfolio in providing services in which they have less expertise. These intermediaries have product and supply chain proficiency and use their existing supplier network to provide varied services. They act as coordinators in the supply chain between the retailer and the manufacturer by properly managing the flow of product, information, and funds through the use of planning. They also use their connections or network of suppliers to achieve economies of scale by collaborating with various retailers and placing combined orders. They act on the behalf of the retailer and check the quality of the products before they are distributed, so the retailer can focus on the core expertise.

Agility can be easily achieved by using intermediaries because they can quickly secure spare capacity from the supplier network. LSPs can also consolidate shipments for various retailers to avoid small volume penalties and can also achieve scale economies in transportation. They also deal with obtaining import and export permits, clearing custom formalities, and taking care of trade regulations. Third party logistics (3PL) and fourth party logistics (4PL) are examples of companies that provide supply chain and logistics services to organizations. 3PL offers many logistics services in an integrated manner to move the product and information. 4PL, which is also called a supply chain integrator, is a concept coined by Accenture in 1996; it integrates and manages its own capacity and that of others to provide comprehensive end-to-end services to retailers (Gattorna, 2006).

15.2.5 Warehousing and distribution

With so many different factories working separately to create a vast amount of garment styles, garments need to be collated into one warehouse. Apparel arrives at one of hundreds of distribution centers located in different regions throughout markets. The garments are unpacked, allocated, and sent to retail stores to ensure "just in time (JIT)" delivery. Garments are then stored, checked, branded, quality assured, and dispatched to retailers from these warehouses, which are positioned in a central location or in close proximity to retail stores.

Distribution centers store the products for short time which is necessary for the fast fashion industry and to stay competitive, as opposed to warehouse where the garments are stored for long time. In some cases, the garments or products are not even stored but are cross-docked after they are received from different manufacturers. The garments are first sorted and then consolidated into small consignments according to individual retailer's needs and are then dispatched. Some value-added activities can take place in distribution centers, such as tagging, packing, and customization, to conform to the principle of agility. Many technological advancements have made distribution center activities efficient, such as pick to voice and pick to light, which save a lot of time in finding the right product (De Vries et al., 2016).

15.2.6 Retailing

The finished garments are sent to retail stores, which may be departmental, discounted, mass merchandise, or specialty stores. When stock arrives at the retail store (from the distribution center), the staff at retail store will unpack the product and commence quality control checks, ensuring that the consumer is purchasing a product without faults (this is especially vital if the purchase is made online). Products are then received into the system and barcoded with the correct recommended retail price after the correct currency conversion (if applicable). Staff will then steam and hang the garments and then will follow a merchandising directive for in-store presentation.

Many supply chain strategies are used to control inventory, such as vendor-managed inventory (VMI), ABC classification of products, and radio-frequency identification (RFID) (Trifilova et al., 2010; Nayak et al., 2015a). The ABC classification of products can be done according to various perspectives such as cost, popular items, and high-margin items, with the sole purpose of avoiding any missed sale owing to nonavailability and controlling stock at least to the minimum to prevent ordering a high inventory and carrying cost. A retail operation is owned by either a company or a franchise, which is a company's strategic decision based on the country where products are to be sold, local government requirements, and the capital required to own the retail stores.

15.3 Contemporary issues in garment supply chain

Ever-changing and volatile demands, the complex global supply chain, more import and export documentation, and custom formalities expose the garment industries to higher risks. Christopher and Lee (2001) suggested that today's fashion industry is considered to be a fast fashion industry, which is inevitably compelled to offer more new product ranges, thus increasing the number of times merchandise is changed in the store. This increased inventory turnover is a modern trend pioneered by Zara in the fashion industry; the company changes its entire stock more than 20 times a year, which is why the industry is called "fast." This surely comes with huge responsibilities to manage the supply chain sensibly, to stay ahead to capture new trends and to translate them into products. Reliance solely on outsourcing may not be able to support this strategy, which is why Zara works on vertical integration, in which capital and technologically intensive works are conducted in-house, such as fabric manufacturing and dyeing, but labor-intensive tasks are outsourced. Even for these outsourced tasks, Zara has more than 300 subcontractors, each specialized in its own production task and working exclusively for Zara. This strategy of conducting processes in-house enhances the opportunity of achieving economies of scale, especially for capital-intensive operations, and realizing cost efficiencies (Christopher et al., 2004).

The growing tendency of the retailers is to rely largely on outsourcing, which in turn increases the complexity of the supply chain. Usually the companies receive much gain in terms of cost if products sent offshore, but it significantly increases the lead time. The geographical distance affects the lead time in addition to the internal problems of both the buyer and seller and also import and export processes. The whole supply chain is thus left with a pipeline inventory that is neither in the hands of the supplier nor the retailer. This has a huge negative impact, especially for the fast fashion strategy, in which fashion changes rapidly. The consequence of this huge inventory is obsolescence, because new trends and fashions hit the market and old ones are either marked down or taken off the shelves.

The lead time becomes even a bigger issue when suppliers are not concentrated in one place and/or multiple suppliers are involved to fulfill the customer's demands. Therefore, outsourcing from multiple suppliers needs to be considered carefully if lead time is a priority of the company for target customers. Conversely, having multiple suppliers gives more options from which to choose and creates a sense of competitiveness among suppliers (Mangan et al., 2008). If the network of suppliers are working exclusively for the retailer, it can reduce the lead time. This is because the subcontractors work in collaboration and make their way toward achieving the same goal of agility that Zara and Benetton follow (Christopher et al., 2004). Many issues affect the performance of the garment supply chain; which are discussed in the following section.

15.3.1 Market restructuring

Garment manufacturing is a labor-intensive industry that provides an opportunity to low-cost labor countries to gain hold on producing garments. A major shift has been observed in garment production moving to low-wage, developing countries. Trade regulations and continually lowered trade barriers have also made this industry global, especially for production (Şen, 2008). This has increased the amount of imports in developed countries for different types of garments. This in turn has led to major market restructuring in various countries including the United Kingdom (Bruce et al., 2004), the rest of the Europe (Keenan et al., 2004), Australia, the United States and Japan (Taplin, 1997).

15.3.2 Third party logistics involvement

The fashion supply chain is rather complex and less adaptable (Davis, 1993; Danese et al., 2004). Dependence on global suppliers and offering short–shelf life products make the fashion industry even more complex. The complexity is also intense as a result of product mass customization to suit different customer needs, diversified skills and the required knowledge, quickly changing designs, and finally, the immense flow of information (Croom et al., 2000).

Working in isolation and not passing on information on time, mistrust, and wrong information may lead to a chaotic situation in the garment supply chain. This leads to holding the wrong stock, which in turn has the potential for supplies to become obsolete or sold at markdowns or discounts, which results in missing threshold profitability (Christopher and Lee, 2001; Handfield and Bechtel, 2002). Another difficulty is that if the entire onus for procuring, manufacturing, and delivering garments is given to the 3PL company, it will lessen the control on the supply chain activities. Thus, heavy reliance on a third party may leave the retailer more susceptible to being exploited. Hence, agile supply chains have ways to alleviate these risks (Masson et al., 2007).

15.3.3 Product or buyer driven supply chain

Fast fashion, to which most fashion retailers claim their competitive advantage, relies fundamentally on speed, which has helped this industry transition from production-driven to market-driven moves; however, both are led by globalization and industrialization (Bhardwaj and Fairhurst, 2010). To stay competitive, many retailers are pushing the boundaries to transit their supply chain from production-driven to market-driven. To achieve this, they have built their supplier network in different markets (Barnes et al., 2006a,b,c).

Production-driven chains are normally large chains steered by big transnational companies because they capture bigger international market than domestic. According to Gereffi (1999), these companies are normally manufacturers and they have a pivotal role in the supply chain by controlling upstream and downstream linkages. Generally, these industries are capitally affluent and technologically savvy, and include manufacturers of automobiles, aircrafts, computers, and heavy machinery. Customer-driven commodity chains include large retailers or branded marketers, which are the focal point, to drive mainly the upstream network by decentralizing production in typically low-wage countries. This concept is common in industries such as garments, toys, footwear, houseware, and handicrafts.

The apparel industry relies heavily on global sourcing in which the production of the garments is generally diverted to third-world countries where contractors and subcontractors specialize in making components because of the difficulty of matching costs if these products were produced in developed countries. In producer-driven commodity chains, products are manufactured before the order is placed. This works on the "make to stock" principle, but the manufacturers/producers control upstream suppliers and downstream distributors and retailers because of the product expertise, patents or intellectual capital (Gereffi, 1999). On the contrary, the competitive advantage of consumer-driven chains lies in the globally spread supply network and the flexibility to customize products to cater to individual needs. The profit in customer-driven chains is generated by offering variety with speed, and not in scale economies, whereas in production-driven chains the sources of profit are mainly intellectual capital, achieving economies of scale in production, and high-value research.

There are different barriers to enter in both production- and customer-driven commodity chains. Gereffi (1999) suggested that the capital-intensive way, technological advancements, and know-how may be the tangible assets in production-driven commodity chains for use as barriers to enter. For example, the Japanese production system diversion from mass production to mass customization, the introduction of JIT, modular production, flexible production, lean manufacturing, etc. are the tangible assets. On the other hand the intangible assets such as brand recognition, marketing skills, relational ties, strategic alliances with suppliers and sharing real-time information are most important in customer-driven commodity chains. In the apparel commodity chain, which is considered to be consumer driven, entry barriers are lesser, owing to less capital-intensive and more labor-intensive, but capital-intensive processes become more important as the chain moves upstream, especially in yarn and fabric manufacturing.

15.4 Contemporary trends in apparel supply chain

A major shift for the manufacturers and retailers was experienced in the 1990s, when low cost became the key to survive in the market. Big market players could afford to manage cost by achieving scale economies in their production processes, procuring large amounts, sea freight transportation, and centralized distribution but they were still unable to match the price of garments produced in the Far East and low-wage countries. Thus companies operating on long cycle times needed to improve to make their supply chains more flexible and shorter (Fernie and Azuma, 2004). This led to the evolution of contemporary supply chain practices such as JIT, total quality management, and total productive maintenance, that helped to manage complexities relating to substantial geographical distances (Bruce and Daly, 2006).

Supply chains are integrated virtually more than ever before because of the use of electronic data interchange (EDI), although some big players in the fashion industry such as Zara and Benetton prefer vertical integration, especially for capital and skills/intellectual-intensive processes (Birtwistle et al., 2003; Bruce et al., 2004). Some contemporary trends that are widely accepted and used in the garment industry are discussed in the following section.

15.4.1 Comanaged inventory

According to Christopher et al. (2004), this particular strategy reduces the transaction cost for both parties involved, the sender and receiver. Comanaged inventory is the process of the supplier collaborating with the retailer to manage product flow to the distribution centers or retail store. The minimum and maximum stock are agreed upon by the supplier and retailer and real-time information is shared by the retailer with the supplier to replenish on time and to generate orders to the upstream suppliers accordingly. This strategy works best when demand is considerably stable and known, which is an option for basic and seasonal products. This strategy may not fit well for fashion products owing to their changing demand and short life cycle (Şen, 2008).

15.4.2 Quick response

The emphasis of quick response (QR) is to compress the supply chain system, to reduce the pipeline inventory, which in turn would reflect the customer demand more explicitly. In QR, the company tends to be flexible and more responsive in providing the goods to the customer in the exact quantity and right quality, on or before time, at the right place and, most importantly, at the promised price. An instantaneous reaction to the QR is a reduction in the cycle time, controlling the inventory at each stage of the supply chain even during a last-minute change and producing near the season (Şen, 2008).

Many technologies such as CAD/computer-aided manufacturing, pick to voice and pick to light technologies at automated distribution centers, point of sale (POS) scanners in retail stores, RFID to control and manage inventory, EDI to share the information instantaneously, etc. support QR at different stages of the apparel supply chain to reduce the cycle time. QR gives the option to be open to last-minute changes and the ability of the process to be flexible to make such changes without keeping extra inventory, adding extra cost, or affecting the lead time (Gunston and Harding, 1986). QR is the most important phenomenon in fashion production owing to its dynamic nature and speed.

Christopher et al. (2004) stated that the traditional practice of manufacturing is pushing the product through the market, in which garments are produced before the season. In QR, a small quantity is produced and sent to the market to get the sense of demand and acceptability through POS data and listening to the cash counter. Data are analyzed regularly to understand changing customer preferences. In the case of high demand, the product can be produced and replenished accordingly, although that would put extra pressure on manufacturing. However, it is more cost-effective than producing the product in bulk that might not get sold.

15.4.3 Collaborative planning and decisions

Competition and collaboration are two opposing forces in which the former focuses on being the best to offer the best product with own capabilities, and the latter brings the best together by sharing and trust. According to Porter (1998), both of these phenomena are important for running a business effectively, especially when the firms work within proximity to each other. He claimed that the companies collaborate and share ideas and resources with their suppliers and customers, but still compete horizontally to stay in the lead.

Collaboration between firms has always been recognized as good practice (Lamming, 1996; Fernie and Azuma, 2004). Collaboration allows the fashion industries to share ideas, exchange information upstream and downstream to have better visibility in the supply chain, and reduce inventory while enhancing responsiveness (Christopher and Lee, 2001). This synchronous activity is even more important when supply chain activities are globally dispersed to reduce the bullwhip effect. Where the retailer and supplier collaborate through the exchange of information, VMI has enormous benefits for both (Mangan et al., 2008).

15.4.4 Responsibility shift from retailer to manufacturer

Horizontal collaboration and retail consolidation have caused a power shift from manufacturers to retailers in terms of driving the supply chain. Large, well-known retailers such as Walmart and Kmart have made conditions favorable to themselves by changing the contracts that actually push their decisions to the manufacturers in terms of cost, speed, holding inventory, service, and delivery (Jones, 1995). Giant retailers want their orders to be placed near the season for the sake of not keeping inventory in their warehouses/distribution centers and to have continuous replenishment, which forces manufacturers to assume the risk of keeping inventory and staying flexible to make last-minute changes.

Alper Şen (2008) showed that traditionally, processes such as tagging and hanging were taken care by the retailer as a part of the manufacturer operation. In the event of noncompliance, financial and operational penalties are tough for the manufacturer. They need to stay in this competitive environment; otherwise retailers have the option to switch easily to other suppliers. This further inflates the cost to the manufacturer with no benefits gained. For example, in the early stage of adopting RFID, Walmart asked its suppliers to implement this technology or they would be left out of the business. Another example is that of Sears, which expected all of its vendors to employ and use EDI and to make products ready to be sold to meet its policy of floor-ready product standards (Magazine, 1998).

15.4.5 Product consolidation (factory gate pricing)

The low-cost trend in the apparel industry has singled out suppliers that are not competitive, are inflexible, and cannot conform to last-minute requirements. Entry in the apparel industry has low barriers, but without cost as a competitive weapon, companies cannot continue for long owing to the globally dispersed garment supply chain in which there is a number of contractors and numerous subcontractors who produce a proportion of the total order. These small orders, which are distributed individually or in isolation from other collaborative partners, increase the price of transportation; this may not be acceptable to the retailer, who competes on cost. These retailers have started using factory gate pricing (EXW), in which products are picked up from the vendor's or manufacturer's premises using their own or contracted transportation, for consolidation into a larger shipment, hence reducing the cost. Many big retailers have initiated this practice: for example, Tesco uses outside work contractors for better visibility and to control the supply chain and still controls cost (Potter et al., 2016).

15.4.6 Lean transformation

The concept of lean manufacturing can be traced back to the Toyota Production System, which was introduced in Japan. It was an important need of that time because of the economic breakdown of Japanese industry as a result of World War II. According to Ohno (1988), the underlying focus was to reduce the cost of manufacturing without using mass production like their American counterparts.

Toyota realized that the main asset of its company was its employees. Even the best technology would not be able to result in good returns if the employees were not trained and motivated. The company introduced the concept of continual improvement, in which it kept working on reducing wastes from its system in any form that lowered profitability. To reduce cost, the company did not depend on achieving economies of scale, like its competitors, but on being flexible and quick to change its models from one to another.

According to Womack et al. (1990), lean manufacturing is a systematic technique to realize more with less effort and infrastructure by reducing waste from the supply chain. They stated that companies work in a synchronized manner although they are individual and separate entities, in the concept of lean enterprise. They proposed that each supply chain activity need to be reviewed for optimization by eliminating wasteful activity. The lean principle works on a customer-centric approach as opposed to the push strategy, which is a supplier approach.

There is a customer–supplier relationship among all processes in-house and an interorganization to satisfy customer demand by providing customers with the best that can be offered by the supplier. The application of the lean concept in the garment industry eliminates non–value added activities from supply chain processes to optimize them. Conventional wisdom says if the shape or characteristics of products are changed by any operation, it is a value-added activity (Womack and Jones, 2010). There are many processes in the supply chain that do not add value but are still necessary for the value to be added by another process. For example, transportation does not add value to the garments while they are being transported, but it is essential to move the product to convert the raw material into the final garment. The kinds of waste in the lean manufacturing culture as discussed by Womack et al. (1990) are:

waiting: a product waiting for the process or a process waiting for the product
overproduction: producing large amounts, as in the batch and queue system, which may not be sold
unnecessary motion: a wasted motion to find the tools and equipment used in the process by the worker
repair/defects: owing to poor quality, if the garments are repaired, that will add extra cost
overprocessing: using overly complicated tools to perform a simple job or processing the garment more than actual specifications
unnecessary inventory: includes raw material, works in progress, or final products. Extra inventory hides defects
transportation: unnecessarily moving the work in progress material for a longer distance to be converted into the final product
improper use of employees: using skilled people in the wrong place

Application of the lean concept is not confined to the manufacturing sector, but is also used in the services sector (Mangan et al., 2008). There are numerous tools to implement the lean culture in companies, including cellular manufacturing, JIT, Andon, Jidoka, Poke Yoke, 5S, the pull system, Kaizen, work standardization, and total productive maintenance, to name a few (Feld, 2000).

15.4.7 Agile supply chain

The global garment supply chain is diverse and has spread across boundaries as a result of high dependence on outsourcing from low-cost countries in the case of basic products and from specialized countries for luxury products. This complexity brings many challenges such as lead time, communication, and responsiveness that need to be taken care of. To mitigate the risk and challenges, the agile model of the supply chain has emerged.

Conventional supply chains are long and complicated; hence they are forecast driven. The reverse is true for modern supply chains. Responsiveness is the key to manage changing demands in the modern supply chains, because the exact demands of fashion products cannot be forecasted and companies that rely solely on forecasting are not able to be sustained (Christopher et al., 2004). Responsiveness also refers to low or no inventory, whereas a forecast-driven supply chain is based on built inventory. To avoid the chaotic conditions that arise in the supply chain owing to the variable demand and the short shelf life, the decision should be made to devise policies that will restructure the supply chain and to source, manufacture, and deliver fashion products on time and in the exact quantity.

According to Childerhouse and Towill (2000), the lean manufacturing principle is well suited for products that can be predicted for their demand. Agile supply chain practices work best for innovative products for which demand cannot be predicted exactly, as in the case of the "fast fashion" industry. Christopher et al. (2004) said that the agile supply chain model is well suited to manage volatile demand. The key reason for the agile supply chain is mass customization.

According to the philosophy of mass customization, which works on the "postponement" principle, most of the products share many components. The shared components are produced in mass to achieve scale economies, and the few final value-added activities are delayed as far downstream as possible. The separate point at which the base product is customized in later processes is called the "decoupling point." Using this strategy, the product can be customized at a later stage of the supply chain, which makes the supply chain flexible enough for last-minute changes. To compete in the market and produce products that suit individual customer's needs, the strategy of mass customization works well. Many fashion retailers use this strategy of postponement; for example, Benetton dyes at the garment stage, which is normally done at the fabric stage. Delaying dyeing until the garment stage gives leverage to Benetton by mass producing the garment and customizing it into different colors only when the company receives orders from its customers (Naylor et al., 1999).

The agility concept has been well used by companies such as Zara, Mango, and H&M, for which the process of adoption of the design on to the final product on shelves can take around 4–5 weeks (Barnes et al., 2006a,b,c). The Spanish fashion giant Zara uses different types of supply chain strategies. Their designers produce more than 40,000 designs a year, from which more than 10,000 are actually picked to be produced into final garments. They produce each design in small batches so as to acknowledge demand first and then to produce the design in masses (if needed); else they continue with their strategy of changing the entire stock quickly to introduce more designs (Lee, 2004).

15.4.8 Electronic commerce and radio-frequency identification

The transformation caused by advancements in information technologies and the Internet has been experienced by almost all types of industry; the apparel industry is no exception to adopt this. Selling products online is the biggest market, superseding the brick and mortar stores and rapidly growing because of various cost–benefits. By offering products online, retailers need to keep the minimal level of inventory to serve customers at a centralized location, and that, too, may be in a low-cost country.

RFID is the most promising upcoming technology used to identify products automatically using radio waves (Dutta et al., 2007; Whitaker et al., 2007). The barcode scanning system has been succeeded by RFID technology as barcode scanning has a lower capability then RFID to store data and RFID offers the additional benefit of scanning products quickly even without being in the line of sight (Nath et al., 2006; Miles et al., 2008). RFID can store up to 1000 bytes of data, whereas the amount of information stored on a barcode is far less.

RFID is also used in the apparel industry at different stages of the supply chain and it is perceived as one of the most significant technological innovations in modern times. RFID is used in retail, distribution, and logistics to track and match products, for theft control, to control and manage inventory, etc. (Gimpel et al., 2004; Liu et al., 2010). Spinning is a capital-intensive process in which different sizes (counts) of yarn are produced and are easily mixed. Even one bobbin of wrongly mixed yarn can display its effects far downstream when the fabric or garment is dyed, owing to different dye pickup by different yarn counts. At the yarn stage, RFID can be used to prevent yarn mixing. At the fabric manufacturing and processing stage, the lots can be segregated with the use of RFID.

A significant benefit is seen at the garment stage, because retail stores and distribution centers may carry thousands of stock-keeping units with which garments can easily be tracked and traced instantaneously, as opposed to using the barcode scanning system, which needs a line of sight and is a time-consuming process. Department stores carry many products and brands under one roof, so keeping track of all items at an individual level is easy and feasible by using RFID (Loebbecke and Huyskens, 2008). Many retailers such as Walmart, Tesco, and Prada have gained benefits by using RFID, and especially the American apparel companies. It has saved many labor hours and reduced out-of-stock products (Nayak et al., 2015a).

Sankei, a Japanese manufacturer, is using RFID at the apparel manufacturing stage to control inventory and trace products easily (Wu et al., 2009). Walmart was among the early adopters of this technology and asked its suppliers to start using this technology if they wanted to stay in business with Walmart. A report published by the American Production and Inventory Control Society said that Walmart persuaded its suppliers to use this technology by giving them access to POS data on a regular basis to stay updated regarding their level of inventory to minimize the cost of overproduction (Weil, 2005). The well-implemented technology by Walmart expedited its corporate strategy of being cost competitive and having a QR (Vowels, 2006).

Fast fashion companies such as Zara, H&M, and Benetton have gained a major share of the market owing to their speed, low cost, and high inventory turnover ratio.

These all are possible only if a company has the technology to support it, which makes RFID an appropriate choice (Nayak et al., 2015a). According to Loebbecke and Huyskens (2008), a well-known German brand, Kaufhof, used RFID technology to suggest clothes and trends to men in the fitting room automatically regarding what suit or accessory would be appropriate using a "smart mirror." An RFID reader attached to the smart mirror scanned the products by scanning the tag attached to the garment brought into the trial room/fitting room. Suggestions were given about matching accessories to upsell on an appropriate interface. Products attached to the RFID also controlled the occurrence of theft by sending out signals and information to the relevant departmental authorities in case the product was taken out of the store without permission or without being scanned.

15.5 Conclusion

The fashion industry is characterized by:

a short shelf life: in which products are transient and swiftly change with changing individual moods, so they need to be changed quickly
fluctuations in demand: demands are not stable and keep changing with the changing season, influenced by trends, film stars, and role models
low predictability: forecasting is extremely formidable because of the highly competitive market with numerous suppliers offering different ranges/trends of products
impulse purchases: the decision to buy the garment is made instantaneously, so the constant availability of products is necessary

Modern trends in the garment industry pose threats to suppliers to stay competitive. The industry runs on keeping a low inventory and a low margin. Thus, companies need to produce products and replenish quickly. Widely used overseas manufacturing creates another issue regarding lengthening the supply chain, adding indirect costs and some hidden problems that may affect the retailer's performance if they are not planned and handled properly. The contemporary paradigms of using lean, agile, VMI, EDI, and technological advancements help to maintain rapport between the supplier and retailer, because upstream players can become more flexible and downstream members can share real-time information.

References

Abernathy, F.H., et al., 1995. The information-integrated channel: a study of the US apparel industry in transition. In: Brookings Papers on Economic Activity. Microeconomics, 1995, pp. 175–246.

Barnes, L., et al., 2006a. Fast fashioning the supply chain: shaping the research agenda. Journal of Fashion Marketing and Management: An International Journal 10 (3), 259–271.

Barnes, L., et al., 2006b. Supplier management in fast moving fashion retailing. Journal of Fashion Marketing and Management: An International Journal 10 (3), 272–281.

Barnes, L., et al., 2006c. Supply chain influences on new product development in fashion clothing. Journal of Fashion Marketing and Management: An International Journal 10 (3), 316–328.

Bhardwaj, V., Fairhurst, A., 2010. Fast fashion: response to changes in the fashion industry. The International Review of Retail, Distribution and Consumer Research 20 (1), 165–173.

Birtwistle, G., et al., 2003. Quick response: perceptions of UK fashion retailers. International Journal of Retail and Distribution Management 31 (2), 118–128.

Bruce, M., Daly, L., 2006. Buyer behaviour for fast fashion. Journal of Fashion Marketing and Management 10 (3), 329–344.

Bruce, M., et al., 2004. Lean or agile: a solution for supply chain management in the textiles and clothing industry? International Journal of Operations and Production Management 24 (2), 151–170.

Childerhouse, P., Towill, D., 2000. Engineering supply chains to match customer requirements. Logistics Information Management 13 (6), 337–346.

Christopher, M., Lee, H.L., 2001. Supply Chain Confidence: The Key to Effective Supply Chains Through Improved Visibility and Reliability. Global Trade Management, vol. 6.

Christopher, M., et al., 2004. Creating agile supply chains in the fashion industry. International Journal of Retail and Distribution Management 32 (8), 367–376.

Croom, S., et al., 2000. Supply chain management: an analytical framework for critical literature review. European Journal of Purchasing and Supply Management 6 (1), 67–83.

Da Silveira, G., et al., 2001. Mass customization: literature review and research directions. International Journal of Production Economics 72 (1), 1–13.

Danese, P., et al., 2004. Managing business processes across supply networks: the role of coordination mechanisms. Journal of Purchasing and Supply Management 10 (4), 165–177.

Davis, T., 1993. Effective supply chain management. Sloan Management Review 34 (4), 35.

De Vries, J., et al., 2016. Exploring the role of picker personality in predicting picking performance with pick by voice, pick to light and RF-terminal picking. International Journal of Production Research 54 (8), 2260–2274.

Djelic, M.-L., Ainamo, A., 1999. The coevolution of new organizational forms in the fashion industry: a historical and comparative study of France, Italy, and the United States. Organization Science 10 (5), 622–637.

Dutta, A., et al., 2007. RFID and operations management: technology, value, and incentives. Production and Operations Management 16 (5), 646–655.

Feld, W.M., 2000. Lean Manufacturing: Tools, Techniques, and How to Use Them. CRC Press.

Fernie, J., Azuma, N., 2004. The changing nature of Japanese fashion: can quick response improve supply chain efficiency? European Journal of Marketing 38 (7), 790–808.

Ford, H., 1926. Mass production. Encyclopaedia Britannica 13, 821–823.

Gattorna, J., 2006. Living Supply Chains. Prentice Hall, London, England.

Gereffi, G., 1999. International trade and industrial upgrading in the apparel commodity chain. Journal of International Economics 48 (1), 37–70.

Gimpel, S., et al., 2004. Textile-based electronic substrate technology. Journal of Industrial Textiles 33 (3), 179–189.

Guercini, S., 2001. Relation between branding and growth of the firm in new quick fashion formulas: analysis of an Italian case. Journal of Fashion Marketing and Management: An International Journal 5 (1), 69–79.

Gunston, R., Harding, P., 1986. Quick response: US and UK experiences. Textile Outlook International 10, 43–51.

Handfield, R.B., Bechtel, C., 2002. The role of trust and relationship structure in improving supply chain responsiveness. Industrial Marketing Management 31 (4), 367–382.

Hodge, G.L., et al., 2011. Adapting lean manufacturing principles to the textile industry. Production Planning and Control 22 (3), 237–247.
Jones, J., 1995. Forces behind restructuring in US apparel retailing and its effect on the US apparel industry. Industry, Trade, and Technology Review 23–27.
Keenan, M., et al., 2004. A dying industry-or not? The future of the European textiles and clothing industry. Foresight 6 (5), 313–322.
Lamming, R., 1996. Squaring lean supply with supply chain management. International Journal of Operations and Production Management 16 (2), 183–196.
Lee, H.L., 2004. The triple-A supply chain. Harvard Business Review 82 (10), 102–113.
Liu, J., et al., 2010. Development of bobbin tracing system based on RFID technology. The Journal of the Textile Institute 101 (10), 925–930.
Loebbecke, C., Huyskens, C., 2008. A competitive perspective on standard-making: Kaufhof's RFID project in fashion retailing. Electronic Markets 18 (1), 30–38.
Magazine, A.I., August 1998. Supply Chain Links Bolster Retailer–Vendor Strengths.
Mangan, J., et al., 2008. Global Logistics and Supply Chain Management. John Wiley & Sons.
Masson, R., et al., 2007. Managing complexity in agile global fashion industry supply chains. The International Journal of Logistics Management 18 (2), 238–254.
Miles, S.B., et al., 2008. RFID Technology and Applications. Cambridge University Press Cambridge.
Nath, B., et al., 2006. RFID technology and applications. IEEE Pervasive Computing 5 (1), 22–24.
Nayak, R., et al., 2015a. RFID in textile and clothing manufacturing: technology and challenges. Fashion and Textiles 2 (1), 1–16.
Nayak, R., et al., 2015b. The role of mass customization in the apparel industry. International Journal of Fashion Design, Technology and Education 8 (2), 162–172.
Naylor, J.B., et al., 1999. Leagility: integrating the lean and agile manufacturing paradigms in the total supply chain. International Journal of Production Economics 62 (1), 107–118.
Ohno, T., 1988. Toyota Production System: Beyond Large-Scale Production. CRC Press.
Pine, B.J., 1993. Mass Customization: The New Frontier in Business Competition. Harvard Business Press.
Porter, M.E., 1998. Cluster and the New Economics of Competition.
Potter, A., et al., 2016. Modelling the Impact of Factory Gate Pricing on Transport and Logistics. Developments in Logistics and Supply Chain ManagementSpringer, pp. 231–239.
Şen, A., 2008. The US fashion industry: a supply chain review. International Journal of Production Economics 114 (2), 571–593.
Sinha, P., et al., 2001. The Mechanics of Fashion. Butterworth-Heinemann, Oxford.
Sparks, L., Fernie, J., 2004. Logistics and Retail Management: Insights into Current Practice and Trends from Leading Experts.
Stengg, W., 2001. The Textile and Clothing Industry in the EU. Enterprise papers(2).
Stevens, G.C., 1989. Integrating the supply chain. International Journal of Physical Distribution and Materials Management 19 (8), 3–8.
Taplin, I.M., 1997. Backwards into the future: new technologies and old work organization in the US clothing industry. Rethinking Global Production. Ashgate, Brookfield, VT.
Trifilova, A., et al., 2010. Study of vendor-managed inventory practices in Indian industries. Journal of Manufacturing Technology Management 21 (8), 1013–1038.
Tyndall, G., et al., 1998. Supercharging Supply Chains: New Ways to Increase Value Through Global Operational Excellence.

Vowels, S.A., 2006. A strategic case for RFID: an examination of Wal-Mart and its supply chain. In: Proceedings of Ninth Annual Conference of the Southern Association for Information Systems. Hyatt Regency Jacksonville, Florida.

Weil, M., 2005. Life after the Deadline: a look at where Wal-Mart's RFID initiative stands. APICS Magazine 15 (3), 51–53.

Whitaker, J., et al., 2007. A field study of RFID deployment and return expectations. Production and Operations Management 16 (5), 599–612.

Womack, J.P., Jones, D.T., 2010. Lean Thinking: Banish Waste and Create Wealth in Your Corporation. Simon and Schuster.

Womack, J.P., et al., 1990. Machine That Changed the World. Simon and Schuster.

Wu, D.-L., et al., 2009. A brief survey on current RFID applications. In: Machine Learning and Cybernetics, 2009 International Conference on, IEEE.

Index

Note: 'Page numbers followed by "f" indicate figures and "t" indicate tables.'

D

Q

T

U

V

W

Y

Z

CPI Antony Rowe
Chippenham, UK
2017-11-22 11:41